Fishes Around Indian Ocean

Fishes Around Indian Ocean

– *by* –

K. P. Biswas

2018

Daya Publishing House®

A Division of

Astral International Pvt. Ltd.

New Delhi - 110 002

First Published, 2009
Reprinted, 2018

ISBN: 978-81-7035-620-2 (Hardbound)

Publisher's note:

Every possible effort has been made to ensure that the information contained in this book is accurate at the time of going to press, and the publisher and author cannot accept responsibility for any errors or omissions, however caused. No responsibility for loss or damage occasioned to any person acting, or refraining from action, as a result of the material in this publication can be accepted by the editor, the publisher or the author. The Publisher is not associated with any product or vendor mentioned in the book. The contents of this work are intended to further general scientific research, understanding and discussion only. Readers should consult with a specialist where appropriate.

Every effort has been made to trace the owners of copyright material used in this book, if any. The author and the publisher will be grateful for any omission brought to their notice for acknowledgement in the future editions of the book.

Published by : **Daya Publishing House**®
A Division of
Astral International Pvt. Ltd.
– ISO 9001:2008 Certified Company –
4736/23, Ansari Road, Darya Ganj
New Delhi-110 002
Ph. 011-43549197, 23278134
E-mail: info@astralint.com
Website: www.astralint.com

Dedicated to

Mrs. Manju Biswas

Preface

Nature has endowed marine animals with acute intelligence which makes them elusive to capture. For instance, a 6-feet long adult bottle nose dolphin male has a brain weight of 48.6 ounces. By contrast, a 6-feet tall man has an average brain weight of 47.2 ounces. Thus the dolphin seems to have a slight edge in length-brain ratio over man. This has tantalised scientists with the thought that in intelligence dolphins may be equal to or greater than man himself.

Gone are the days when man thought he only had to drop the net in the sea to catch fish. A battle of wits has to be waged between the fisherman and his prey. In this battle the most potent armament in man's armoury is to acquire the right type of intelligence about the movement and occurrence of fish to select right type of craft and gears.

In order to be successful at the job of locating schools of fish the fisherman must acquire knowledge and experience to the point where he is intimately familiar with the distribution and migrations patterns of the fish in the different seasons and in different sea conditions.

The fisherman first choose the fishing ground where the fish is likely to be found based on the weather and sea conditions, his own recent fishing success, and informations gathered from other fishermen. The fisherman must have some basic informations about the availability of his target species. Upon reaching his chosen fishing ground, he must then search for fish schools by means of signs, such as, current rips (the feeding ground of yellow tail) or water temperature (which determines the migrating lanes for fish like tuna).

Rocky reef areas around islands tend to be feeding grounds for migrating fishes which make excellent fishing grounds. When trolling in such a fishing ground, it is best to search for the schools of fish by starting near the shore and following the isobath lines one after another out into the deeper waters.

Because of the fact that fish change their migrating depth depending on factors such as the time of day, water temperature and their degree of appetite, inorder to achieve good fishing results, it is necessary to be able to troll at a number of different depths from the surface to the middle. To make the line and hooks troll at the desired depth, devices such as submarine board or a surface-planning device are required, to be used. The device is designed in such a way so that its smooth surface receives the waves and rides up over them easily. In addition to this function of riding on the surface of the water, this device also function as a vibrator to give a jerking motion to the baited hooks behind it of the sort that will stimulate the feeding instincts of the fish.

After a fish school is located, the boat is required to approach the school in a way that will not disturb their movement, and begin the trolling process, sweeping in front of the school at a speed of 3 or 4 knots. During this process, three factors must be considered; the direction of the sun, the movement of the boat and the movement of the school of fish.

In the case of both live bait and lures, however, it is considered extremely important that the bait be one that appeals strongly to the fish and stimulates its feeding instincts by, (a) moving through the water with a quick "swimming" motion similar to that of a real fish, (b) that it provides a strong visual stimulus by means of bright colours and intermittent flashings of reflected light and (c) that it has a slight sound as it "swims" like that of a real group of migrating fish.

For catching fish with net gear, informations on the assemblege of target species, their location in fishing grounds is essential. Fish move seasonally in search of water of the proper temperature, in search of food and to breeding spots. This habit is called migration. The study of these migrations is extremely important to the fisherman. For example tuna migrate to areas where the water temperature is 14^0-15^0 C. Therefore they will not be found outside such areas. Salmon always return to their home area for breeding. At this time salmon can easily be taken in shallow waters without extensive equipment. Many fishing methods make use of these habits and migrations.

Fish eggs which hatch in sea weed areas migrate gradually to deeper sea as they mature, usually returning to their original spawning area.

Fish migrate looking for more food at same depths. Deep sea fish remain in deep area in the day time, but come up to the surface at night in search of food.

All these useful informations along with major commercial fish species available in Indian Oceans have been incorporated in the book with a hope that the book will he beneficial to the fishery operators, students, teachers and researchers working in the field of fisheries and marine sciences.

K.P. Biswas

Contents

Introduction

Indian Ocean covers over 14 per cent of the earth's surface. Indian Ocean bed forms a depression. The Central Indian Ridge stretches along 65 to 75 degrees east, dividing the ocean into a western and eastern part. Inorganic deposits, formed from erosion of the shore and bottom, make up the bulk of bottom composition of the continent shelf and slope. Sand and terrigenous mud predominate on the slopes, especially in the west. On the shelf the bottoms consist of sand, pebbles, stones and coquina, and are on the whole suitable for trawling. Rocks and coral formations occur on some sectors of the shelf, on banks and around islands.

The geographical position of the Indian Ocean determines the main features of its hydrology. A tropical climate prevails over most of the ocean with little fluctuations of water and air temperatures and a high humidity.

The land masses in the northern area bring about the development of a monsoon region which embraces the entire ocean expanse north from 10 degrees south.

The Indian Ocean is sometimes described as an ocean of warmed waters, as it lies almost entirely in a region that is thoroughly penetrated by the sun. The only source of cold water is the Antarctic region. The influence of cold west Australia current is negligible. The other permanent currents of the Indian Ocean are the Equatorial countercurrent, the South Equatorial current and the Mozambique Agulhas and Somalia currents, which wash the east coast of Africa. All these currents bring very warm waters whose surface temperature is rarely below 20 degrees celcius.

Heat exchange in the Indian Ocean is largely conditioned by the warm waters of the Red Sea and Persian Gulf that enter the Arabian sea. The sun penetrates down to 100 to 150 metres and in this layer the temperature is uniform and very high.

Salinity is relatively high in the Indian Ocean and fluctuations occur only at the surface.

Circulation in the Indian Ocean involves the waters of the upper and lower layers. That in the surface layer is most important as it creates the winds prevailing over the ocean.

The following divergences have the greatest effect in summer on the distribution of plankton and corresponding distribution of fish. The divergence of the waters of the western part of the Somalia current and the coastal waters. The equatorial divergence between the North Equatorial current and the counter-current. The divergence of the Monsoon current west of the Maldive islands and the divergence between the West Australia current and the coastal waters of Australia.

A tropical and a temperate fish fauna exist in correspondence with two zoogeographical zones, tropical and south temperate in the Indian Ocean. Their range of distribution can be arbitrarily defined by the 20 degrees surface layer isotherm in February and by the 25 degrees isotherm in August.

The waters of the southern region of the Indian Ocean possess 460 species of fish, of the central region, 900 species. The northern region also possesses abundant fish.

The waters of the Arabian Sea, Persian gulf, Bay of Bengal and Andaman Sea harbour about 1000 species of fish.

The Indian Ocean pelagic abounds in large fish. The formulation of accumulations, their movements and distribution of the young depend on oceanographic conditions like temperature, currents and forage supply, that is, plankton and small fish.

An important feature of tropical fish is year round spawning.

All the evidence establish the abundance of fishes available in the Indian Ocean, the Arabian Sea and the Bay of Bengal.

The three main taxocenes, the aggregate of a species of a certain taxonomic group, of marine fishes in the Indian Oceans are, (a) littoral, (b) epipelagic and (c) deep water. They have sharply differentiated compositions which overlap to a considerable degree. Such overlappings exist as a result of active and passive migrations.

The ocean contains 400 families of the Teleostei group (bony fishes) and about 40 families of the Elasmobranchii groups. These groups contain profusion of species that run into several thousands. These are distributed in the littoral, epipelagic and deep water ichthyocenes of the ocean.

The western part of the Indian Ocean demarcates into three regions, southern, central and northern. These regions possess more than 1500 species of fish. To name a few are, tuna, billfish, marlin, sword fish, sail fish, pelamid, dolphin, sardine, herring, anchovy, hake, mackerel, scad, snock, gilthead, emperor, snapper, dog fish, bass, bream, sole, shark etc.

The waters of the Arabian Sea, Persian Gulf, Bay of Bengal and Andaman Sea harbour about 1000 species of fish.

The eastern regions of the Indian Ocean also possess a rich variety of fishes. More than 2000 distinct species of fish are known to exist in Indian waters. Of these, about 500 species classified into 16 groups, easily qualify as the commercially important fishes. Indian waters also possess five chief commercial edible shell fish; (i) prawns, (ii) lobsters, (iii) crabs, (iv) clams and (v) oysters.

Nine species of deep water prawns, besides 5 species of penaeid prawns of shallow waters, have been discovered in depths of 200 to 400 metres and 190 to 760 metres. Evidence is also on record of the *Nemato carcinus*, a typical deep sea shrimp. Its very long legs and antennae help it to keep afloat and find food. This shrimp is found in the Indian Ocean at a depth of 900 meters.

Closely related to shrimp and looking very much like them are the handsome euphausids, commonly called krill. The name which means "shining light" refers to their remarkable luminescence. One or more species is found in every ocean. Although most live between the surface and 1000 meters, some range at depths of 2166 meters.

Those that live in warmer, deeper water are meat eaters, but most others strain floating plants out of the sea.

Sir Francis Day (1889) described 1418 species of fish under 342 genera from British India. P. K. Talwar in 1991 reported 2546 species of fish belonging to 969 genera, 254 families and 40 orders. Their description were mainly to give fish species an identity for an universal nomenclature, taxonomic affinities and status among related species based on the morphological and meristic characters. Knowing the species, thus light have been thrown on their distribution in different geographical areas and habitats. The distribution of marine fishes is rather wide and some genera are common to the Indo-Pacific and Atlantic regions. Fifty seven per cent of the Indian marine fish genera are common to the Indian Ocean and to the Atlantic and Mediterranean.

Much of the world's wealth of fish diversity is found in highly diverse marine and coastal habitats. These habitats range from shallow coral reefs to the dark ocean floor's hard rock and soft sediments at abyssal depth. While the total number of described marine fish species are exactly not known, scientists are continually discovering new concentrations of diversity. Coral reefs are already known to be among the richest habitats in fish diversity. The deep-sea bottom, dark and subjected to tremendous pressure is now thought to be a dwelling place for a number of luminescent fish species.

Coastal ecosystems play a vital role in the economy of countries bordering Indian Ocean by virtue of their resources, productive habitats and rich bio-diversity. The fish constitute the dominant taxon in the nekton, an important group among the pelagic organisms. About 4000 species of fish are known to occur in Indian Ocean, of which about 50 per cent occur in Indian seas. A majority of these species occur in coastal waters supporting valuable fisheries.

Marine fisheries constitute a highly productive sector, a source of valuable food and employment. For India with strong fisheries interest, the largest fish production comes from the coastal capture fisheries, which contribute on an average, 62 per cent of the total fish production. The marine jurisdictional area, the Exclusive Economic Zone (EEZ) is extensive, spanning 2.02 million sq km which is 38 per cent of the total marine, fresh water and land areas of the country. On 3651 fishing villages situated along the 8129 km coast line, about one million people are employed full-time in marine capture fisheries. The marine fishing sector which is dominated by small-scale and semi-industrial operators supports several ancillary industries such as boat-building, yards, processing plants etc. Marine fisheries production of India which was only 0.5 million tonnes in 1950 increased through time and peaked to 2.7 mt in 1997. In 1997 the production (2.2 mt) from inshore waters (<50 m depth) has reached the maximum harvest potential (2.2 mt) and hence scope for further increase in production from inshore waters is limited.

Examining the history of fishing and fisheries makes it abundantly clear that humans have had for thousand years, a major impact on target species and their supporting ecosystems. Indeed, the archaeological literature contains many example of ancient human fishing associated with gradual shifts, through time, to smaller sizes and the serial depletion of species that we now recognize, as symptoms of overfishing. According to State of World Fisheries and Aquaculture's reports, only an estimated 25 per cent of the major marine fish stocks or species groups are under-exploited or moderately exploited, while the rest 75 per cent have been fully exploited (47 per cent over exploited and 28 per cent significantly depleted). The fishery wealth of ocean, once deemed inexhaustible, has proven finite and fish once thought "the poor man's protein" has become a resource coveted and fought over by nations.

Less than 7 per cent of world's ocean are continental shelves, shallower than 200 m. Shelves generate biological production supporting 90 per cent of global fish catches. The overwhelming majority of shelves are now sheltered within the Exclusive Economic Zone (EEZ) of maritime countries, which also include all coral reefs and their fisheries. According to the 1982 United Nations Convention of the Law of Sea, any country that can not fully utilize the fisheries resources of its EEZ must make this surplus available to the fleet of other countries. This along with eagerness for foreign exchange, political pressure and illegal fishing has led to all of the world's shelves being trawled for bottom fish, purse seined for pelagic fishes and illuminated to attract and catch squids and cuttle fishes.

Indian EEZ has become a much coveted fish poaching avenue not only for Taiwanese tuna long liners (some of them are regularised through Letters of Permission) but also of other countries. Thai fishing vessels are poaching in Indian EEZ and their presence is noticed in the Bay of Bengal and Andaman Sea. The latest addition to the poacher's list is Sri Lanka. The tuna longlining vessels of this country have been seen fishing for tuna in Indian waters as north as Visakhapatnam. The author, during his tenure as Director Fisheries in Andaman and Nicobar Islands, in 1982 inspected a Taiwanese long liner, apprehended by Indian Coast Guards on the charge of poaching fish from the territorial seas of India. About 28 Thai fishing vessels were caught and seized by then from Andaman Sea on the same charge.

Small boats, traditional fishermen, twelve millions in numbers, yet they catch only about half the world's fish. In some poor countries, a small boat with a basket of line may be the last resort for a fisherman to survive. Although technology helped quadruple the world's catch since 1950, a nearly empty basket is all a fisherman in Cochin, India has to show for several hours of toil, a complaint heard from all the traditional fishermen around the world. A fishing industry functions on the basis of continuing interaction of replenishment capacity of natural resources, intensity of fishing and effective use of the catch. For fishing to emerge as an independent industry, development of fishermen communities linking the development of techniques for catching, preserving and processing of marine products must be fulfilled. In the neighbouring South-East Asian country there are a dozen of basic fishing methods operated basing on the knowledge of a number of fish species about their habitats, behaviour and migration in the marine ecosystem. Fishermen involved in small scale coastal fishery generally choose three or four of these fishing methods which they engage in on a rotating schedule round the year.

With this backdrop, the identification, occurrence and distribution, food and feeding habits, age and growth, maturity and breeding, migration and population dynamics of major group of fish forming commercial fishery in maritime nations bordering Indian Ocean have been described in the subsequent chapters.

Chapter 1
Oceanic Regime

Physical Habitat

Seventy one per cent of the Earth's surface is covered by oceans having an average depth of 3800 m. The oceanic habitat is defined as saline water portion that extends over the continental slope and the abyssal plain. The portion of waters over the shelf, which conventionally extends out to a depth of 200 meters is usually referred to as the neritic environment.

Out of a total ocean surface of about 360 million Km^2, the neritic environment over the continental shelves covers almost 32 million Km^2 and hence the oceanic region accounts for over 91 per cent of the world oceans. Indian Ocean, have surface area of 60.06 million Km^2, 16.76 per cent of total world ocean. Its neritic habitat over the continental shelves covers around 4.27 million Km^2; leaving almost 93 per cent for the oceanic region.

Although mean productivity per unit area is much lower in the oceans than on land, their very large surface means that the oceans still account for at least a third of the annual global carbon fixation. For this reason, oceanic communities contribute significantly to global processes.

Oceanic phytoplankton is responsible for the primary production of the oceans and constitutes the basis of the food chain in the high seas. Primary production is restricted to the euphotic layer, or upper part of the photic zone, where sufficient light penetrates to allow photosynthesis. The depth of the euphotic zone depends on the amount of suspension and detritus present in the water and can vary from 40-50 m in turbid waters to over 100 m where the waters are particularly clear. Production is also limited by the availability to inorganic nutrients.

Below the photic layer there is the aphotic zone, where no light arrives and primary production is absent. Organisms living in this zone which are not performing vertical migration to upper waters are exclusively carnivores, suspension and detritus feeders.

Large quantities of nutrients are continually lost to the aphotic zone and are not available for photosynthesis. However, large-scale ocean circulation linked to Earth's rotation climatic cycles (seasons) and topography of the ocean basins allow periodical or semi-permanent (in some areas) mixing of superficial and nutrient rich deep waters. This phenomenon, called upwelling, is the cause of extremely high productivity of some fishing areas.

Trend in World Oceans

Global catches of oceanic species have been steadily increasing during 1950-1999 and reached about 8.6 million metric tons in 1999, except a small decrease in early 1980s. The share of oceanic catches in global marine catches ranged between 4 and 8 per cent from 1950 to 1989. In recent years the contribution of oceanic catches to total catches increased and exceeded 10 per cent in 1998 and 1999.

Until 1975, catches of deep water species were relatively small, ranging 2 and 10 per cent of the total oceanic catches, but since the late 1970's their contribution has consistently been greater than 20 per cent, reaching 33 per cent of the total oceanic catches in the last two years.

Among the epipelagic species, catches of tuna and tuna-like species have been increasing dramatically throughout the years. Since the mid-1960's the rate of increase of tuna and tuna like catches has been much higher in comparison to other epipelagic species and tuna catches are still growing at a rapid pace, while those of the other species have decreased in recent years. The deep water group, dominated by a single Gadiformes species, *Micromesistius poutassou*, blue whiting, do not show increasing catch trend, as that of tuna species and are low valued fishes, which are mostly processed into fish meal.

Total marine catches by Distant Water Fleets (DWFs) increased from less than one million tonnes in early 1950s to about 8 million tonnes in 1972, fluctuated this value until 1991 and then declined rapidly to about 4.5 million tonnes, remaining stable in the most recent years. As a proportion of total marine captures, DWFs catch reached a maximum of 15.5 per cent in 1972 and then declined to about 5 per cent, a level at which they have stabilised since 1993. The starting points of the two marked decreasing trends of DWF catches coincides with the oil price hike (1973) and the dissolution of the Former USSR (1991), whose fleets were actively fishing in all oceans.

Following the declarations by an increasing number of countries of the Exclusive Economic Zones (EEZs), after the United Nations Convention on Law of the Sea (UNCLOS) of 1982, distant water fishing nations had to negotiate access to the marine resources living within the 200 miles limit. This new situation, together with the increasing price of fuel oil led to an overall increase of costs for DWFs that progressively shifted to oceanic species which are both highly valuable (tunas) and can be often caught in the high seas, outside areas of national jurisdictions.

However, in terms of quantities, the increase of oceanic catches in recent years is entirely due to the contribution of bordering countries whose catches of oceanic species have been steadily increasing since the early 1980s.

Western Indian Ocean

Marine captures in this area increased continuously from 1950 onwards and have stabilized around 3.9 million tonnes since 1997. For the whole time series, the oceanic share has always been significant, reaching 20 per cent for the first time in 1995 and a maximum of 22 per cent in 1999. Since 1983, more than 75 per cent of the oceanic catches have been from tuna species, while deep-water catches are Indian catches of hairtails (Trichiuridae), a group of fishes which could also be considered as epipelagic because it shows vertical feeding migration. A deep-water fishery that could be developed in future years is that for lanternfishes (Myctophidae) in the Arabian Sea.

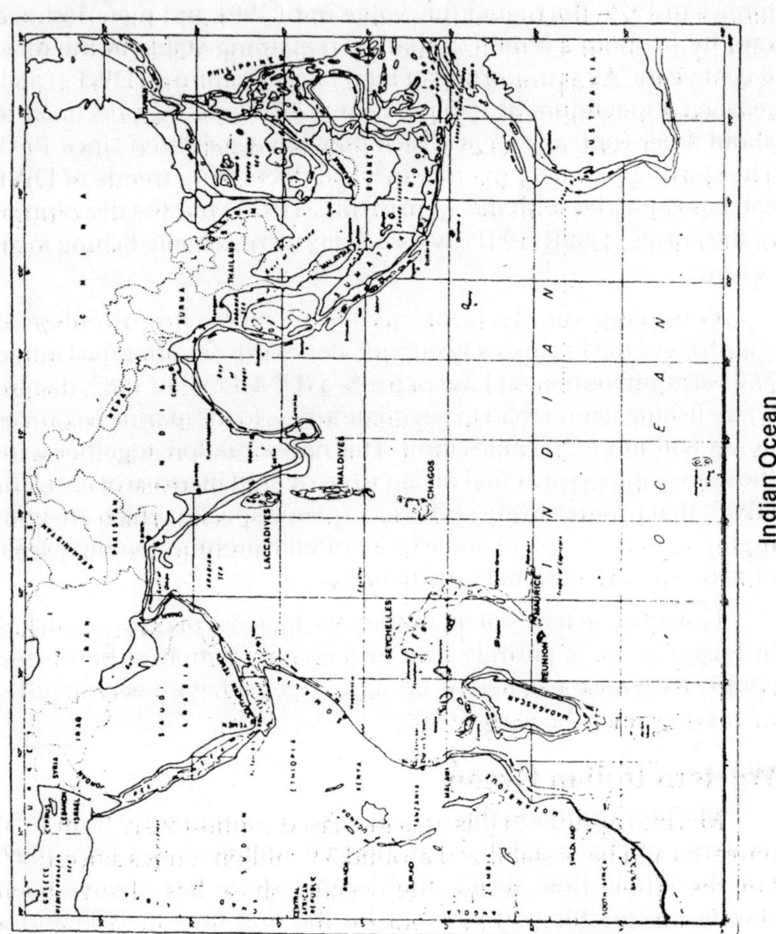

Indian Ocean

Since 1984, catches of oceanic tuna in Western Indian Ocean have been increasing steeply and they exceeded 700000 tonnes in 1999. About two thirds of these catches are harvested by European (Spain and France) and East Asian (Japan and Taiwan Province of China) fleets. Main tuna species caught are skipjack, yellowfin and bigeye tunas. Epipelagic species other than tuna, represented, are dolphin fish and sharks of the family Carcharhinidae.

Eastern Indian Ocean

Trends in both total marine and oceanic captures in the Eastern Indian Ocean are very similar to those in Western Indian Ocean. In both areas total catches have been progressively increasing along the entire time series and tuna catches constitute the bulk of oceanic catches. The major differences are that the steep increase of tuna catches in the Eastern Indian Ocean took place about ten years later than in the Western Indian Ocean (in 1993 instead of 1984) and that distant water fleets have a more limited role in the Eastern area.

Skipjack, yellowfin and bigeye tunas are also the main target species in Eastern Indian Ocean. Bordering nations with important tuna fisheries are Sri Lanka and Indonesia, while Japan and Taiwan Province of China are the main distant water fishing fleets. From 1960 onwards, Sri Lanka has been reporting considerable catches of silky shark (*Carcharhinus falciformis*). Since 1980, catches of this species have ranged between 10,000 and 25,000 tonnes.

The share of deep-water catches in this area is slightly higher (on average 18 per cent of the oceanic catches) in comparison to the Western Indian Ocean, with the highest quantities (mostly hairtails catches by India and Indonesia) recorded in 1976 and in 1999. Significant catches of orange roughy (*Hoplostethus atlanticus*) in the Eastern area were reported for 1998 and 1999 (4857 and 7553 tonnes respectively) by Australia, while in previous years catches of this species were mostly concentrated in Southwest Pacific.

The main distinguishing features of Indian Ocean marine ecosystem is the high catch percentages for miscellaneous coastal fishes and miscellaneous pelagic fishes. The catches of herrings, sardines and anchovies and of crustaceans in the Indian Ocean ecosystem exceeds 10 per cent on average. This ecosystem is characterized by fishing activities mainly concentrated on the coastal areas and this explains high percentages of miscellaneous coastal fish catches.

Primary production ranges from moderately (Bay of Bengal) to high (Arabian Sea). Three of the Indian Ocean ecosystems (Somali Coastal Current, Arabian Sea and Bay of Bengal) are influenced by monsoons. In the Somali Coastal Current and in the Arabian Sea, the southwest monsoon from May to October cause seasonal upwelling phenomena that are on the other hand lacking in Bay of Bengal. In the Arabian Sea about 65 per cent of fish landings derive from artisanal fisheries and this would explain the prevalence of coastal species catches, but it may also be influenced by the presence of low-oxygen water, which restricts productivity at depths of 200 m and more.

Indian Ocean as a Habitat

The Indian Ocean, covering an area of 49 million square kilometer is the world's smallest ocean basin. It can be divided into two main areas from east to west, along 72°E, both sides of Maldives and Chagos Islands. The western areas comprises the Gulf of Oman, the Arabian Sea, the Somali Basin, the Mascareignes Basin as well as the Mozambique channel. The more extensive eastern area comprises the Bay of Bengal, the central Indian Ocean Basin, the Cocos Islands Basin, the Wharton and North Australian Basins between Australia and the Sunda Islands.

Indian Ocean, hemmed in to the north, east and west is characterized by well defined meteorological and oceanographic phenomena which distinguish it from the Atlantic and Pacific Oceans. The Indo-Asian continent prevents any temperate marine influences from the north, and the strong interaction between ocean and continent results in the seasonal monsoon.

Climatic Conditions

The monsoon regime occurs when the direction of the average prevailing winds changes more than 90° from summer to winter. The regions affected by the monsoons are exclusively limited to the northern hemisphere. The south of the Equator, the east and west extremes of the ocean are more influenced by the monsoon than the centre.

Throughout the year, anticyclonic high pressure prevails in the South Indian Ocean. It shifts seasonally from south to north. In January, the low atmospheric pressure present in the South-west Indian Ocean extends from the southern anticyclonic high pressure areas of 30°-40°S to those of the Asian continent. The low atmospheric pressure present in the Eastern Indian Ocean extends from northern Australia to Malaysia.

In July, the southern anticyclone reaches its most northerly position at 25°S. It is stronger than in January. Low pressure areas are found in the Arabian Sea (Gulf of Oman) and in the Bay of Bengal, Atmospheric pressure drops steadily from north Madagascar to the Asian continent.

In the tropical region of the Indian Ocean, out of the monsoon regime, south-easterly trade winds prevail all the year round. These become strongest during the southern winter when the south anticyclone reaches maximum intensity, regularly exceeding 20 knots.

In monsoon affected areas, the northeast monsoon blows from December to March following the north-south pressure gradient. Moderate northeast wind at 4-6 knots on average, collects moisture over the Bay of Bengal and the Arabian Sea and meets the south-east flux at the intertropical convergence zone (ITCZ). Crossing the Equator, the north-easterly winds change to north-westerly and blow in a southwesterly direction, the predominent westerly winds remain very light. This equatorial calm can however be interrupted by strong winds lasting several days. Cyclonic gyres are observed in the area. The northeast Indian Ocean is dominated by easterly winds blowing off the land. The average wind force on the west coast of Thailand ranged between force 3 and 4 of the Beaufort scale.

As an extension of south-east trade winds, the southwest monsoon, north of the Equator lasts from June to September following a steep pressure gradient situated between the southern anticyclone and the low pressure areas of the Northern Indian Ocean. Atmospheric circulation is particularly strong along the coast of Somalia. The Somalian jet is an important component of this monsoon as it carries humidity towards the Arabian Sea and induces a coastal upwelling which saps heat from the surrounding atmosphere. The seasonal flux changes direction during the winter monsoon carrying humidity towards the southern hemisphere.

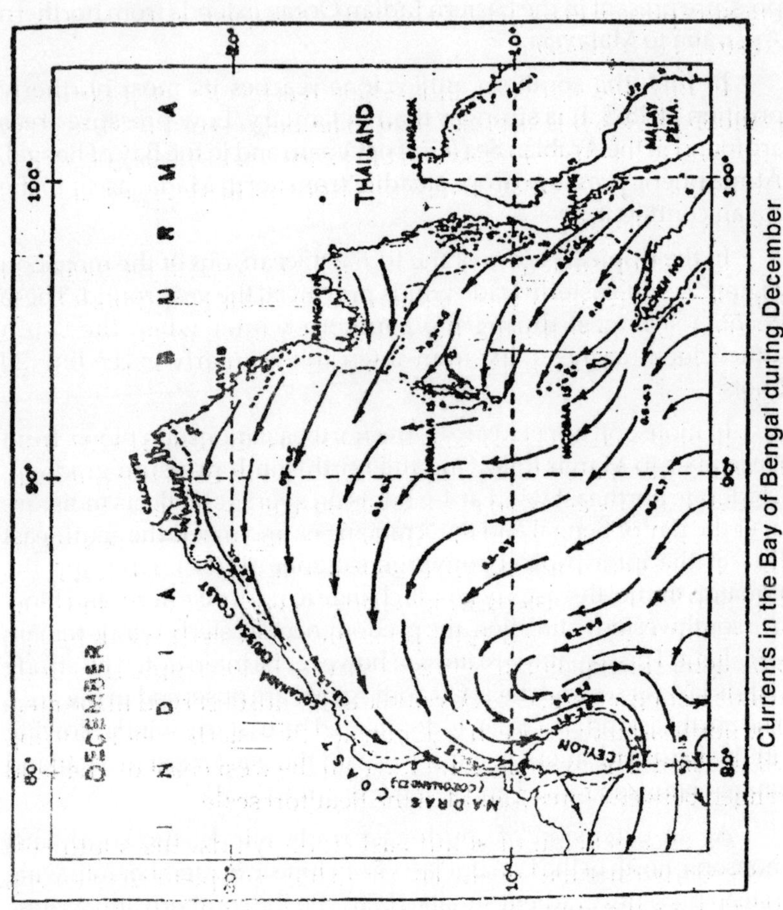

Currents in the Bay of Bengal during December

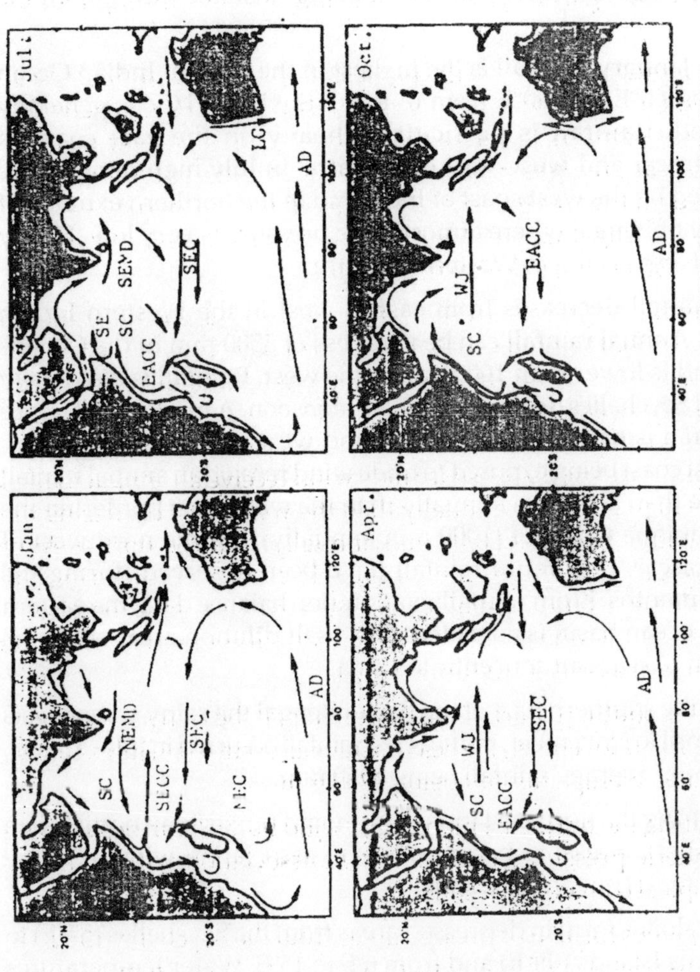

Surface Currents in January, April, July and October (Fieux, 1985)

SC: Somali current; SE: Somali eddy; NEMD: Northeast monsoon drift; SEMD: Southeast monsoon drift; SECC: South equatorial counter current; EACC: East African coastal current; SEC: South equatorial current; WJ: Equatorial jet; JC: Java current; MEC: Madagascar east current; LC: Leuwin current; AD: Antarctic drift

During the transitions of inter-monsoons range from April to May and from October to November, a strong westerly flux dominates in the equatorial areas. There is no prevailing wind in the Bay of Bengal, wind direction being unstable throughout the area.

In January, rainfall is the highest in the Central Indian Ocean between 60°E and 90°E from 0° to 10° S where ITCZ is generally situated. Rainfall is particularly heavy on the east coast to Madagascar and west coast of Sumatra. In July high rainfalls are recorded on the west coast of India and in the northern extreme of the Bay of Bengal where atmospheric pressure is very low. Heavy rainfall also occurs in Western Sumatra.

Rainfall decreases from east to west in the Western Indian Ocean. Annual rainfall can be in excess of 3500 mm in the Chagos area and is lower than 1000 mm to the west. Rainfall is also heavy around Seychelles during the winter monsoon. A marked difference in rainfall is noticed between east and west coast of Madagascar. The east coast being exposed to trade wind receive an annual rainfall of more than 2000 mm annually than the west coast bordering the Mozambique Channel (1000 mm annually). On the northwest of Madagascar, maximum rainfall have been observed during the winter months. From rainfall-evaporation balance data, the eastern part of ocean basin is considered as a salt dilution area, while the western part as salt concentrated area.

In the southern part of the Bay of Bengal the rainy season lasts from April to November, the heaviest rainfall occurred in July-August, the annual average rainfall being 2250 mm.

During the northeast monsoon, "wind bursts" can occur when atmospheric pressure becomes low in association with an active intertropical convergence zone.

Cyclones form in depressed areas from the Scychelles (55°E) to the Cocos Islands (95°E) and from 6°S to 15°S. Water temperatures are high (28°-30°C) and there is a dense layer of humidity in the atmosphere. Active cyclone formation lasts from December to April, with a maximum around January-February. Cyclones generally travel westwards south of 8°S. The vulnerable areas are the islands of the Mozambique Channel, Madagascar, Reunion and Mauritius, whereas the northern Seychelles Islands and Chagos Islands are off the usual trajectories. The movement of these tropical depressions sweeps southward.

Hydrological Conditions

The warm waters of the Red Sea and Gulf of Oman are high in saline content, while the eastern Indian Ocean shows the warm waters and low salinity bordering Indonesia where rainfall is heavy. Surface waters with temperature below 22°C range from 34.5 per cent to 36 per cent and are situated principally south of the Equator. The upwelling of Somali coast (8°N-12°N; 50°E-54°E) is the only exception, whose characteristics are similar to those found in the southwest of the Indian Ocean.

The different water masses are present in the western Indian Ocean. Large seasonal variations occur in mixed layer.

(*a*) From the Equator to 6°S, the equatorial surface water has an average salt content of 35.3ppt

(*b*) From 6°S to 22°S, east of Madagascar, the south equatorial surface water comprises a layer of water 80 to 110m thick. In the northern section of this area, between 6°S and 11°S, salinity is the lowest in comparison to the rest of the entire western basin of the Indian Ocean, minimum being 34.5ppt to 34.8ppt.

(*c*) In west of Madagascar, the Mozambique Channel surface water varies in salinity during the seasonal changes, in relation to the alternation between a rainy season (November-April) and a dry season (May-October).

(*d*) Around latitude 22°S, the Mozambique Channel water mixes with tropical surface water of higher salinity (35.1ppt) coming from the south.

Sub-surface waters occupy a layer from 100 to 200 m thick generally at the level of sharp thermal gradient, between 15°C and 23°C. Salinity is high, more than 35.2ppt, oxygen level minimum and nutrients content is maximum. At the Equator, the same properties are found at depths from 60 to 150 m and 300 m at 17°S.

(a) North of 10°S, the north equatorial sub-surface water originates in the Arabian Sea.

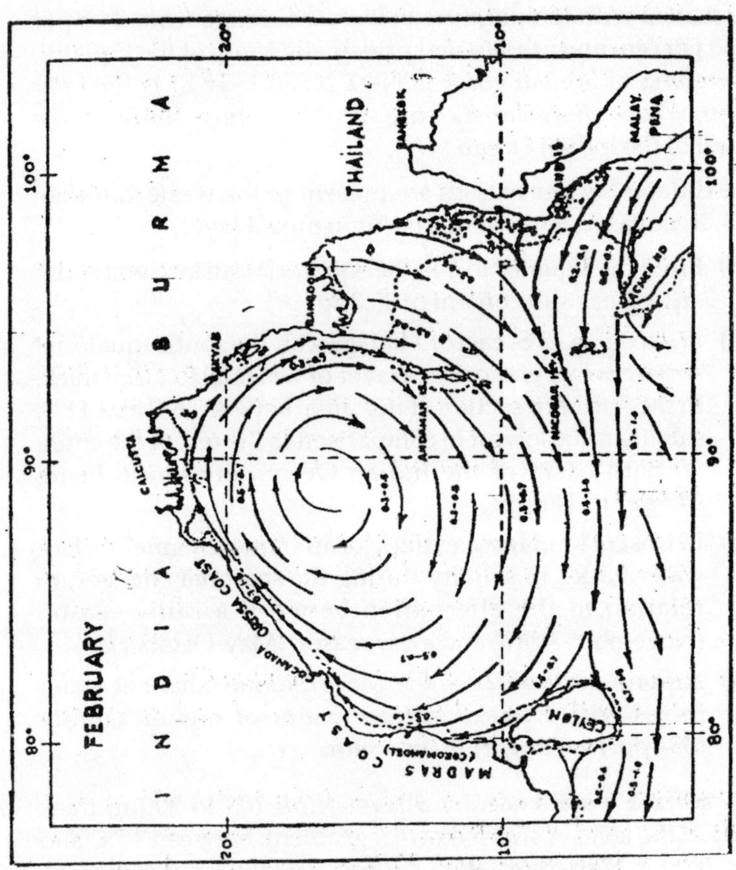

Currents in the Bay of Bengal during February

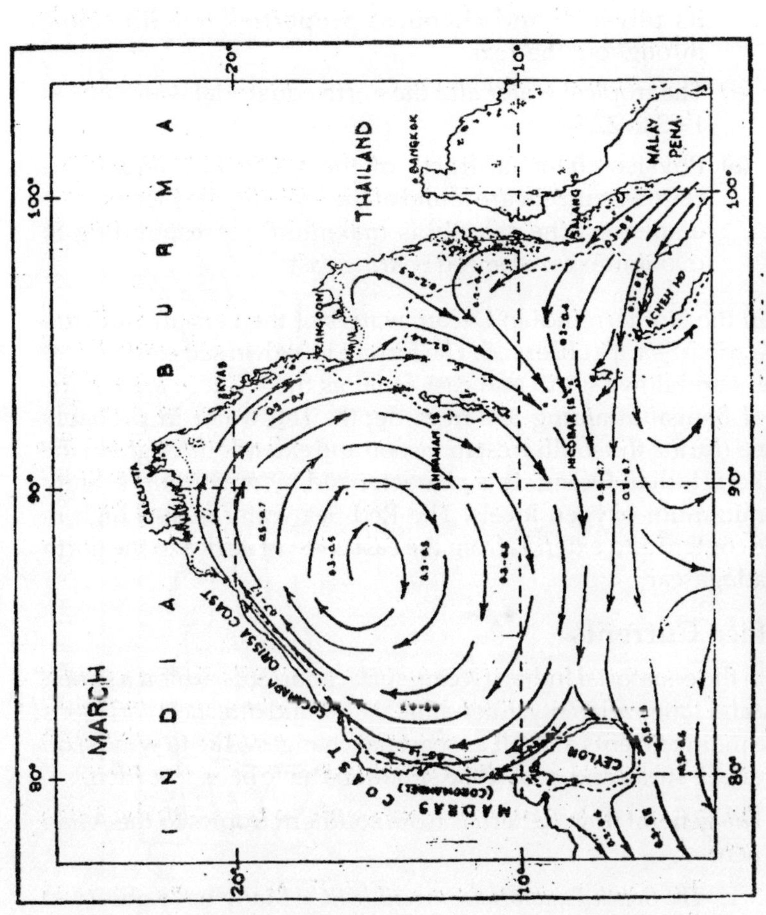

Currents in the Bay of Bengal during March

(b) The east Madagascar tropical water is found in the Northern Mozambique Channel at 11°S-12°S, after passing around Cape d' Ambre to meet the water systems of the north Mozambique Channel.

(c) The north equatorial water from the Northern Mozambique Channel is situated between 12°S and 17°S, its physical and chemical properties remain stable throughout the year.

(d) The tropical water and the north equatorial water mix at 19°S to 22°S.

(e) Physico-chemical traces of the subsurface equatorial counter-current are found at the equator, the thermocline dispenses, the salinity is maximum corresponding to maximum zonal speed bearing east.

In the Nothern Indian Ocean, waters of the Persian Gulf and the Arabian Sea are observed. The former is high in salt content and density and flows south and east as far as the western area of the Bay of Bengal attaining 350 m in depth. The water in probably formed during the south west monsoon and extends throughout the Northern Indian Ocean. It is characterized by maximum salinity and minimum oxygen levels. The Red Sea water is also high in saline content and extends from the east coast of Africa to the north of Madagascar.

Surface Currents

In the equatorial Indian Ocean, surface currents exhibit a strong space and time variability. Inter-annual flow and direction variations of the main currents as well as seasonal changes (due to winds) on the surface strata and eddy type situations have been recorded.

The general flow pattern is from southern tropics to the Asian continent

1. *The South Equatorial Current (SEC)*: Many have observed this west-flowing stream. In the Eastern Basin, the south equatorial current extends from 9°S to 14°S at a speed of approximately 1 knot. During its westerly course, the SEC expands from 10° to 20°S at 60°E and divides into two main streams at about 16°S along the east coast of Madagascar; whereas the northern stream flows round

Cape d' Ambre as it accelerates to 3-5 knots. To the north of Madagascar, the SEC carries with it a 300 m deep layer and 200 km wide. When the SEC reaches the east coast of Africa, north of the Mozambique Channel, it divides into two.

The south-bound stream starts a wide anticyclonic gyre in the north basin of Mozambique Channel while simultaneously feeding the Mozambique current flowing south along the African coast.

The north-bounds stream generates the coastal East African current. From June to September, the SEC spreads out northward (2°S to 5°S) due to the influence of the southeast trade winds.

2. The East African coastal current flows in a northerly direction reaching the Equator, and sometimes crossing it. During the southwest monsoon it is prolonged by the Somali current, whereas during the northeast monsoon, a confluence with the seasonal monsoon current which flows in a southerly direction along the east coast of Africa. The meeting point varies from year to year, and throughout the season according to the strength of the northeast monsoon.
3. The south equatorial counter-current (SECC) flows in an easterly direction from December to April, between the SEC and the northeast monsoon surface drift. It is essentially a seasonal current which corresponds to the northeast monsoon period. Its latitudinal extension varies from one year to the next as well as during the season. At its widest point (end of the northeast monsoon) it can spread from 2° to 9°S. The SECC travels at 1.5 knots. In December and April, the flux is about $1.10^6 m^3/sec$. During the southwest monsoon, the SECC disappears leaving cyclonic eddy movements which develop at the Equator between the SEC and the southwest monsoon drift.
4. The equatorial jet is exclusively seasonal. It moves in an easterly direction along the equatorial belt, reaching a speed of 2-3 knots. It can be observed only during the inter-monsoon periods, between April-May and October-November. The stream flows along the Equator attaining

a width of 500 km. The principal cause for the existence of the current are the westerly winds which blow on the equator at the end of the monsoons during the transition periods. The ocean quickly responds and the jet becomes strongest in the central part of the ocean (between the Maldives and the Chagos Islands), before slowing down after a month, due to internal waves moving from east to west.

5. The north equatorial current (NEC) also known as the "northeast monsoon drift" flows in a westerly direction and is exclusively dependent on the northeast monsoon. The NEC moves through most of the northern hemisphere, extending as far as 2°S in the southern summers. This current is capable of forming whirl type movements even 200 km wide, as observed in the south Arabian Sea.
6. The Somali current is the most pronounced and the most spectacular oceanic response to the onset of the southwest monsoon. The atmospheric jet off Somalia from July until September is strong, reaching 35-40 knots. The resulting circulation along the coasts induces a reaction of unique intensity. The water mass it conveys comprises a layer 200m deep estimated at 40.10^6 to 50.10^6 m^3/sec., which can attain a speed of 7.4 knots at latitude 4°S. The water mass is related to the strong coastal upwelling. There are existence of two anticyclonic eddies associated with the Somali current.

The intensity of the south eddy shows large variations from one year to the next. It is centred on the Equator west of 50°E. The north eddy spreads from 4°-9°N, east of 50°E, and persists after the southwest monsoon ends. Traces of it remain until the beginning of the following northeast monsoon. The mass flux moving eastwards drops from $26.10^6 m^3$/sec in October to $7.10^6 m^3$/sec in December.

During the northeast monsoon the current flows southward along the Somali coast. There is a complete inversion of the direction of the flux which modifies the hydrological physionomy of the whole region and ensures the arrival of warm water towards the coast in the very area of the summer upwelling.

Vertical profiles of the currents made on the Equator indicated the existence of an east-flowing under current. It is active only after the northeast monsoon comes to an end. It makes up a layer accessible to surface gears (purse seines) operating in depths of 50-120m. It seems that the under current is affected by a strong interannual variability, as its existence is associated principally with the winds blowing around the Equator during the previous months. Vertical profiles have illustrated in the deeper layers the existence of various jets moving in opposite directions. These subsurface jets recorded from 2° N to 2° S are stronger and last longer in the Western Indian Ocean than at the centre or the east.

Sub Surface Currents

Circulation in the Mozambique Channel is strongly influenced by the topography of the sea bed. The anticyclonic gyre situated in the north basin, is associated with a water convergence in the centre. The average depth of the mixed layer is 100 m, from July to September, during the cool season of the northern winter monsoon. Isotherm of 20°C is found at a depth of 180–200 m. In the channel's southern basin, a converging eddy was located. At 16°S–18°S, an inverted circulation (cyclonic) would seperate the two eddies during the southern summer. In the southern winter, the north eddy would spread towards the southern basin, as the cyclonic circulation was disappearing. Upwellings have been recorded west of Grande Comore Island, west of Cape d' Ambre (Madagascar), Cape Delgado (Mozambique) and Juan de Nova. This results in nutrients being brought to the surface waters. Similar upwellings around 25°S and 40°S were noticed in the south Mozambique Channel. Upwelling can also be found off Fort-Dauphin (Madagascar).

Hydrologically contrasted, the Seychelles and neighbouring banks are situated in a belt where monsoons are alternating. The continental slope extends along 1500 km. This area can be crossed by the south equatorial current and the south equatorial countercurrent. Resulting topographical effects cause seasonal changes in the surface layer. Upwelling was recorded in the southeast Mahe Bank in April-May, at the end of the northeast monsoon, while the south equatorial countercurrent is flowing strong. During the southwest monsoon (southeasterly winds blowing on the Seychelles) a similar phenomenon can be found off Saya de Malha banks, from 8° to 11° S and 59° to 60°E.

Two major characteristics of the area are; in the south part the divergence between the SEC and the SECC and the convergence, in the north, between the SECC and the NEC.

The divergence is observed at the boundary of two currents flowing in opposite direction (the SEC flowing west, while the SECC is flowing in an easterly direction). In fact there is no real form of upwelling which would cause cold spots of water on the surface. In the best case, a slight drop of in water temperature may occur. Usually, the warm mixed layer reduces to 20-30 m. The thermocline becomes dome-shaped, and very noticeable, the vertical thermal gradient is in excess of 4°C/10m.

The position of divergence varies from one year to the next according to the intensity of the northeast monsoon (north-westerly winds around the Seychelles) and also during a same season. It was located at 6°S-7°S in April-May 1984. In May 1983 it was placed at 8°S, south of the Seychelles, and at 7°S around the Chagos Islands. In April 1983, it was again recorded at 8°S, towards 55°E, whereas in May 1981 it was recorded further south, at 9°S. Thus the divergence can be considered to occur between 6°S and 9°S.

The convergence zone is situated north of the Seychelles. It is characterized by a thick mixed layer. In February, 1982 the thermocline was located at approximately 100 m depth and showed a vertical thermal gradient in excess of 4°C/10m. The convergences are the areas of concentration of forage biomass for high trophic level species, as well as drifting debris which have an aggregative effect on tuna species. Hence the period and meeting points of the surface current convergences become of great interest. The convergence of the SECC and the NEC takes place during the northeast monsoon around 2°S. However, during the rare periods that the northeast monsoon is particularly strong (as in April 1972) the convergence area may shift south as far as 4°S-5°S, north of Mahe Island, Seychelles.

Somali region is subject to a quick oceanic response to seasonal climatic changes. The most spectacular event is a strong coastal current, starting in May-June, which later flows away from the coast forming two eddies, one to the north of the Somali Basin, the other to the south. The consequent upwelling is visible soon after the beginning of the southwest monsoon. The surface water temperature changes suddenly from one week to the next during this period.

This occurrence may be observed on the weekly satellite maps. Coastal temperatures drop by 3°C. This phenomenon occurs generally during the first three weeks of June and lasts until September.

Rotary Currents in Bay of Bengal

Long bands of alternate smooth and rough surface water were observed in many places along the coast of Eastern and North-eastern India. In some cases these bands extended to the horizon. They varied in number from two or three to at least ten. Single band varied in width from an estimated 25 to 200 metres. The orientation was always parallel to the coast which is in the direction of the prevailing drift. By repeated bathythermograph observations and by noting the surface current, it was concluded that these are "Rotary currents" their flow is in the form of interlocking spirals with their axis in the direction of the prevailing current.

A similar rotary system called "convection cells" are caused by the wind blowing on the surface and are oriented in the direction in which the wind is blowing. Frequently sea weed and flotsom orient themselves along lines of convergence. Rotary currents cannot be the same as the convection cells, since they occur during the absence of wind, and have also been observed to be oriented at right angles to the wind. The speed of flow in rotary currents is estimated to be as high as 5 knots which is greater than that attained in convection currents. Both circulations on the other hand, appear to be similar interlocking cells in the surface layers of the sea.

This creates lines of convergence where the interlocking surface flows are towards each other and lines of divergence where the flows are away from each other. The vertical thermal structure shows deeper isotherms under lines of convergence and shallower isotherms under lines of divergence. The vertical thermal gradients have developed more layers depths and stronger gradients in areas of convergence than in areas of divergence. Rotary currents are believed to be developed in the turbulent region where off-shore tidal currents come in contact with the prevailing coastal currents.

In the region off the numerous mouths of the rivers, strong sets to the north and south caused by the tide running in and out of the river estuaries have been experienced. This set is felt as far as 50 kilometers from land due to the numerous river estuaries and particularly the constricting configuration of the upper reaches of the Bay of Bengal.

It is believed that the motion is to the north throughout most of the north band and to the south throughout most of the south band. The southern part of the rough area may be partially caused by churning, hence the higher wavelets. It is also believed that the surface features are displaced a little to the south due to wind and current. In any case the band circulation is alternately north and south.

In addition to the rotary cells there is a deeper effect. A slow displacement takes place to deeper levels which extends to about 125 feet in this area.

The entire surface layer is moving towards the west. The particles of water in the rotary cells will then describe as clock wise and counter–clock wise spiral.

Upwelling and Sinking Off the East Coast of India

The phenomenon in the sea of subsurface layers of water moving upwards towards the surface is known as upwelling. Such a process may occur anywhere in the sea but does so most commonly in regions where the prevailing winds blow at such a angle with the coast that surface water is displaced away from the shore, and low temperature subsurface water moves up from the depths to take its place. Upwelling is easily detected in a vertical temperature section, as the isotherms are titled upwards in the zone of maximum upwelling.

In some regions the prevailing wind blows in such a direction that the surface water is concentrated or piled up, usually against the coast. This displaces the subsurface water downward, and the process is called sinking. Here the isotherms of a vertical temperature section are bent downward.

In the east coast of India off Visakhapatnam, September and October are months of sinking. The surface manifestations of sinking are relatively high temperature, low salinity and high sea level. In October the current flow is south-west with a speed of 2.5 knots. The low salinity is caused by discharges from the large rivers (Ganges and Brahamaputra) which dilute the surface layer and flow down the coast. High temperatures are due to summer heating and the warm rivers during this early fall season. The wind at this time of the year deflects the surface water towards the shore, and piles it up there, this displacing the thermocline downward and causing sinking. As a result, the surface water of lowest salinity is largely restricted to the coast. Its relatively low density also causes it to remain at the surface. This is the season of highest sea level which is further indicative of warm low density water and is one of the most important factors in increasing sea level in this locality.

In November and December the surface temperature falls off sharply, due to the beginning of winter cooling at the surface, and the salinity increases due to the decrease in river runoff and consequent dilution. When the surface cools and becomes more dense than the water below, some convective mixing with subsurface layers also take place. Thus the surface water is combined with water of higher salinity. There appears to be a lesser concentration of low salinity surface water next to the beach, with isohalines running at right angles to the coast, rather than parallel as in the preceding months.

In January the surface temperature is at a minimum, although the salinity increases. These changes are caused principally by vertical mixing. At the end of January and throughout February both salinity and temperature increase markedly. This is attributed largely to the reversal in current direction at this season of the year (with current now to the north-east) which brings warm high saline water from the equatorial region. At the same time the SSW wind causes as offshore drift of surface water. As colder subsurface water rises to take its place, upwelling starts. The denser water is reflected in a lowering of sea level.

The north-east current and the offshore drift continue in this season. By the end of March the surface layer has become reduced. Inshore it disappears entirely, and the thermocline extends upwards to the surface, indicating the development of the upwelling process. This upwelling continue throughout April and part of May, whereas it stopped or appear to stop in April next year. In this season with the maximum amount of dense water occurring near the shore, the sea level is minimum.

In May and June the currents are still north-east but flow at a slower speed. Now the summer heating has offset the upwelling and the surface again becomes heated near shore. The salinity remains high. Although no subsurface temperature data are available for this time of the year, it is reasonable to assume that the upwelling has been reduced, because sea level also rises. This is evidence of the presence of warm low density water.

In July the surface temperature falls, reaching a minimum around the first of August. The salinity remains high, although this is the season of south west monsoon. The cause of this low temperature period is not apparent. It is associated with a seasonal low in sea level, maximum rain and a change in current direction. It seems likely that the cold water at this season is not from the rain but from sub-surface water again coming to the surface.

This is substantiated by the offshore direction of currents which, though allegedly flowing from the north and south are deflected out to sea at this time of the year. Further evidence of possible near shore upwelling is the shallow depth of the thermocline on the western side of Bay of Bengal, observed in August.

Environmental Factors Affecting Vertical Temperature Structure of the Sea

Off the east coast as with other seas, the temperature of the sea is not uniform, but varies in space as well as in time. The resulting changes of temperature with depth are known as "vertical temperature gradients". Major gradients are most common in the upper layers where external environmental influences predominate. Temperature gradients in the sea are important in explaining many physical, chemical and other ecological phenomena.

The numerous environmental factors which contribute to the production, or elimination of vertical temperature gradients in the sea may be divided into two groups (a) energy transfer processes at the sea surface and (b) advective transfer processes below the sea surface. Those factors concerned with the transfer of solar radiation to the sea result, by selective absorption, in highest temperatures nearest the surface, thus producing "negative" temperature gradients. Conversely the removal of heat from the sea tends to eliminate negative gradients and produce slightly "positive" gradients with higher temperatures below the surface.

The second group of factors producing temperature gradients in the sea is concerned with the external and internal forces which cause movement and displacement in the water without a transfer of heat to or from the atmosphere. These forces produce tide, current, wind etc. which transport water from one place to another. If the advected water differ in temperature from the adjacent water in its new environment, a temperature gradient, vertical or horizontal will be produced. Certain geographical and density boundaries will influence the path of the transported water, and thus influence the gradients.

Temperature Gradients Resulting from Energy Transfer at Sea Surface

The most common factor in the production of thermal gradients near the surface is simply the transfer of heat at the sea surface. The processes of heating the sea surface are the absorption of radiation from the sun and the sky Qs; convection of sensible heat from atmosphere Qh; and condensation of water vapour Qe. These processes produce negative gradients near the surface. On the other hand, cooling or reverse processes, have the opposite sign. Back radiation from sea surface, Qb, convection of sensible heat to the atmosphere, and evaporation reduce negative gradients and sometimes produce slightly positive gradients near the sea surface.

A combination of these factors gives;

$$Qs - Qb - Qh = Qr$$

where, Qr is the heat transferred to the water at the surface to produce temperature gradients.

Radiation

The radiation available to heat the sea surface (total radiation minus back radiation) depends upon the altitude of the sun, absorption in the atmosphere and cloudiness. Under condition of high radiation, *i.e.,* during the sunlight hours or in summer, heat is transferred over a period of time from the atmosphere to the water. This heat added to the water is absorbed mainly at the surface and consequently develops negative temperature gradients. The magnitude and depth of gradients are proportional to the amount of heat added, mixing etc.

When the solar radiation reaching the sea surface is low *i.e.* at night or in the winter, negative gradients are so reduced that slight positive gradients are frequently formed. At these times heat has been transferred from the water to the atmosphere.

Heavy cloudiness in the sky reduces the solar radiation reaching the sea, and consequently interferes with the development of temperature gradients. When a light cloud coverage is present, relatively more heat is transmitted, resulting in stronger negative gradients. Of course, the largest surface gradients are produced by this effect when the sky is clear.

The cloud factor in the formation of temperature gradient is mainly confined to the surface layer. However, when long periods of clear or cloudy weather persist, the effect will extend, by absorption and diffusion to deeper layers.

Clouds not only reduce the amount of solar radiation reaching the sea but also act as a blanket, or "green house" effect at night to reduce cooling of the sea surface.

Cloud types vary in thickness and density and thus in the amount of heat they reflect and transmit. Low clouds, such as stratocumulus, are denser, and reduce the production of surface temperature gradients more than do the same amount of high cirrus clouds.

Convection

The convection of sensible heat to and from the sea surface depends largely upon temperature difference; sea surface temperature minus air temperature. This temperature difference between the sea and the air produces temperature gradients by convection. When the sea is warmer than the air, weak positive gradients are formed near the sea surface. These gradients are usually 0.1 to 0.3° F. When the air is warmer than the sea surface, negative temperature gradients develop at the surface. The magnitude of gradients produced by convection, over a period of time, is proportional to the temperature difference between the sea and air, other variables being constant. However, the total heat transferred to the sea is generally small, due to low specific heat of the air and the stability of both air and sea which develops under these conditions.

Evaporation

Evaporation, or the reverse condensation depends on the vapour pressure difference, that is, vapour pressure at the sea surface minus vapour pressure of the air. When the vapour pressure at the sea surface is greater than the vapour of the air above, heat will be removed from the sea to the atmosphere by the process of evaporation, resulting in a slight positive gradients in the surface. Conversely, if the vapour pressure above surface is greater than at the sea surface, heat will be added to the water by condensation. This results in negative temperature gradients at the surface. It is possible for convection and evaporation to be acting in opposition to each other.

Rains

Water warmer or colder than the sea is transferred from the atmosphere to the sea surface through rain. Positive gradients may occur during rainy weather when a heavy rain is at a lower temperature than the sea. However, if the heavy rain is warmer than the sea, negative gradients may occur. In all cases the effect is believed to be only temporary and minor and is confined to intermediate surface layer.

Temperature Gradients Resulting from Advective Processes Below Sea Surface

When variation in currents causes lateral or vertical displacement in the sea, the resulting water masses are redistributed and temperature gradients are affected. These water movements or advective factors may be further broken down into wind characteristics, current types, boundary influences etc.

Wind Force

Surface temperature gradients vary inversely with the wind force. When the wind force is "O" Beaufort, heating may occur at the surface, resulting in large negative gradients. As the wind force increases, the surface layers will become mechanically mixed, thus producing a homogeneous layer which is devoid of temperature gradients to the depth of mixing. With increasing wind force the mixing will extend deeper. The depth of mixing will also depend on the duration of the wind, its fetch, and upon the density gradients in the water. With the wind force of "4" to "5" Beaufort (12 to 21 miles per hour) the mixing usually extends to at least 30 feet.

When a wind mixed water coloumn is heated at the surface and is again remixed by a weaker wind, steps will be formed in the temperature depth curve. These successive mixings and heatings are common in spring time, and result in a series of negative gradients and isothermal layers. When the wind produced discontinuities are first formed they are sharp. Over a period of time diffusion reduces the sharpness of these gradients, and the seasonal thermocline thus developed appears as a continuous gradient.

Turbulence

Temperature gradients may be influenced by surface and subsurface turbulent mixing of the water column. At the surface, wind and waves produce an isothermal layer. However, below the surface, mixing can also take place where currents flow over rough sea bottom. The magnitude of the turbulent layer depends upon speed of currents and roughness of bottom.

Currents relative to some depth where motion is assumed to be negligible are controlled by the horizontal change in average vertical density of the water and coriolis force. In the northern hemisphere the water on the right-hand side of the current is lighter, and, consequently usually has a deeper surface layer than the water on the left-hand side of the current. In the southern hemisphere the lighter water is on the left-hand side of relative currents. These are characteristic of the California and Humbolt currents respectively.

In regions where one type of water flows over and displaces another type, gradients will be formed at the boundary between the two waters where cold water of low salinity overrides warmer water of higher density, a positive temperature gradient will occur at the boundary. The converse occurs when the warmer water flows over and displaces colder water. These types of displacement are common along the southern boundary of the cold Oyashio Current off the Kuril Islands and the eastern terminus of the Equatorial counter

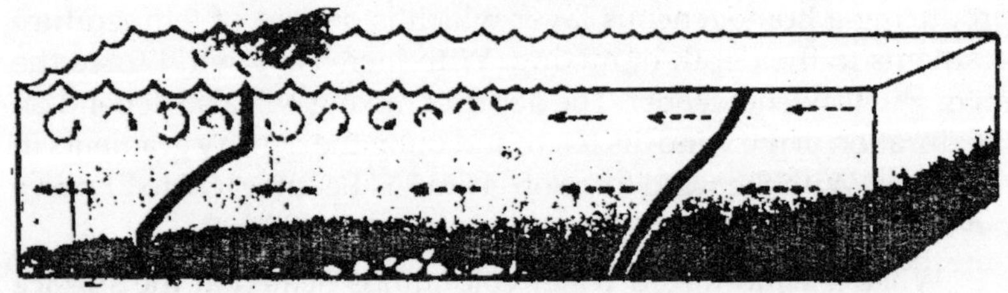

Turbulence

current off Central America. Gradients thus produced are usually extremely sharp.

When the wind blows over the surface of the sea for long periods the surface water, in the northern hemisphere, drifts slightly to the right of the direction in which the wind is blowing. This results in the piling up of a deeper layer of light surface water more or less in the direction towards which the water is moving. A shallower surface layer develops in the opposite direction. The best example of drift currents is in the equatorial region where the north-east and south-east trade winds transport surface water from east to west. In this region the transported water forms a progressively deeper surface layer in the direction of the current.

Current Boundaries

When adjacent currents of different temperature characteristics flow in opposite directions or at different speeds, the boundary between the currents will be made up of numerous eddies. In the boundary layer many irregular and complex gradients will be produced. A good example is the boundary of the Gulf Stream and Slope current. Here the surface layer is shallow at the current boundary, and increases with depth to the right of the current. The current boundary consists of complicated positive and negative gradients.

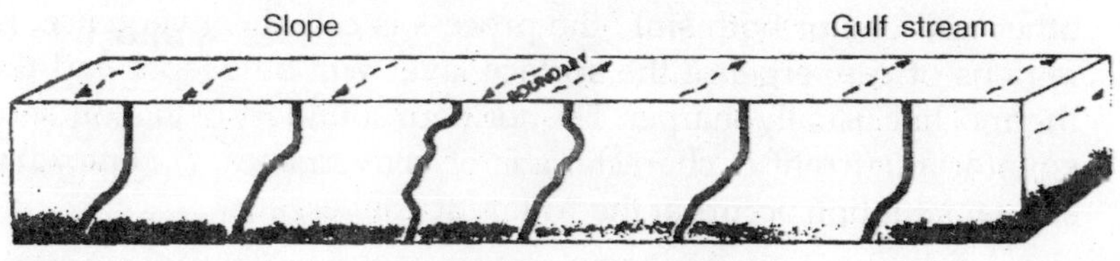

Current Boundaries

Divergence

When surface water is blown or flows away from an area, and allows subsurface water to ascend to the surface, the process is called divergence. It is characterized by a surface layer more shallow than adjacent water, and temperature gradients are usually weaker in the thermocline. Divergences are frequently found between opposite flowing currents, on their left side in the northern hemisphere, right

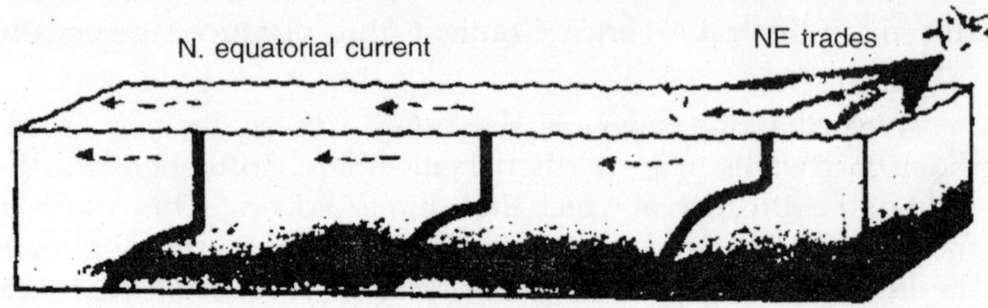

Drift Current

Divergence

in the southern. The southern boundary of the north equatorial current is a region of divergence.

Convergence

When two water masses of surface water flow towards each other and one or both sink, the process is called convergence. In regions of convergence the surface layer will be deeper and the thermocline usually sharper. The northern boundary of the southern equatorial current is characteristic of convergence. A somewhat similar situation occurs at the Antarctic convergence.

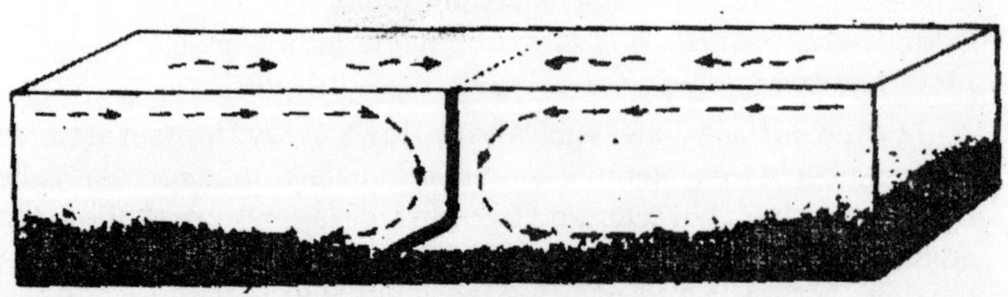

Convergence

Convection Cells

Convection cells are believed to be a series of convergences and divergences on a smaller scale. These cells are occasionally produced by strong winds during a period of surface cooling. The axis of the cell will be parallel with the direction of the wind. The diameters of the cells are approximately equal to the depth of the mixed layer, but the depth of the mixed layer may vary on either side of the cell. The thermocline will be deeper when the circulation is downward, shallower where it is upward. Regions of convergence is characterized by surface slicks. A similar long interlocking series of pronounced cellular circulations is found in the Bay of Bengal. There also, the thermal structure below lines of convergence gives deeper surface layers with sharper gradients than the structure beneath lines of divergence.

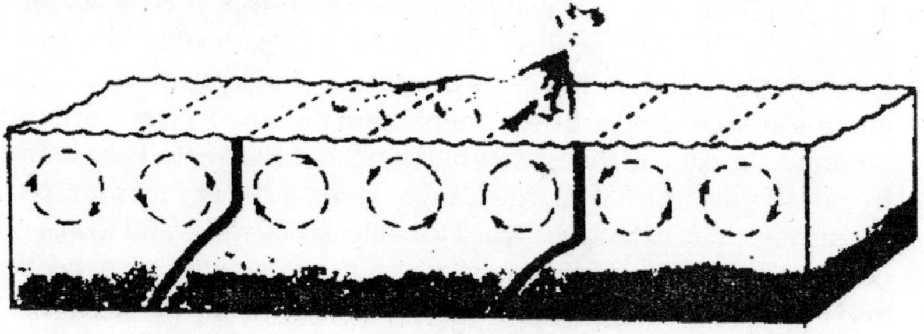

Convection Cells

Internal Waves

Internal waves occur at the boundary between density layers. These waves are universal and vary widely as to height and period. Their amplitude may be as great as several hundred feet, and their period from a few seconds to many hours or longer. The layer depth of an internal wave is deeper in the trough than in the crest.

Tide

Layer depth is influenced by the tide. A semi-lunar "internal tide" has been observed. The phase of the internal tide varies with both time and location. In the majority of cases a shallow surface layer, or internal high tide, frequently occurs three to five hours after high tide at the surface, and a deeper surface layer internal low tide occurs three to five hours after low tide at the surface.

Land Boundaries

The proximity to shore influences the vertical temperature structure. This is believed due to the presence of weaker winds and the greater absorption of heat near shore. The waters near shore are more turbid, contain more micro-organisms and consequently absorb more radiation. This results in stronger negative gradients in summer. The phenomenon is more pronounced in lower latitudes and where discharges enter the sea. The reduced winds, as well as, greater thermal gradients near shore, are believed to result in less mixing in winter. Consequently the layer depth near shore is usually less than the layer depth further out at this time.

River Run Off or Surface Dilution

The fresh water discharged into the sea from rivers is most always lighter than sea water and flows on top. If river water is colder than the sea, a positive gradients will result at the boundary, if warmer, a negative gradient is produced. The depth of the gradient and the area covered are dependent upon the size of river discharge, amount of mixing in the estuary and longshore currents. For example in the fall season, off the large Ganges river, a 10 per cent dilution of sea surface may extend sea ward for at least 500 miles and to a depth of over 150 feet. The diluted surface water may be subjected to heating and cooling causing negative and positive gradients in the surface layer, but still retain a vertical stability.

Upwelling

When the surface water is carried away from shore by wind drift, deeper water rises to take its place. This type of vertical motion of subsurface water towards the surface is known as upwelling. The

Upwelling

phenomenon is similar to divergence. However, upwelling usually takes place near a land boundary, and the upwelling water flows in one direction when it reaches the surface. Since the upwelling water comes from lower depths, it will most always be colder, with weaker gradients than in adjacent regions where no upwelling occurs.

Sinking

The opposite of upwelling is known as sinking. This phenomenon occurs when the wind blows in such a direction as to cause the surface water to be transported towards a coast. In this case the deeper surface layer is built up nearer the coast and immediately off shore. A sinking region is characterized by deep and usually sharp thermoclines. Such a temperature structure is typical of the east coast of India during the north-east monsoon.

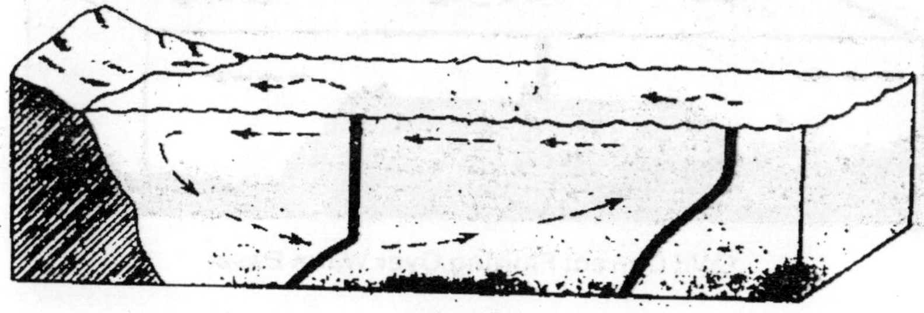

Sinking

Islands

Variations in the temperature gradients occur in the vicinity of islands. This land boundary will act as a barrier resulting in the "piling up" of surface water on the windward side and the "removal"

Islands

of surface water on the leeward side. On the north-east, or windward side of the Hawaiian Islands, for example, the layer depth has been found to be deeper than that on the opposite side.

Basins

Sea water temperatures in a sub-surface basin remain relatively constant. However, the water above sill depth may change with tidal or other currents. A sharp temperature gradient may be formed at the sill depth if the water which flows over the sill has a different temperature than the basin water. These gradients may be either positive or negative, depending upon the type of water transported across the basin.

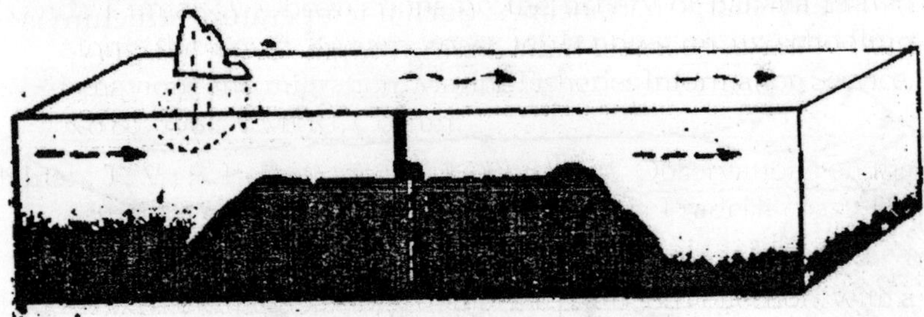

Cold Current Flowing Over Warm Basin

Warm Current Flowing Over Cold Basin

Time Cycles of Temperature Structure in the Sea

In addition to internal waves and tidal influences, longer period diurnal and annual cycles occur. Temperature structure here is a periodic function of time. The factors causing these cyclic changes are, however, numerious.

Three-hour periods are used for the diurnal temperature cycle diagram. Since diurnal heating and cooling are primarily confined to the upper 30 feet, enlargements of this area are shown. The illustrated case is one in which no residual heat is added to or subtracted from the water after the 24 hours diurnal cycle.

The periods chosen for an annual cycle are three months intervals, corresponding to the seasons. Winter is taken as late as early February, spring as early May, summer as early August and fall as early November. The extreme maximum and minimum values of temperature in winter and summer, respectively occur a little later in higher latitudes than in the low.

Diurnal Changes

The characteristics of diurnal heating near the surface depend upon the diurnal changes in the factors which lead to the heating of the water. The most pronounced diurnal heating occurs at mid-latitudes during the summer. Here the water is frequently isothermal near the surface at midnight (0000), becoming slightly positive by 0600 in the morning. By 0900 heating has again returned the water to an isothermal condition, by noon negative gradients appear at the surface and by 1500 the surface heating reaches a maximum. After this the surface starts to cool, resulting in an isothermal layer over a weak negative gradient. The subsurface temperature maximum is reached a few hours after the surface maximum. Cooling continues until, by midnight, the upper layer is again isothermal. Variations of this cycle depend upon location, season, and local meteorological conditions. During windy weather, of course, the surface layer becomes mixed and such a cycle cannot occur.

Low Latitude Seasonal Changes

Throughout the year, in the low latitudes, trade winds maintain a deep surface layer of high temperature water over a permanent thermocline. Although solar radiation is high, the heated water is carried away by currents as fast as it becomes warm. On the equator in mid-Pacific the sun radiation undergoes a semiannual cycle with the greatest intensity in spring and fall. This and probably current variations, produce deeper layers of warm water shortly after the sun passes through the vernal and autumnal equinox.

Mid-Latitude Seasonal Changes

The combination of factors controlling the vertical temperature structure in mid-latitude, around 30°N, causes the upper 500 feet of water to become isothermal in winter. By spring small steps develop at varying depths during this period of heating and mixing. However, the deeper water is actually colder, in spring, due to an apparent advection of colder water from below. In summer, large negative gradients occur at the surface, caused by more heating, diffusion and milder winds. Cooling and increased winds in fall produce the characteristic mixed layer over a sharp thermocline.

High Latitude Seasonal Changes

In the high latitudes, just north of the Aleutian Islands the temperature structure during winter is virtually isothermal to depths greater than 400 feet. Slightly higher temperatures around 500 are attributed to subsurface advection which is characteristic of this region. By spring the temperature depth curve begins to show steps, indicative of successive heating and mixing of the water. By summer these steps become more numerous and by diffusion, merge into the seasonal thermocline. Due to frequent winds in this region, the intermediate surface layer usually remains isothermal even though heat is being added to the water. In the fall the isothermal layer becomes deeper, caused by increased mixing and cooling.

Arctic Seasonal Changes

In the Arctic, north of the Bering Strait, the surface is covered by ice in winter. The cooled surface water creates vertical convection currents which produce isothermal structure. By spring the ice becomes thicker and the water remains virtually isothermal, the minor gradients are caused by subsurface advection. In the summer the surface layer is heated. In addition, warmer water is advected into the region. By fall the surface again freezes, the cooling and mixing processes create an isothermal layer over a slight positive gradient.

In 1979, the maps indicated the presence of two cooler areas, each of them representing a deviation of the existing water flow now heading away from the coast, approximately at 4ºN and 8º N which was often observed at these latitudes. The lowest surface temperatures (17ºC in 1979 and 13.2º in 1964) were recorded for the north Somali coast. These values are usually found between 100 and 200 m. depth.

Convergences are created at the centre of the eddies. The thermocline is pushed down to around 150m depth at latitude 9ºN–

10°N and to about 100m depth in the south eddy. This thick mixed layer lasts as long as the southwest monsoon. Traces of it can be found also at the beginning of the northeast monsoon in December.

The inversion of the Somali current that is influenced by the northeast monsoon generates advection of water towards the coast.

From May to September the current flows along the Soudi Arabian coast and starts a gyre in the northernmost part of the Arabian Sea, along the Pakistan coast. This generates an upwelling alongside Oman which is felt 200 to 400 km away. But on the other hand, the fall in temperature is slight off Pakistan. These two upwellings are not as important as the Somali one. However the situation observed in the Arabian Sea, namely an upwelling reaching far out to sea and moderate currents, would be more favourable to biological production than off Somalia. There seems to be an eddy with a convergence (thermocline at 80 m depth) between the two areas of upwelling (Oman and Pakistan).

Around Minicoy (8°N, 73°E) an upwelling is caused by a divergence of currents near the island, in November.

An upwelling in conjunction with the southwest monsoon whose maximum intensity would be reached in July-August was observed along the southwest coast of India.

An upwelling is formed in the Bay of Bengal on the east coast of India, mainly along the Waltair coast (16°N–17°N) between January and July, while a coastal current flows northeast. An inverse situation arises in September December, the current bears southwest, surface water is driven inshore. Another upwelling, much weaker than that described above, occurs before the summer monsoon, off Tamil Nadu (13°N) in southeast India.

An upwelling was reported to occur along the west coast of Thailand in the Andaman Sea when the east wind blows from December to April. Since the upwelled cold water remains at subsurface levels, a strong vertical stratification of the water occurs (9°C difference between 0 and 100m depth). When the westerly winds blow from July onwards, the warm waters reach the coast and at that point, the difference in temperature between the surface and the 100m depth is only 3°C.

The mixed layer gradually thickens along the Equator between 80° and 100°E. Given that the currents flowing east in the equatorial region are noticeable only during the inter-monsoon periods. It is during these and the two following months that the thermocline's maximum depths were observed. In May-June, the thermocline exceeds 100m depth over a large area (2°N to 2°S and 80°E to 100°E). In November–December it is along the west coast of Sumatra where the thermocline deepens significantly. The vertical thermal gradient exceeds 5°C/10 m from May to August in this area.

In July, convergences were observed along the Equator and at 5°S. A divergence was created in the middle position (2°30′S). In August, a single convergence remained along 2°S. During these two months, the thermocline was located around 100m depth with a strong vertical thermal gradient in excess of 8°C/10m.

The two currents originating in the northwest coast of Australia are the Leuwin current which flows along the coast in a southerly direction and soon leaves the tropical zone and the south equatorial current.

A west bound flow probably originate in the north (Indonesian archipelago) before being diverted westward by the Australian continent.

An upwelling was recorded at 12°S (between Australia and Sunda Islands) in winter. This is the most noticeable seasonal variation in a region which generally shows a relatively constant hydrological pattern.

Surface Temperature

Surface water temperature remains stable over a large area of the tropical Indian Ocean and particularly during the northeast monsoon. Surface temperatures near the Equator often exceed 28°C. The upwelling along the 0 man and Somali coasts causes water temperatures to drop considerably. Thermal fronts delimit these upwelled waters. Beginning in June, upwelling last until August in Oman and September in Somalia. They are sufficiently strong and regular from year to year to appear on maps averaging 60 year of observations. Other areas may cool through wind action and heat exchange processes with atmosphere. During the north east monsoon, the thermal Equator can be found between 5° and 15°S and from 0° to 10°N during the southwest monsoon.

Surface temperature analysis on a shorter time basis (one week for example), show great movement for some regional patterns. From January to April, from the Equator to Madagascar, warm water pools (more than 29°C) appear and thermal fronts are created at the boundary. However, it seems unlikely that these quick developing warm water pools can generate a trophic chain due to their short duration and to the high temperatures registered. From June to September, during the Somali upwelling and winter cooling process which affects the southern hemisphere, regular frontal zones are found from 5°S to 5°N, north of the Seychelles (55°E) and with more important inter-annual variations, similar structures ranging from 5°S to 10°S between 65°E and 75°E. The relevant temperatures (25° to 27°C) are a border line between a warm water pool (28°-30°C) between 0° and 5°N and colder waters originating along the east African coast or in subtropical regions.

Temperature Variations in Western Indian Ocean

The seasonal temperature variation decreases from the coastal sections to the sections situated further offshore. The highest temperatures are similar in every section (28.5°-29°C), while the lowest temperatures are recorded near the African continent (less than 23°C along the Tanzanian and Kenyan coasts). The waters comprised in the sections situated in the centre of the Western Indian Ocean basin remain warm all year long. The more southern waters reach maximum surface temperature from January to May and minimum temperature in August.

The section bounded by 0°-4°N and 40°-70°E is affected by two cooling events (double cycle of cooling heating). In August, the drop in temperature could be attributed to wind and to an upwelling water dispersion mix which taxes place to the north of this section.

Surface Salinity

Surface salinity helps identify the origin of different water masses. The strongest haline discontinuities are observed in the Northern Indian Ocean. The Arabian Sea encompasses highly saline waters (from 35.5ppt to more than 36.6ppt increasing towards the north) due to the excess of evaporation over rainfall. On the other hand, in the Bay of Bengal where heavy rainfall is recorded in summer and where the water volume from bordering rivers is also high, salinity is low (32ppt in the most northern part of the Bay). Surface currents drive these water masses and haline fronts are then created. In January-February the north-east monsoon drift appears as a low salinity tongue-shaped band (less than 35ppt) as far as 60°E. The south equatorial current disperses its low saline content waters from west Sumatra to the African coast in March-April. Low surface salinity is maintained by heavy rainfall occuring in the intertropical convergence zone (ITCZ). In November-December, the waters flowing from the Arabian Sea reach their most southerly point, towards the Equator. Another tounge-shaped band with higher saline content is found along the Somali coast in January-February in the Somali current.

Dissolved Oxygen

The surface waters are slightly oversaturated in the tropical part of the Indian Ocean (from 4.7 to more than 5.0 ml/litre). Lower levels are found along the north-west coast of Sumatra (less than 4ml/litre). On the other hand, from a depth of 100m, the levels vary greatly from the Northern Indian Ocean (Arabian Sea, Bay of Bengal) where less than 2 ml/litre are registered to the Southern Indian Ocean where the content increases from 3ml/litre at 10°S to 5 ml/litre at 20°S. The minimum oxygen layer (2.8 ml/litre on the Equator) becomes deeper and thickens from 60 to 150 m on the Equator to 100–300 m at 20°S.

These variations in dissolved oxygen underline the difference between the Southern Indian Ocean, which is an "open" ocean where levels are greater than 4ml/litre, and the Northern Indian Ocean (above 10°S) which is "closed" by the African and Asian continents, and where waters are renewed slowly (approximately 50 years for a layer ranging between 75 and 1200m). Dissolved oxygen levels are lower than 3 ml/litre except for the Somali coast where the water is renewed in oxygen by the upwelling occurring during the south-west monsoon. In the Northern Arabian Sea, in winter, surface waters sink and help restore oxygen to subsurface layers. The low dissolved oxygen level found in the Arabian Sea are due to excessive consumption, principally because of organic debris oxidation, and to the loss of oxygen along the way of the water flow responsible for oxygen regeneration. Currents play therefore an important role in the redistribution of dissolved oxygen. The south equatorial currents brings water relatively rich in dissolved oxygen to the African coast, then, the Somali current drives that water northwards while regenerating in oxygen, during the south-west monsoon. The sudden drop of dissolved oxygen levels in the Bay of Bengal largely depends on the thermohaline stratification which reduces the vertical exchange processes.

Surface Salinity and Temperature Variations in Bay of Bengal

The waters of the Bay of Bengal is of great interest owing to such opposite influences as the south-west and north-east monsoons and the north-east coast is of particular interest as most of the large rivers of Indian, such as, the Ganges, Mahanadi, Godavari and Krishna open into the sea on this coast. As a result of the monsoons and the influx of large volumes of fresh water brought down by the great rivers there are wide fluctuations of both surface temperature and salinity in the different months of the year. Added to the influences of the monsoons and the rivers, there are two opposite strong currents skirting the coast, a southwesterly current from August to November and a north-easterly current from middle of December till July. These two currents also play important roles in the distribution of surface salinity and temperature fluctuations.

Observations of surface temperature and salinity estimations from 48 stations collected from the inshore and off shore waters off Visakhapatnam, extending between Lat. 17°03' N and 17°14' N and Long 83°18.2'E and 83°57.6'E reveals that, October is the transition month between the south-west and north-east monsoon. In the northern and central portions of the Bay the south-west monsoon diminishes rapidly. Northerly winds of north-east monsoon first appearing at the head of the Bay extends further to the south. There are two seperate counter-clockwise currents one off the Orissa coast and the other off the Coromandel coast confined to the western half of the Bay. The current off the Orissa coast moves in a southerly direction along the Visakhapatnam coast and merges with the other off the Coromandel coast. The strong southerly current has a maximum flow around 15 miles out.

The average salinity for the month in the different 5 miles sections from the inshore to the off shore waters upto a distance of about 50 miles ranges from 18.26 to 28.59ppt. The minimum salinity of 17.83ppt is in the 5 to 10 mile zone. There is a steep rise in salinity from the 20 mile line located 50 miles offshore, the salinity is 28.59ppt. The temperature of the surface water is also lowest in the 5 to 10 mile section and there is a steep rise in the 10 to 20 mile section after which it remains steady within the 82°F to 83°F range. The low salinity and temperature in the 5 to 10 mile zone off the coast is due to the prevailing strong southerly current which brings from the north large volumes of fresh water from the Brahmaputra, Ganges and Mahanadi, which are still full after the south-west monsoon. It has also been observed that during the month there is a general sinking of surface water on the continental shelf.

The north-east monsoon has fully become established over the Bay during December. The currents are the same as in October with an increase in the westerly sector to about $2^1/_2$ knots within the 14 miles from the coast, beyond which it decreased to 1 knot. The low salinity waters ranging from 26 to 26.36ppt are restricted to a coastal belt extending to about 7 miles offshore and after this zone there is a steep rise in salinity to about 32ppt in waters 20 miles offshore. The southerly current has considerably weakened in the offshore waters. The surface waters have considerably cooled owing to the general cooling by the atmosphere at this season. From the inshore to the offshore waters the temperatures ranges from 75.9° to 81°F. This may be partly due to cooler water discharged by the northern rivers at this season.

The month February is characterised by the retreat of the north-east monsoon and the weather is more stable than in the previous months. The current during the month is in a north-easterly direction at a short distance from the land between Sacramento shoal and Bhimilipatnam. From January to April both inclusive, it attains its greatest rate which is sometimes as much as 4 knots in the vicinity of Sacramento shoal and about 5 miles from the land. Between Kakinada and Bhimilipatam it extends farther offshore with a less rate; inshore of this current there is generally slack water.

The salinity ranges from 32 to 33ppt within the 5 miles zone which is maintained steady upto a distance of 30 miles offshore. There is an increase of both surface temperature and salinity which may be due to the inflow of high saline waters from the mouth of the Bay and from the equatorial regions along the eastern coast.

During March the retreat of the north-east monsoon continues and the currents are the same as in February. A general upwelling has been noticed all along the upper shelf especially upto 10 miles from the coast. There is a north-easterly current with a maximum flow between 7 and 22 miles offshore.

There is a lowering of salinity and increase of temperature with distance from the coast upto 10 miles range. Beyond the 10 mile line both salinity and temperature remain steady.

General Features of the Malabar Coast

The Malabar Coast, which is more or less straight without any large bays, is characterized by a number of short rivers joining the sea after running through the comparatively narrow belt of low land seperating the sea from the Western Ghats. The plains over which these rivers flow consist geologically of laterite formation fringed on the seaward side by a narrow belt of recent alluvium. These rivers discharge into the sea every year immense volumes of water which bring with it large quantities of organic and inorganic substances collected along its course.

The shore along this coast is mainly sandy, but there are some low cliffs or reefs of laterite here and there as at Cannanore, Tellicherry, Elathur and Beypore. The submerged sea bottom is predominantly muddy and devoid of weeds with very few rocky and sandy patches. With a few unimportant exceptions the bottom from about a mile from the shore out to 30 fathoms from Ponnani northwards to Mangalore is composed of soft dark grey mud with small dead shells, chiefly bivalves, more or less sparsely distributed through it. As a rule the sedentary fauna of this area outside the 10 fathom line is extremely scanty and algae are of course entirely absent.

A special feature of the Malabar and Cochin inshore sea bottom as distinct from other mud bottoms is the presence of what are called "mud banks" scattered at various places along the coast, namely, Alleppey Narakal, Calicut, Quilandi, Pantalayani, Beypore, Munambam, Chellanam and off Cochin. These have been known for a very long time as areas where the sea remains calm even when the roughest weather prevails and the water is very rough in the neighbouring areas. The fine mud of such banks is in an unconsolidated state and the banks themselves have been known to shift from place to place now and then. A great agitation of the bottom mud was noticed during the rough weather season in many parts along the coast and judging from the regularity of the south-west monsoon here, such on agitation seems to occur every year though variable in intensity and extent.

Climatically, Malabar has four distinct periods in the year. From June to the beginning of September is the characteristic south-west monsoon season with strong westerly winds, rough weather and heavy surf on the sea shore. There is heavy rainfall, especially in June and July. September and October constitute a transitional period with a fairly dry weather but occasional rains. Next comes the cool north-east monsoon period from November to February with land winds and a calm sea. In February the wind veers to north and from March to May come the hot, dry months with continued calm sea and light breezes from the north-west. By the end of May transitional conditions set in and in June the south-west monsoon is in full swing again.

The features of the sea bottom, physico-chemical as well as biological form aspects of the environment of marine fishes, the fauna of the inshore sea bottom of the Malabar coast reveal certain marked features. The most important of these is that the animals are not uniformly distributed at all the depths even within the limited inshore fishing areas, there being a zonation in vertical distribution both qualitatively and quantitatively although the nature of the bottom is practically the same. The reasons for the existence of such zonation may be due to the differences in the complex of environmental conditions, such as, salinity, temperature and pH ranges, the incidence of light, the incidence of currents and turbulence effects on the bottom, between the deeper and the shallower zones. The deeper region has a more stable environment in that the range of variation of these factors is more limited there than in the shallower regions.

The 10, 8 and 6 fathom levels formed one zone, while the 2 fathom and the near low water levels formed another zone, the intervening 4 fathom region being very poor in fauna.

The zone in the shallower region is rich in fauna during the premonsoon months and is dominated by polychaetes and phoronids. During the monsoon months this rich belt disappear altogether while the fauna of the deeper levels also decline.

Recolonization of the shallow region start when the post monsoon conditions are fully established but is very slow and the density of the fauna is very low even by the end of November. Among the factors determining the behaviour of the fauna, the nature of the bottom material is probably the most important. The formation of the fauna being regulated and cyclical wherever there is a relatively stable mud at the bottom but showing considerable fluctuations wherever there are mud banks.

Chapter 2
Biological Resources

Oceanic resources include species that inhabit beyond the continental shelf, although they may spend part of their life cycles in the coastal areas. Oceanic resources are marine animals living in the epipelagic, mesopelagic and bathypelagic zones in the oceanic region. Exploitable species living in these zones are fishes, crustaceans, cephalopods and marine mammals. Fishes have the greatest importance both in number of species and in terms of fishery revenues. Out of approximately 25000 species of fishes about 325 are epipelagic, representing 1.3 per cent of the total. Meso-pelagic and bathy-pelagic fishes comprise about 1250 species, corresponding to 5 per cent of the total. Rest 92.7 per cent are distributed in the neritic environment over the continental shelves.

Though fish species living in the oceanic region have been classified as either epipelagic or deep water species inhabiting the meso- and bathy-pelagic zones, it is difficult to assign species to one of the two categories, since several species undergo vertical migrations in relation to feeding, reproductive season and circadian rhythms. In such cases, species have been classified on the basis of the zone in which they are usually caught by commercial fisheries.

The reason for categorizing epipelagic and deep water species among the oceanic resources is that the fisheries targeting the two groups of species are often different in terms of importance, technology, history and value. The valuable and still developing fisheries for tuna and tuna-like species constitute the bulk of the fisheries targeting epipelagic species, although other epipelagic resources, such as, cephalopods, short lived with a rapid turn-over, might sustain expanding oceanic fisheries, being able to respond promptly to favourable environmental changes.

On the other hand, most oceanic deep-water resources are very dispersed and difficult to harvest and several fisheries on these resources have been discontinued because they were not economically viable. Catches of some deep-water species, in particular, blue whiting, *Micromesistius poutassou*, which constitutes almost half of the deep-water catches in the last twenty years, are mostly meant for reduction into fish meal because of the rapid deterioration of their flesh, the presence of parasites, and the low market value for the fresh or processed product. In addition, the lack of sound biological information is often the reason of uncertainty on the long-term sustainability of such fisheries. Deep-water species are in general characterized by slow growth rates and late age at first maturity (25 years for orange roughly, *Hoplostethus atlanticus*), which lead to weak biological compensation of fishing mortality.

Oceanic resources are usually exploited by long-range fleets operating in areas where target species concentrate for feeding or reproduction. The more rapid increase of world fishery fleet sizes as compared to catches and the consequent depletions of some coastal resources have contributed to the increase of fishing effort in oceanic areas.

The complex interrelations between economic and political factors and the scarce knowledge of oceanic stocks, the issue of oceanic resources management is increasingly coming to international attention in the light of a growing world human population and limited food fish supplies. Also, considering that oceanic species live in a virtually boundless environment and exhibit extensive migratory behaviour amongst high seas and national jurisdictions, their management necessitates international cooperation.

Nutrients

Nearly all the marine biology activity is concentrated in the upper layers. Photosynthetic processes, the first step ending with the large predator species all along the trophic levels, require the mineral substances concentrated in deep offshore waters. Upwellings mainly recirculate these vital substances. Another important source of enrichment, particularly significant in the North-east Indian Ocean comes from rivers (3000 km representing 9 per cent of the world's total in only 3 per cent of the total ocean surface).

Due to the seasonal nature of the coastal upwellings in the Indian Ocean, seasonal variability of the nutrient distribution in the surface layers can be expected. The richest areas of phosphate distribution are in the Northern Indian Ocean. Minimum quantities were observed east of 90°E during the northeast monsoon (less than 0.10 uatg/litre), where as the contents are in excess of 0.25 uatg/litre west of 60°E. During the south-west monsoon, upwellings developing in the west cause significant increases of nutrient concentration in the surface waters, from 1 to 1.5 uatg/litre of PO_4, and more than 10 uatg/litre of NO_3. The south west monsoon drift distributes these contents towards the Eastern Indian Ocean.

The divergence located south of the Seychelles is also an important enrichment factor. A potential high production area which seems to be the most extended in the Indian Ocean (10^6 km^2). A rise in nitrate content along the divergence line from February to April with values ranging from less than 100 matg/m^2 at 4°S–5°S (also at 16°S) to 600 m atg/m^2 at 9°S-10°S were observed. These results when compared to those observed in the Pacific on the same latitudes show a similar structure. In the Indian Ocean they show twice the values observed in the Pacific at 10°S. However, it seems there can be a large interannual variation of the nutrient contents found in this part of the Indian Ocean.

The South-west Indian Ocean has high enrichment potentialities of the surface layers due to the seasonal movements of the divergence (5° to 10°S) and particularly when vertical mixing is induced by the wind or by eddies occurring near islands and banks. The most likely season for the enrichment processes to occur is when the southwest monsoon blows. Nevertheless, strong winds (even cyclones around 10°S) during the northeast monsoon when the thermocline is shallow (particularly within the area crossed by the south equatorial counter current) can help to enrich surface layers.

Environment and Production

The nutrient enrichment in surface waters initiates a food chain starting with vegetal micro-organisms, the phytoplankton, and evolves temporally and spatially before ending with the large predators. This chain is made up of three levels;

1. A primary level consisting of phytoplankton which synthesizes organic matter from nutrients through the use of light energy.
2. A secondary level, including the vegetarian animals (zooplankton) which feed on the micro-organisms in the preceding level.
3. A tertiary level, which is very wide spread, beginning with the animals feeding on zooplankton (first order carnivores) to the large predators (second order carnivores) of which the tuna, barracuda form part.

Based on the environmental conditions, an attempt can be made to identify most suitable areas in which large fish species are most likely to concentrate.

Primary Producers in Ocean Environment

Marine algae are important to life on earth, because they (a) control atmospheric carbon dioxide (CO_2), (b) form the basis of marine food web and (c) produce marine biotoxins.

At low concentrations, blue-green algae is a part of the natural environment. But under favourable conditions, phytoplankton can sometimes grow and reproduce at such a high rate that they create dense highly coloured patches on the water, popularly known as algal bloom and red tides in the marine environment.

Phytoplankton blooms normally occur in response to either an increased supply of nutrients or increased exposure to sun light. Most of the phytoplankton blooms occurring in the oceans are induced by natural causes and seasonal cycles. The North Atlantic Bloom is the largest seasonal bloom observed every year. There are also large blooms in the Arabian Sea due to the monsoon cycle and increase in productivity associated with the rainy season in Bay of Bengal.

In the Arabian Sea blooms of diatoms, dinoflagellates and cyanobacteria are frequently found. Following species contribute to the diatom blooms. They are *Coscinodiscus* spp, *Rhizosolenia* spp, *Skeletonema* spp, *Melosira* spp, *Thalassiosira* spp, *Chaetoceros* spp, *Thalassiothrix* spp and *Nitzschia* spp. An extensive bloom of *Navicula* spp was recorded in the Arabian Sea during November and found to be most important due to its non-toxic character.

Dinoflagellate blooms in Arabian Sea appear as response to monsoon upwelling and winter cooling in central and northeast Arabian Sea. The bloom of *Noctiluca scintillans* appear annually. The flagellum present does not propell the cell, and considered to be non-motile. It can regulate its vertical position in the water column by controlling its buoyancy. Colour of *Noctiluca* bloom is intense green. Its harmful effect on flora and fauna was not noticed. Other dinoflagellates blooms are, *Gymnodinium* spp, *Gyrodinium* spp, *Peridinium* spp, *Prorocentrum* spp, *Alexandrium* spp, *Ceratium* spp and *Dinophysis* spp.

Cyanobacterial blooms (*Trichodesmium* spp) in the Arabian Sea were recorded during November to May. *Trichodesmium* is a colonial cyanobacterium that is large enough to be visible to the naked eye. Each colony is about 5 mm in length and 1 mm across and the colony concentration varies from one to many colonies. They grow in clean optically transparent and pristine environment with warm temperature of more that 25°C. It grows when other phytoplankton does not grow due to paucity of nutrients.

Other organisms like *Phaeocystis globosa* from Arabian Sea and holococcolithophore from southern Malabar coast also form blooms in the Arabian Sea. During the bloom of holococcolithophore fish kills were recorded.

The anoxic bottom or "Dead zones" where the sea floor water is with a very low (zero) concentration of dissolved oxygen. Very few organisms can tolerate the lack of oxygen in the dead zone area and thus the habitat of numerous bottom dwelling organisms is destroyed spreading over larger areas of the sea floor. The organic matter produce by phytoplankton at the surface of the ocean sinks to the bottom, where it is subjected to break down by the action of bacteria. Bacteria use oxygen and give off carbon dioxide. The oxygen used by the bacteria is the oxygen dissolved in water and that is the same oxygen that all other oxygen respiring animals on the sea bottom (crabs, clams, shrimp and a host of mud-loving creatures) and swimming in the water (zooplankton and fish) require for life to continue. Most of the time oxygen is renewed by wind mixing the water especially in winter. In summer when the winds are weak and sun-warmed water at the surface increases stratification oxygen becomes depleted.

Dead zones grow if phytoplankton produce more organic matter to feed the bacteria. Agricultural fertilizers, runoff from cities and outflow from sewage when reach the ocean, supplement the natural nutrients available to the phytoplankton to grow and multiply which leads to more organic matter reaching the bottom, more bacteria and more dead zones.

In the Arabian Sea, the rivers and the monsoon wind driven upwellings regularly bring nitrogen and other nutrients from the sea floor surface, stimulating the rapid reproduction and growth of microscopic algae, phytoplankton. These algal blooms are natural events that benefit life in the Arabian Sea by generating tonnes of phytoplankton–a major source of food for larger organisms. But some phytoplankton species produce harmful blooms which release and can synthesize toxins that can poison molluscs and fish. Excessively large blooms can also overwhelm a marine ecosystem depleting oxygen in the water. Many harmful blooms are artificially fueled by fertilizer runoff from farms, mainly by excess uses of chemical fertilizers, which dumps tonnes of excess nitrogen into the rivers during south west and north-east monsoons that eventually flow into the sea.

The algal bloom due to *Trichodesmium erythrium*, a blue green algae is well known on the north west coast of India. Initiation takes place sometime in the end of February reaches a peak by the third week of March and generally declines by the end of April. Every year the bloom is reported to occur near Goa, on the north west coast of India, besides other localized areas on the west and east coast. The role played by the sudden appearance of the bloom is significant from the point of view of biological productivity, carbon dioxide flux (acting as a sink) and nutrient enrichment (nitrogen fixation) in the ocean.

First Link of Food Chain in Indian Oceans

Excessive production of minute plants are known to occur under various conditions in the inshore regions. This may take place often with cyclic regularity in any particular area when some optimum conditions of temperature, salinity, sun light, nutrients etc. prevail in marine environment. The bloom usually takes place rather suddenly and may spread with amazing speed, sometimes changing the colour of the surface waters into red, green or hay colour. The blooms are normally monotypic.

The most important factor for the phytoplankton bloom to appear is the presence of high quantities of nutrients in the surface waters which may be due to the process of upwelling or run off from land. The majority of blooms, thus, appear during monsoons or soon after that.

The south-west coast of India is known for intense upwelling during the southwest monsoon period. During this period, the nutrient rich subsurface water is brought to the surface, which provides congenial condition for high productivity at the primary level, sometimes leading to the flowering of certain species of phytoplankters.

Noctiluca miliaris, Coscinodiscus sp. and *Fragilaria oceanica* are the three common species which appear in blooms regularity. Some other phytoplankters also bloom occasionally at restricted areas. Among them are *Ceratium, Chaetoceros, Rhizosolenia, Biddulphia, Skeletonema, Thalassionema, Thalassiosira subtilis, Thalassiothrix* and *Trichodesmium.*

Noctiluca miliaris appear in blooms in some place or other in every year. The period of blooming of the species is found to be between August and April of every year. *Noctiluca,* as a rule, appears in blooms during post monsoon and premonsoon season when the water temperature is comparatively high and required amounts of nutrients are present in the surface waters especially during the post-monsoon period.

Blooms of *Coscinodiscus* sp has been reported from the northern parts of the south-west coast of India, namely, Karwar, Mangalore and Tamil Nadu on the east coast. In general, November-March is the period of bloom for this species. During January and December, 1982 it appeared in blooms at Karwar and continued upto March,

1983. On the east coast the *Coscinodiscus* bloom take place in April, September and December and even in January in 1985 in TamilNadu. The period of bloom for *Coscinodiscus* is almost the same as that of *Noctiluca miliaris* though they are not seen in blooms simultaneously in any year.

Fragilaria oceanica is a typical phytoplankton which appears in blooms during the monsoon months especially in the southern parts of the southwest coast of India. The species occur in blooms in one of the months of June, July or August.

Chaetoceros sp. was found in blooms in February at Karwar and August in Vizhinjam. *Rhizosolemia* sp was in bloom in November at Mangalore, while at Vizhinjam it was in June of the same year. Blooms of *Skeletonema* sp was observed in February at Mangalore, but in June at Tamil Nadu during 1982.

Biddulphia sp. was in blooms at Mangalore in November, *Thalassionema* sp and *Thalassiosira subtilis* at Vizhinjam in June, *Thalassiothrix* sp in blooms in Tamil Nadu in June. *Trichodesmium* sp was reported only from lower east coast. The period of bloom vary from year to year. In Tuticorin it appear in August and October. In Tamil Nadu the bloom appeared in April and June.

Noctiluca blooms appear with a drastic reduction of salinity values, 19.34 ppt. Blooms of *Coscinodiscus* and *Chaetoceros* appear in salinities ranging from 32.4 to 34.43 ppt. The surface temperature was relatively low and between 25.6 and 27.5°C when the blooms occurred. The dissolved oxygen content was well above optimum level and ranged between 3.79 and 4.25 ml/l.

In general the salinity values were considerably less when blooms occurred. The surface temperature was low when the bloom of *Noctiluca miliaris* appear. When *Coscinodiscus* was in blooms, the surface temperature and salinity showed lower than normal value. *Rhizosolemia, Skeletonema* and *Thalassiothrix* were also in blooms in low saline cool water.

The occurrence of phytoplankton blooms is a regular phenomenon in the coastal waters, than incidental. While it is possible that the interaction of environmental parameters act as triggers for the sudden outbursts of blooms, it is difficult to pin point any one parameter as responsible for this. May be a combination of various parameters is necessary to create an optimum condition for the flowering of any one species. Sudden increase or decrease in the values of certain environmental features that act as triggers for an organism to multiply at an explosive rate.

Trophic Relationship and Energy Transfer

The entire population of the oceanic waters is interconnected by a complex relationships mainly trophic and forms a complicated self-regulating system whose various components ensure the transformation of a considerable portion of the energy in the ocean.

The transmission of energy from certain trophic levels to others from the surface layers to the depths of the ocean is largely determined by the character of the vertical distribution of plankton and its amount and composition at various depths.

The cycle of chemicals in the ocean, the migration of chemicals through the water and the processes of sedimentation are also largely determined by biogenous factors and depend upon the vertical distribution of plankton.

The behaviour of commercial planktivorous fish is closely connected with the vertical distribution of plankton and its variations in time and space. The depths where the fish live and the regions where they concentrate, as well as the density of their schools, often depend upon the depth at which concentrations of edible plankton occur.

The regular variations in the depths of concentration of many pelagic animals affect the sound transmission of various water layers and the intensity of bioluminescence at various depths.

Trophic relationships at various depths are a principal factor in determining the quantitative distribution and qualitative composition of plankton.

The upper layers of the pelagic division are inhabited by three major types of pelagic fauna, boreal, tropical and antartic. The ranges of these fauna correspond to the main climatic zones, arctic-boreal, tropical and antarctic, which represent the three principal zoogeographic regions.

The most clear-cut distinction in the distribution of the plankton is the division into cold-water (polar and sub-polar) and tropical areas. The major criterion of this classification is the presence or absence of marked seasonal changes during the year, that is, the phytoplankton either vegetates more or less uniformly or exhibits distinctly cyclical growth. Differences in the vertical distribution and migrations, of the zooplankton of the high and low latitudes represent a tendency towards more efficient utilisation of the phytoplankton.

In the cold-water (polar and subpolar) areas, the nutrient elements required for primary production rise to the euphotic surface layers from the depths mainly with seasonal (winter) turbulence of the water. Their utilisation by the developing plankton begins only in the spring when the increased stability of the surface zone permits the formation of a seasonal pycnocline and the increasing intensity of solar radiation widens the euphotic layer.

In the tropical zone, illumination remains at a sufficiently high level, so that the critical depth is always below the limit of the upper mixed layer. Therefore, tropical phytoplankton can develop the year round in all layers from the surface to the depth of the basic pycnocline.

In tropical areas the nutrient elements enter and are utilised in the surface steadily the year-round, or almost so, compared with only a few months in polar and sub polar waters.

The oceanic regions of the tropical zone show a variety of water structure and consequently, various environmental conditions for the plankton. This zone can be divided into waters of equatorial and neighbouring tropical and subtropical structures.

A more or less permanent stratification of the water prevails throughout the year over most of the tropical water area. Solar radiation here is sufficiently high to support photosynthesis in the whole layer above the thermocline, while the supply of nutrient elements in the surface layer shows no seasonal fluctuations as a limiting factor.

The large pelagic animals (macroplankton and nekton) exert a major influence on the whole pelagic population and the overall food balance of the ocean. Most species of the macroplankton are carnivorous and feed mainly on the mesoplankton.

The uneven changes in planktonic composition with depth, the presence of water layers with an especially sharp faunistic succession caused both by physical and biological factors, the possibility of distinguishing zones whose fauna has certain common morphological straits, the change with depth of the dominant trophic groupings and other ecological features of the population-all make it possible to divide the pelagic division into biological zones.

The surface zone (0–200 m), consisting of waters from the surface to the boundary between the cold intermediate layer and warm intermediate layer, is characterised by the development of phytoplankton and the mass accumulation of a few epiplanktonic and interzonal species.

The intermediate zone (200-500 m) corresponding to the waters of the upper part of the warm intermediate layer, has a sharply increased diversity of fauna. The main mass of plankton living in this layer consists of interzonal forms also living in the surface zone. However, deep-sea species not ascending to the surface also appear.

In the deep sea zone (500-6000 m) corresponding to the waters of the lower part of the warm intermediate layer and deep waters representatives of the deep-sea fauna are numerically predominant, but interzonal species remain important especially in the upper part of the zone. The diurnal vertical migrations, of the interzonal species cover only the upper layers of the zone. Most of the deep sea species are distributed throughout the whole zone. According to the composition of the fauna, the deeps sea zone can be sub-divided into two sub-zones, the upper (500-2000m) and lower (2000-6000 m). Depth zones in the ocean are divided into inshore waters, which extend upto a pressure range of 2 atmosphere *i.e.* 2066 grams pressure per square centimeter that stretches upto 20 meters. Off shore waters which extend upto a pressure range of 6 atmospheres *i.e.* 6198 grams pressure per square centimeter that stretch upto 60 metres and deep sea which represents depths from 60 metres beyond.

The depth of 80 metres is reached well outside the territorial waters of India from any point along the coast. The 80 metre depth line falls in international waters.

Fishes inhabiting the surface layers of the open ocean form a natural community, the taxocene of epipelagic fishes or the epipelagic ichthyocene within the biocenosis of the ocean epipelagic.

The biotope of the epipelagic fishes is the upper mixed layer of the ocean, demarcated below by the main thermocline. Environmental characteristics, such as, the thickness of the upper isothermic layer, the temperature of the water, the system of oceanic currents, the zones of division of dynamic structures, the biological productivity of the water and the degree of equilibrium of the trophic cycles of the pelagic communities play an important role in the distribution and mode of life of the fish.

Adaptation of fishes of life in the ocean pelagic has led to their classification in one or other of the two main life forms of pelagic organisms, planktonic and nektonic.

The epipelagic ichthyocene includes quite varied elements that are distinguished according to their relationship to the biotope. The holoepipelagic fishes inhabit the surface layers of the open ocean through all stages of their individual life cycle. As distinct from these the microepipelagic fishes spend only part of their lives in the epipelagic. The third group constitutes the zeno-epipelagic fishes which normally live outside the epipelagic but form in it pseudopopulations whose individuals are not capable of completing the life cycle.

Of key importance in the distribution of planktonic organisms including plankton fishes, is the passive transport effected by the currents. Nekton fishes move actively throughout their life cycle, migrating within their ranges. The boundaries of their distribution areas are determined chiefly by the temperature of the water.

On the basis of the distribution of fish in the ocean epipelagic, the following ichthyogeographical regions may be singled out; arctic, north boreal, south boreal, north sub tropical, tropical proper, south sub tropical, notalian, low antarctic and high antarctic. These regions are grouped together in three main biogeographical regions, arctic-boreal, tropical and antarctic-notalian.

An important feature of the life of epipelagic fishes is active and passive migration whose main purpose consists in somehow guaranteeing energy gains for the migrants. Vertical and horizontal migrations, have a significant influence on the process of redistribution of organic matter in the ocean.

Primary and Secondary Production

Maximum chlorophyll *a* values (indicative of photosynthetic activity) are attained quickly in areas of upwelling or divergence. It has been found that the process of vegetal production occur with depth in the divergence areas of the western Indian Ocean. The highest values were found at 11°S at depth ranging from upto 80 m, in the thermocline and in the layer where highest nitrate concentrations were recorded. In spite of the potentially available nutrients, chlorophyll *a* levels averaged only 20 mg/m^2 in the first 100m, which led to assume that the zoo plankton, whose dry weight averaged only less than half of that measured in an eutrophic area near Madagascar, was grazing heavily.

In the Arabian Sea, phytoplanktonic blooms of *Trichodes mium* were found during the south-west monsoon due to greater coastal upwellings. On the western half of the Arabian Sea some of the highest production levels ever recorded in the entire Western Indian Ocean were measured. These exceeds 1 gram of assimilated carbon/m^2/day.

In the Mozambique channel, the most outstanding primary production areas are situated along the coast of Mozambique off cape St. Andre (Madagascar) as well as in Juan de Nova, an island which is rich in phosphates, derived from seabird guano. Along the African coast line, river flow is responsible for regeneration of the environment with chlorophyll a values exceeding 1mg/litre. In front of cape St. Andre where another current divergence is located, chlorophyll *a* values average 0.4mg/litre. In a general sense there is a higher plankton biomass in neritic waters than in oceanic waters. A smaller or larger extension of the continental shelf is sufficient to explain the differences in phytoplanktonic concentrations; the values measured in Tulear (South Madagascar) are generally lower than those measure in Nossi-Be (north-west of Madagascar) where a large planteau stretches off the coast.

Around the Seychelles, during the more suitable periods for the formation of upwellings along the plateau, chlorophyll a values measured in the first 100m deep layer, were greater than 40mg/m^2 compared to 20 mg/m^2 measured offshore. In the same way, zooplankton dry weight ranged from 3 gram/m^2 on the continental slope to 0.5 g/m^2 offshore out at sea.

Chlorophyll *a* concentrations to the north-west of Australia are higher in winters a period in which there is an upwelling near 12°S. Chlorophyll *a* values between 10 to 30 mg/m^2 were registered in the water columns between 0 and 150 m. The areas richest in zooplanktonic biomass coincide with those where photosynthetic activity is high, namely along the northwest coast of Australia and in the upwelling area. Highest production is attained in August-September and February-March.

Tertiary Production

Plankton proliferates in moving water masses. This means that from the initial enrichment zone, water masses will be subject to a maturation process, during which a succession takes place of the various links of the food chain. This would seem to indicate that concentrations of predators and tuna like species in particular, will occur outside the upwelling areas along the nearest convergences.

The trophic level attained at a point in time *t* will be proportional to the distance covered by the water mass, and the area of potential concentration is related to the speed and the direction of the current which has moved this water mass. This is purely theoretical. The need for a regular supply of mineral nutrients in the enrichment area for a sufficiently long time (several weeks) would eliminate ephemeral phenomena. Moreover, if a direct relationship between the available mineral substances and phytoplankton production is observed, it is impossible to predict the amount of potential biomass available at the end of the food chain due to a complex succession of events which take place during the maturation process. The effects of predation are multiple and difficult to understand. Furthermore, the future of a given water mass which has been enriched in organic matter is dependent on the physico-chemical processes to which it would be subjected during transport.

Fish Related Plankton

Information on standing crops of plankton in an area often facilitates the detection of fish shoals and hence the location of rich fishing grounds. The majority of plankton serves directly or indirectly as food for economic marine life. Although some of the planktonic species are harmful and repellent to fishes, it is probable that the occurrence of plankton is almost always associated in some way with the presence of economic fishes. For this reason, the distribution of plankton and its quantitative, qualitative, seasonal, diurnal and geographical occurrence has much to do with the distribution, migration and the detection of the existence of marine animals.

The relationship between the distribution of aquatic animals and the food on which they depend can be made use of in finding fish congregating areas. Centain types of commercial fisheries are making use of this technique.

An animal in a higher order may depend indirectly on the presence of other, lower organisms even though they do not enter directly into its food. Plankton in such an association with economic fish can be used as an indirect indicator of fishable waters.

There are some plankton and economic fishes occur in a common environment although there is apparently not a food relationship. In such cases, proper judgement on the periodical occurrence, amount and geographical distribution of fish may sometimes be determined from the presence of this type of plankton.

In contrast with direct and indirect food planktons, there is another type of plankton which is either detrimental or unrelated to fish life. Some of these may be actually harmful to fishes and often drive them away. Others may live in an environment altogether different from that suitable for fishes. Fishes will therefore not be available in an area where these types of plankton are found.

Movements of fish through the water sometimes stimulate a phosphorescent glow caused by certain luminous plankton. The size and location of a shoal of fish can thus be observed at night.

Swarm of some types of plankton eat up the long line catch operating for cod and similar species.

Role of Plankton in Fisheries Production

To secure information on the nature of the fish stocks as well as their quantitative distribution in space and time, the method of trial and error is the time-honoured method. Through the ages, a great deal of information on the nature of fish stocks in different parts of the world oceans as well as their seasonal fluctuations in abundance has been accumulated. This knowledge has played an important part in indicating the influences which the basic physical and chemical elements such as sun-shine, wind, currents, temperature salinity etc. exert on the biotic elements such as plankton and fish life. Investigations have revealed the complex mosaic showing the integration of these physical, chemical and biological elements of the marine ecosystem. The physical chemical and biotic elements of the ecosystem are closely integrated. It was found that the physical elements such as, rainfall, windforce, temperature, salinity and phosphate are inter-related.

Rainfall is correlated with phosphate, salinity, number of copepods and the amount of total catch. Windforce is correlated with temperature and salinity. There are inter-relationships between the different biotic elements of the marine ecosystem, namely, phytoplankton, zooplankton and fish.

In the sea, the prime relationship between organisms is that associated with nutrition. A study of the food of different species of fish all over the world has shown that many pelagic fish and the very young stages of almost all species of fish feed on plankton.

Since plankton is such as important item in the diet of fish, the quantitative relationship between plankton and fish is very important. It has been observed at widely seperated locations all over the world oceans that there is a positive correlation between the density of plankton and the density of the plankton-feeding fish population. It has been found that the occurrence of plankton feeding whales is positively correlated with the distribution of planktonic life upon which they are known to subsist. Positive correlations, between the distribution of blue and fin whales and the abundance of zooplankton have been observed, especially on the abundance of their favourite food *Euphausia superba*.

Using plankton indicator and in correlating the numbers of Calanid copepods in the plankton with the amount of herring caught in the same areas, it was found that in most instances the greatest number of adult herring were caught in water rich in *Calanus*. Along the west coast of India, the landings of fish were directly proportional to the quantity of plankton produced in the different months of the year.

In Singapore straits there is a statistically significant positive correlation between the number of copepods and the amount of fish caught in the same areas. At the same time, the correlation coefficient between phytoplankton and *Stolephorus* spp is statistically significant and in turn, the catch of *Stolephorus* spp is correlated with the catch of *Scomberomorus* spp. In simple terms, where there is an abundance of plankton, an abundance of plankton-feeding fish may be expected. Where these voracious plankton feeder, such as, large stocks of

Stolephorus spp, *Clupea* spp and Carangids are abundant it may be expected that species of *Scomberomorus* and *Chirocentrus* will be abundant.

A study of the distribution of plankton in space and time will serve as a means of indicating the spatial and temporal distribution of the probable centres of aggregation of plankton feeders and their predators.

Chapter 3
Fishery Resources

Indian oceans, smallest of the three major world ocean, which is bordered by 35 countries and 6 island groups harbouring about 1450 million people, a third of the earth's population.

Prasad, Banerjee and Nair (1970) estimated the potential yield of fish of the Indian ocean based on the data on primary production and zooplankton biomass. The area of 51 million sq. km. considered by them lies between 20°-120° E. longitude and from the Asian land mass to 45°S latitude. They estimated the primary production as 3.9×10^9 tonnes of carbon per annum. By assuming three variants, conversion ecological efficiency from primary to tertiary level between the production of different organisms they estimated the fish production for the western and eastern Indian oceans seperately.

From the estimated zooplankton biomass of 5.19×10^8 tonnes, the potential yield of fish was estimated to be 29 million tonnes by assuming 10 per cent conversion coefficient and 45 million tonnes at 15 per cent conversion coefficient. They inferred that 55 per cent of the total stock can be exploited safely.

Based on the yield ratio of carbon production of 0.005 per cent as well as the estimated potential yield derived from the results of the fishery exploratory surveys, they estimated the potential yield of fish to be 11 million tonnes.

Gulland (1971) estimated the exploitable fish stocks of pelagic, demersal, crustaceans and tuna and skipjack seperately at 7.5, 6.0, 0.25 and 0.30 respectively totalling 14.05 million tons based on the data on catch/effort, length and age composition, while the unexploited stocks have been estimated from exploratory fishing, eggs and larval studies and echo-sounding survey data.

According to Qusim (1977) the area of the Indian ocean lies between 25° and 120° E latitude and 25°N to 40°S longitude amounting to about 46.8 million sq. km. He estimated the potential fish yield of the Indian Ocean to be 17 million tonnes.

Based on 934 observations, available from C_{14} technique, column primary production of the Indian Ocean was estimated to be 4.42×10^9 tonnes of carbon per annum. This production has been converted to tertiary production by taking 0.1 per cent as the transfer co-efficient. The tertiary production thus calculated has been converted to live weight by multiplying by a factor of 10 and thereafter the potential production of fish has been estimated at 15 million tonnes by considering the yield ratio of 0.035 per cent of carbon.

An estimate of 10.2 million tonnes, of which 4.5 million tonnes are demersal and 5 million tonnes are pelagic has been reported by FAO (1979).

The marine fishery production system is always described in very general terms as an ecosystem with a linear transfer of energy from phytoplankton to zooplankton to fish. The estimates also differ primarily in the productivity weightage given to various trophic levels and the yield ratio of carbon to fish. The calculations are not supported by proven conversion coefficient for animal products from one trophic level to another, so that in the final stage, Dalal (1987) made arbitrary corrections to obtain reality. In his opinion, Gulland's (1971) estimates, which are based on catch effort and exploratory fish surveys data forms another category of assessment of potential resource, which lack ecological considerations.

In view of all these facts and some recent findings of the role and magnitude of picoplankton and benthos to the final end product, the fish, the concept of trophic levels, combined together with the available fishery production estimates are not more than guesses based on certain assumptions.

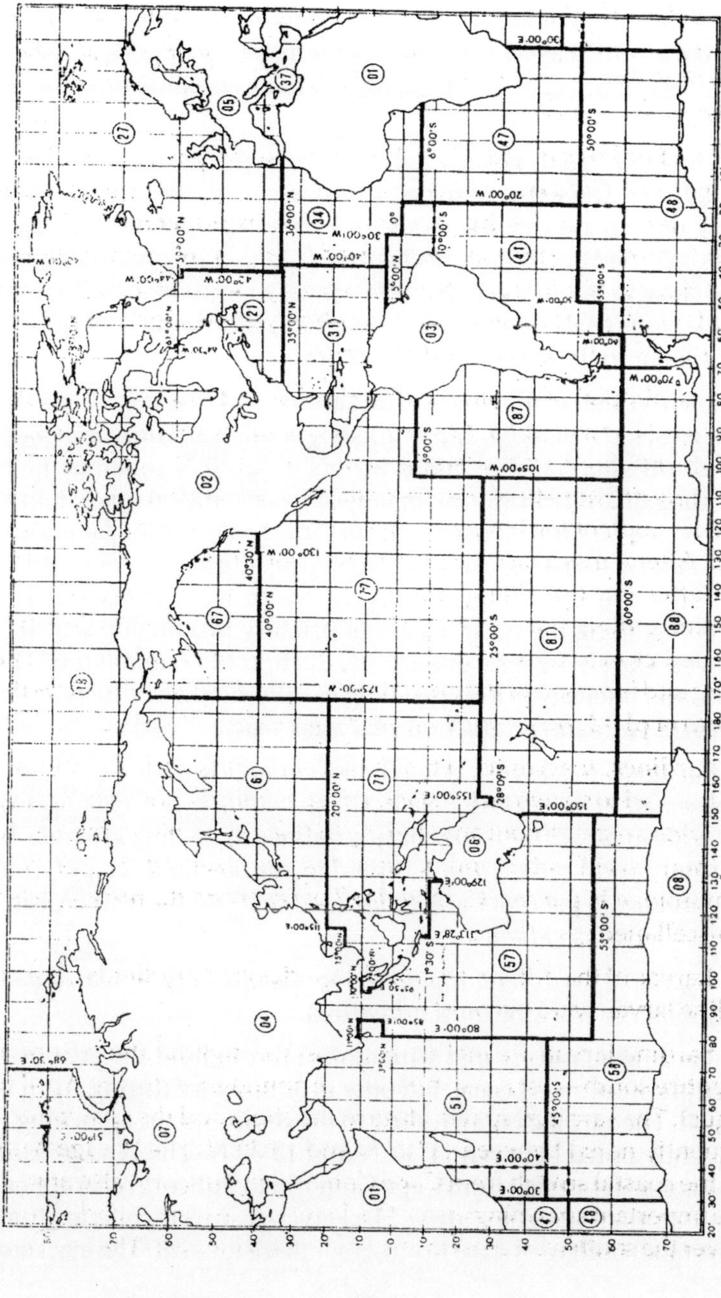

Major Fishing Areas

Pelagic Fishery

Of the approximately 2.0 million tonnes of marine fish catches in India, as much as 55 to 60 per cent is made up of pelagic fishes. More than 50 per cent of this pelagic varieties is constituted by oil sardine and mackerel.

The landings of pelagic fishes have, in the past, shown large fluctuations. The existing fisheries are seasonal, the season being clearly demarcated by the monsoons. The exploitation of pelagic shoals is carried out by non-mechanised fishing craft when the shoals move close to the shore. In some years very large shoals of fish are abundant in coastal waters, while in other years no such large scale migration into the coastal waters occurs.

Distribution of the stocks of oil sardine and mackerel along the south west coast of India, approximately south ward from Ratnagiri in both off shore and in shore waters at lat. 5°N to about 17°N extending offshore to approximately 250-300 nautical miles, reveal that the zooplankton biomass is at very low levels during January to April. Afterwards it increases to reach a peak during July-September soon after the upwelling during the south-west monsoon. The upwelling along the coast trigger of primary production which is followed closely by a secondary production of zooplankton. The timing and intensity of the upwelling to a great extent influences the pattern of plankton production year after year.

Sardines, mackerels, white baits, carangids and coastal tuna are observed to spawn on the shelf. The spawning is, however, spread over wide areas without any sharply defined spawning grounds. In the total larval collection, Clupeidae represented 21 per cent Scombroidae 16 per cent Carangidae 7 per cent and the rest consisted of miscellaneous varieties.

Larvae of the different *Sardinella* species are very similar, but oil sardine larvae were the most numerous.

Sardine larvae are met with almost throughout the year over the entire south-west coast, but only in abundance during April to August. The sardines spawn close to the shore and the spawning is frequently noted between 11°30' N and 15°30′N. The Wadge Bank and the coastal stretch from Cape Comorin to Tuticorin also appear to be important spawning areas. Mackerel larvae were collected from all over the south west coast in most months of the year. The mackerel spawning occurs over the inner area of the shelf, but generally more off shore than in the case of oil sardine. The whole coastal area between Cape Comorin and Karwar appears to be the spawning area, particularly the southern part.

The period from April to September is the main spawning season for white bait with a main peak before the monsoon and a secondary peak afterwards. The white bait stock consists of several *Anchoviella* species, all having extended spawning seasons. White bait larvae are observed during most of the year all over the south west coast and form the major component of the clupeoid larval population. The March-July period is however, observed to be the season with greatest larval abundance of the species.

The most important species of Carangidae in the south west coast are golden scad, *Caranx kalla*, horse mackerel, *Megalaspis cordyla* and scad, Decapterus spp, Larvae of Carangid fishes are found throughout the year, but spawning seems to be most frequent in the February-September period. The Cochin-Cape Comorin region is the major spawning area for these species.

Coastal tuna, especially, *Auxis* spawn mainly between Cape Comorin and Calicut, the most of these intense spawning season being May to August.

Mackerel

Mackerel shoals are recorded throughout the year, but surface shoals are mostly seen after the south west monsoon. This seasonal formation of surface shoals is evidently related not only to upwelling but also to the rich bloom of planktons, resultant to the upwelling. When the upwelling ceases in October-November. the fish is no longer restricted to the surface and a considerable portion of the stock especially the juveniles move into the close shore conventional fishing zone. Mackerel shoals occur throughout the year mainly in a belt all along the coast between 10 and 25 nautical miles offshore. Only a part of the stock moves closer to the shore after the upwelling. The juveniles also are distributed to a great extent exactly in a manner similar to the adults. Maximum concentrations of juveniles are found in the southern region between Quilon and Calicut. The mackerel spawns on the inner shelf over an extended period, particularly from March to October, the most intensive spawning being at the beginning and end of the period. Mackerel is short live and the stock is affected by a high natural mortality. The fish spawns when a year old and a few appear to survive for a second spawning. The vessel data and arial surveys in 1973, 1974 and 1975 with an extensive coverage estimates an average stock of 450000, 100000 and 300000 tonnes respectively. Adult mackerels are fast swimmers and they avoid trawls. Because of its shoaling behaviour, the fish can most successfully be caught in purse-seines. Small and shallow purse seines caught only the 0-year group mostly. Commercial sized catches is possible in turbid waters.

Oil Sardine

The distribution of oil sardine is more or less similar to that of the mackerel, occuring about 10 to 25 nautical miles off shore. After the upwelling, part of the stock moves close to the shore and contribute to the conventional fishery. Sardine shoals are, however, observed further out on the shelf throughout the year, the juveniles found dispersed in the shelf water south of Karwar. Some occasional dense patches occur between Karwar and Alleppey, Annual fluctuations are seen similar to those of the mackerel stock. One factor of significance appears to be that towards the tail end of the upwelling, a major part of the sardine stock, as much as 75 to 80 per cent is found in the zone of 10°N to 13°N. The oil sardine spawn close to the shore over several months, the intense spawning period being May to July. They are fast growing and spawn by the end of first year of life. They are short lived and second time spanners are seldom observed. The surveys in 1973, 1974 and 1975 shown an estimate stock of 400000, 500000 and 700000 tonnes respectively. Oil sardines are difficult to be caught by trawl but they are much more amenable to capture by purae-seine than mackerel. Appreciable quantities of sardines were located from 40-110m depth during January-March along the east coast. Sardines constituted 18.5 per cent of the catch taken by demersal trawling. Sardines were caught for the first time in such abundance (Average catch 160kg/hour) from deeper waters of Andhra coast. During the same time large surface shoals of sardines could not be observed Maximum catches of sardines were obtained from the depth range of 80-90m. 60 per cent of the catch in February accounted for *Sardinella* spp. A maximum catch of 1433 kg. per hour for this variety was recorded during the same period. The common size as per length frequency study was 19 cm.

The investigations showed that during the period of intense upwelling in May-September, the mackerel and oil sardine stock move outside the conventional close shore fishing zone and occur in a narrow belt usually some where between 16 to 25 km offshore. These shoals at that time are very emenable to purse, seining, catches as high as 10 tonnes per set having been taken by 54' and 57' vessels. With medium to slightly bigger vessels, these can be significantly stretched outside the conventional coastal zone in August–October and April-May.

White-bait

The next important resources, the white-bait stock consists of several species of which *Anchoviella heteroloba, A. bataviensis* and *A. zollingeri* are the most important. White-bait occurs most frequently between 25 and 35 metres depth. In the southern region they occur in much shallower regions. Beyond 50 metres they occured rarely. During the day time, the white-bait are found in dense shoals at or near the bottom. During night they come up higher in the sea and remain dispersed. Throughout most of the year, that is, from October to May, the white-bait concentrations are spread out along the south-west coast. Before the south west monsoon season, there is a southward migration, and during the monsoon they get concentrated in the Gulf of Mannar, particularly on the shelf off Cape Comorin and eastward to Mnapad.

This migration is apparently related to the oceanographic conditions. The south bound migration follow the current which flows southward from March or April. In shifting southwards the white-bait also avoid the oxygen deficient water which starts ascending the shelf north of Quilon at that time.

After the monsoon the white-bait migrate northward along the west coast, again following the current which now gets reversed in its direction of flow. The white-bait attains their northern most distribution in the January-March period.

Juvenile white-bait are distributed more or less along the whole shelf from Tuticorin to Ratnagiri. They are most abundant from April to July and from October to February and the best concentrations are found between Cape Comorin and Mangalore.

White-bait spawns during most of the months with one premonsoon and one post-monsoon peaks. They grow to 8 to 9 cm in six months and spawns in the first year itself and suffers high mortality afterwards. The stock estimated during 1973, 1974 and 1975 range between 208-520 thousand, 18-809 thousand and 80-1054 thousand tonnes respectively. White-bait are easy to catch with both mid-water and high opening bottom trawls and with purse seines. With light attraction the fish can be chummed easily and caught with small purse-seines.

The main species of horse mackerel, namely, *Megalaspis cordyla, Decapterus kurra D. russelli* and *Caranx* spp occur in characteristic schools mainly on the outer and middle shelf but occassionally in dispersed condition. The biggest concentrations are seen between 7° and 9° N in regions of 30 to 100 metres depth. On an average 77000 and 176000 tonnes are estimated potential of these species. Horse mackerels are very difficult to catch by trawl. Better catchability is with purse seines, especially using light to attract them and chum them at the surface.

Catfish and ribbon fish form about 20 per cent of the total fish biomass in south west coast of India. The group consist of several species of cat fish, *Tachysurus* spp, and ribbon fish, *Trichiurus* spp. They are distributed all along the south west coast, particularly around the 50 metres depth. The group has a pronounced southerly distribution. The stock of this group was estimated at 294-709 thousand tonnes with an average of 475 thousand tonnes. Both pelagic and wide mouthed bottom trawls are very suitable gears to catch this group.

The inshore shallow water group of fish mainly consists of silver bellies, *Leiognathus* spp and *Gazza minuta*, golden scad, *Caranx kalla*, butter fish, *Lactarius lactarius* and glass perch, *Ambassis* spp. The proportion of different species in the groups varies from place to place and from period to period.

During the day they are located near the bottom and they disperse in the water column during the night. For most of the year the species of this group dominate in the near shore zone along the entire coastal stretch of south west coast. During the south west monsoon season they remain in the area where upwelling occurs, distributed in the upper layers further offshore.

The average biomass of shallow water fish groups was assessed to be 50000 tonnes and the maximum stock of 185000 tonnes was recorded in May, 1997.

These species are easily caught with various types of trawl and catch rates upto one tonne per trawl hour were obtained during day time as well as during night time.

A great variety of miscellaneous fish contribute greatly to the biomass of the shelf area. The most important of these were pomfret, *Stromateus* spp and seer fish, *Scomberomorus* spp, which are pelagic species often occurring near the sea surface. They are caught with trawl as stray individuals. The other pelagic and semi-pelagic species that contribute to the biomass are lizard fish, *Saurida* spp., threadfin bream, *Nemipterus* spp, Indian anchovy, *Thrissocles mystex*, sharks and rays.

These species were mostly recorded on the middle and outer shelf, and their distribution pattern was more or less uniform throughout the period. They are found in abundance on the northern shelf during post monsoon months and are concentrated on the southern shelf during the monsoon period. Random estimates of these miscellaneous varieties stock is about 390000 tonnes.

The lantern fish group, consisting mainly of the two genera *Myctophum* and *Vinciguerria* may become commercially important because of their relatively greater abundance. Echo recordings of this type were made in oceanic waters, particularly in a belt just beyond the edge of the shelf and north of the Laccadives. This group exhibit pronounced diurnal vertical migrations rising from deep scattering layers during night time and forming a layer near the surface. Swimming crabs, *Charybdis* spp were often recorded in considerable quantities together with the lantern fish. The typical demersal fish which contributed to the echo recordings were snapper, *Lutjanus* spp, rock cod, *Serranus* spp and pig faced bream, *Lethrinus* spp.

Demersal Fish

Bulls eye (*Priacanthus* spp.), a potentially rich deep water fish resources was located from Goa to Mandapam along the west coast and point Calimere to Visakhaptnam along the east coast. The species was recorded from 50-400m. depth along the north Kerala and Karnataka coast with peak occurrence in 100-150 m. depth belt. Its abundance was more during April to June in this region. The highest catch rate was 616 kg per hour; while the average was 72.8 kg. It formed 11.26 per cent of 314 tonnes of fish caught by two fishing vessels operated from Mangalore. The predominant size group was 13-21 cm. The species was available between 90-250 m depth in the Gulf of Mannar and Wadge Bank and the highest catch per hour of 32.04 kg was recorded.

Priacanthids were recorded from 120-200 m depth along the Andhra Pradesh coast with a peak occurrence from 150-170 m. depth belt. While the average catch per hour was about 60 kg., the highest catch was 1225 kg in January. It formed about 5 per cent of the total catch of about 176 tonnes of fish. The common size group in the catch was 16-18 cm. The species was also caught from Tamil Nadu coast between the depth of 50–200 m. While the average catch was about 200 kg per hour, the highest catch rate was about 1700 kg per hour during December. The species appears to migrate both across the shelf and parallel to the shelf depending on cold water current.

The distribution of *Priacanthus* in the Gulf of Mannar was traced in 100 to 250 m depth belt. About 2.5 tonnes of Bulls eye were caught while surveying deeper waters of Gulf of Mannar with an average catch rate of 10 kg per hour.

Priacanthus spp. in Andhra coast accounted for 4.76 per cent of the entire catch. The maximum catch of 6145 kg of this species (21 per cent) was obtained during January 1984. The maximum catch per hour obtained was 1425 kg from a depth of 150 m. From deeper water (151-170 m depth range) the maximum catch obtained was 358.79kg per hour. This is followed by 66.81 kg per hour from 131-150 m depth. The common size of fish in the catch was 16 cm. in length.

In Karnataka and North Kerala coast beyond 200 m depth *Priacauthus* spp. is one of the important varieties forming 11.26 per cent of the total catch. The species is distributed between 10°-11° N with a highest catch rate of 72.38 kg per hour. The species though occured in all the three bathymetric zones, was abundant in 100-200 m depth range. Among deep sea varieties, *Priacanthus* spp. was abundant during April to June. Their length ranged from 8-29 m with dominant size group of 13-21 cm.

Indian drift fish (*Psenes indicus*) is another demersal fish available along both east and west coast. Its occurrence appears to be more predominant in areas south of Mangalore than northern side. It formed 6.2 per cent of about 314 tonnes of fish caught off Mangalore.

The average catch rate was about 20 kg per hour, while the highest catch rate was about 1400 kg. It was found to occur between 200-500 m depth. The species is also found in appreciable quantities along the Wadge Bank and in the Gulf of Mannar in identical depth. While the average catch rate was 16 kg per hour, catch rate as high as 170 kg per hour was also recorded from a depth zone of 100-150 m. On several occassions the species was caught in sizeable quantities with Barracuda. The predominant size group along the west coast was 16-20 cm.

Psenes indicus formed about 3 per cent of the catch (176 tonnes) from Andhra Pradesh coast. It was found to occur in appreciable quantities during months of November, December and March between 50-130 m depth. The average catch was 10 kg per hour as against a highest catch rate of about 400 kg. The species was recorded mainly from 110-400 m depth from the lower east coast. While the average catch per hour was about 17 kg there, the highest catch rate was 1500 kg. The common size group was 14-18 cm.

In Gulf of Mannar, *Psenes indicus* is found to be distributed in 100-200 m depth zone chiefly. The highest catch rate recorded in respect of this species in 157 kg per hour. It could be said that the catch rate to the tune of 16 kg per hour and above.during all the months excepting December is a sign of noteworthy resources. *P. indicus* varies in size from 11-25 cm and majority of them were in 16-20 cm (58 per cent) and 21-25 cm (32 per cent) size groups.

Psenes indicus comprised 1.69 per cent of the total catch along Andhra coast. The maximum catch rate of 400 kg per hour was obtained in the month of March. The maximum catch rate of 49.58 kg per hour was obtained from a depth of 90-100 m. Beyond 130 m depth the species was not recorded. This variety has a good market and largely consumed by the local population. The common size in the catch as recorded by length frequency study was 16 cm.

Along Karnataka–North Kerala coast, between 11°-12° N highest catch rate of 29.99 kg per hour was obtained comprising 6.17 per cent of the total catch.

Barracuda

In the Gulf of Mannar, commercially exploitable quantities of Barracuda (*Sphyraena* spp) have been discovered in the depth range of 100-200 m. The catch consisted of five species belong to the genera

Sphyraena. The most predominant species was *S. obtusatta*. The average catch rate was 92 kg per hour, while the highest catch rate was 1350 kg per hour. The dominant size groups were 21-25 cm and 36-40 cm. Barracuda formed 22 per cent of the total catch in the area. during the period.

All the five species namely, *Sphyraena jello, S. obtusatta, S. langsar, S. picuda* and *S. commersoni* occur in Indian Oceans. But *S. obtusatta* was found to contribute substantially to the demersal catches in Gulf of Mannar. The catch of the species were poor during October and November. But high catch rates were recorded from December. Though most of the areas in the Gulf record rather stray catches of Barracuda especially in shallow waters, a good fishery for Barracuda exists in 100-200 m depth belt. The Barracuda fishery commence during December and had consistent record of good catch till March. An aggregate catch rate of 72 kg per hour was observed, though the catch rate of 1350 kg per hour was the highest yield from the area. The catches were dominated by the individuals in two size groups 21-25 cm (36 per cent) and 36-40 cm (23 per cent).

Perches

Perches constituted 21 per cent of the catch from Gulf of Mannar during the five months survey from October to March. This group consists mainly of *Epinephelus diacanthus* (Kalava), *Lutjanus argentimaculatus* (Red Snapper) and *Lethrinus ramak* (Velameen) besides an assortment of other species. The average catch per hour was about 60 kg while the highest catch per hour was about 90 kg. It was abundant in the depth belt of 30-120 m.

Survey in May at Wadge Bank indicated that perches of identical species constituted about 37 per cent of the catch from 20-225 m. depth. The species was found to occur 25-100 kg per hour of trawling. The depth range of 25-46 m and 183-225 m were found to be most productive for this group. The highest catch per hour was observed during the month of August followed by April-September. It formed about 75 per cent of the catch during March-April. In Mangalore coast perches constituted only 2 per cent of the catch.

From areas, north of Tamil Nadu, perches formed 7 per cent of the total catch. An average catch of 300 kg per hour was obtained during November from 50-60 m depth. Perches mainly belonging to the genera *Pomadsys* formed about 4 per cent of the catch. Its peak occurrence, 234 kg per hour, was observed in the month of March from 50-70 m. depth. It was available throughout the year from 40-170 m depth along the east coast.

In aggregate perches contribute about 38 per cent of the catches from the Gulf of Mannar within 125 m. depth with 60 kg yield per hour. The highest catch was 90 kg per hour followed by catch rate of 55 and 54 kg per hour. In order of abundance, the varieties among the perch group were red snapper (29 per cent), kalava (27 per cent), velameen (14 per cent), *Gatrin* spp (9 per cent), *Acanthurus* spp (11 per cent), *Pomadasys rasta* (7 per cent), *Scolopsis* spp (2 per cent) and *Lutjanus vitta* (1 per cent). The perch fishery in the Gulf during six months period was supported by individuals dominating in 56-70cm with 68 per cent kalava, 76-80 cm with 43 per cent red snapper and 51-60 cm with 37 per cent velameen.

Among the perch group *Pomadasys* spp is the most important variety commonly available in the demersal catch along Andhra Pradesh Coast. A maximum catch rate of 234 kg per hour was obtained in the month of March. This variety accounted for 4 per cent of the total catch, 7 per cent of catch in the month of March. Maximum catch rate was 39.17 kg per hour at a depth range of 51-70m. In the depth range of 110-130 m. a catch rate of 17.92 kg per hour for this variety was recorded. The frequent size was 35cm.

In the Karnataka and north Kerala coast between 14°-15°N, a highest catch of 4.29 kg per hour was obtained constituting 1.61 per cent of total catch in the area from a depth range of 100-200 m.

Black ruff (*Centrolophus niger*), one of the most abundant species of deep sea, was caught from Mangalore coast. It formed 16.1 per cent of the total catch of 314 tonnes and found to occur in 250-500 m depth with highest catch rate from the depth zone of 350-500 m. The average catch rate was about 53 kg per hour, while the highest catch rate was about 2100 kg. It was more abundant during January–February than other months. The common size group was 12-14 cm. Black ruff was also found to occur in Wadge Bank and in Gulf of Mannar in lesser quantities. No appreciable quantity was caught from east coast.

Centrolophus niger was caught between Lat. 11°-12°N along the Karnataka and north Kerala coast with highest catch rate of 52.79

kg per hour and constituted 16.14 per cent of the catch. The species occured beyond 250 m depth with high catch rate between 350-500 m depth range. 7 to 18 cm. was the common length range in the catch, but the dominant size group was between 12-14 cm. The species is rich in protein (14.90 per cent) and the fat content was 5.80 per cent.

Threadfin bream (*Nemipterus* spp) is found to be the most predominently available fish in deeper waters from Wadge Bank to Karwar along the west coast. It formed a little over 17 per cent of the total catch along Kerala–Karnataka coast, while it formed about 24 per cent of the total catch from Wadge Bank. The species was found to occur throughout the year with a peak catch rate of about 181 kg per hour during June-July in the Wadge Bank. The peak catch was observed during February-May along the Kerala-Karnataka coast. Threadfin bream was found to occur from shallow waters to 225m. depth all along the south west coast. An average highest catch rate of about 34 kg was obtained from 100-200 m. depth belt from Karnataka coast. The dominant size range was 14-26 cm.

Threadfin bream was also caught in appreciable quantities from 50-200m depth from the east coast. A highest catch rate of 264 kg per hour was recorded from 150m. depth. However the average catch per hour was found to be 20 kg, less than that is available along the west coast. Catch per hour upto 640 kg was recorded. Its distribution in the deeper waters along both east and west coast in such quantities appears to be very promising to the Indian fishing industry.

Along Andhra coast the species accounted for 3.74 per cent of the entire catch. Maximum catch rate of 3200 kg was obtained from depth range of 131-150 m.

From Karnataka and north Kerala coast *Nemipterus japonicus* was caught between Lat. 11°-12° N with highest catch rate of 31.53 kg per hour. The species formed 17.08 per cent of the total catch along the coast. The species were abundant during February-May. The length range of the catch was 11-29 cm with a dominant size group of 14-26 cm.

Sciaenids

Large sciaenids consisting mainly of *Pseudosciaena diacanthus* (ghol) comprised 8.47 per cent of the total catch in Andhra Pradesh coast. A maximum catch rate of 2666 kg per hour was obtained around a depth of 70 m between lat. 16°N to 16°22' N and long 81°59' E to 82°13' E. Maximum catch of this variety is observed from 51-70 m depth. From the length frequency studies, the most common size of the fish was 75-80 cm.

Cat Fish

Cat fish formed about 12 per cent of the catch from Karnataka–Kerala coast; while it formed about 3 per cent of the catch from Wadge Bank. The average catch rate was about 30 kg per hour, while the highest catch rate was 1200 kg. It formed about 5 per cent of the catch with an average catch rate of 16.4 kg per hour from Andhra coast. Both along the east and west coast its predominance is more between 50-90 m depth. Highest catch rate of 29.98 kg per hour was obtained between lat. 12°-13°N in Karnataka–Kerala coast.

Carangids

The group comprising of *Caranx* spp., *Decapterus* spp and *Chorinemus* spp constituted about 6 per cent of the catch from Wadge Bank and Gulf of Mannar. It was available upto 225 m depth but with high catch rate of 64 kg per hour during January from 40-90 m depth. An average highest catch rate of 90 kg per hour was obtained from Wadge Bank. Carangid formed 2 per cent of the catch from Karnataka region. The group constituted about 7 per cent of the catch from the east coast. October-November yielded catch rates upto 320 kg per hour from areas of Tamil Nadu from a depth range of 50-125 m.

Large growing individuals of the genus *Caranx* is an esteemed food item. Other allied species included in this group is King fish, *Rachieentrus canadus*. Carangids are distributed in all the areas in the Gulf of Mannar upto 250 m depth.

A higher catch rate of 63.96 kg per hour was recorded in January followed by October (18 kg per hour).

Carangids including horse mackerel, *Decapterus* spp contributed 16 per cent of the total catch in Andhra Pradesh coast. *Decapterus* spp was caught at the maximum rates from Kakinada-Pentakota, where a maximum catch rate of 2000 kg per hour was netted. The species is abundant in 71-90 m depth range. It is available upto 150 m. depth in appreciable quantity. The common size in the catch was 19 cm.

Off Karnataka and north Kerala coast carangids were the important varieties within 200 m depth. In the coast *Caranx* spp constituted 4.20 per cent, while *Decapterus* spp contributed 3.26 per cent of the total catch. The highest catch rate of 16.73 kg per hour was obtained from lat. 13°-14° N. They were abundant in 50–100 m depth range. *Decapterms* spp were of 12-25 cm with dominant size of 14-22 cm.

Pomfrets

Some potentially rich fishing grounds in relatively deeper waters were located for pomfrets, an esteemed table fish and an increasingly important export commodity in the east coast. It was found to occur south of Visakhapatnam between 50-90 m depth. There the species formed 4 per cent of the catch and was found to be available throughout the period with peak abundance in the month of March. Pomfrets constituted 2 per cent of the catch from the depth range of 50-110 m. The areas along the lower east coast appear to be very productive fishing areas for pomfret. In the Gulf of Mannar, pomfret contributed 3 per cent of the catch mainly from the depth range of 30-60m. Catches upto 113 kg per hour was obtained during December. The catch consisted predominantly of white pomfret.

The silver pomfret, *Pampus argenteus* is the principal constituent of the catches from the Gulf of Mannar, whereas the black pomfret, *Parastromateus niger* is occasionally caught. The main area producing this variety of fish are shallow waters upto a depth of 50 m. Their month wise catch rate (kg per hour) was 19.30 in October, 12.60 in November, 112.90 in December, 28.66 in February and 42.90 in March. The maximum yield recorded was 113 kg per hour in December followed by 43 kg per hour in March. The aggregate catch rate of 24 kg per hour indicates resource of significant magnitude of this much sought table fish variety. Both silver and black pomfret caught were in their prime size 21-30 cm. (94 per cent and 85 per cent of the catch respectively).

Pomfret occupies 3.70 per cent of the entire catch of Andhra Pradesh coast from April. A maximum catch rate of 800 kg per hour was netted during the month of March. The best fishing grounds for pomfrets are found between Narasapur and False Devi point areas from where maximum catch was obtained at 29.44 kg per hour at a depth range of 71-90m. This is followed by 51-70 m depth range where a catch rate of 13.61 kg per hour was obtained. The most frequent size was 28-31 cm.

On the south west coast of India, especially from Colachel to Alleppy, fishing for the perches with hand lines from the indigenous canoes and catamarans is an annual feature during the months of December to March. 120-140 metres provide good fishing grounds.

Approximately a stretch of 200 nautical miles of the shelf extending from south west of Cannanore to south west of Quilon, between lat. 8°30' N and 10°15' N indicates that the shelf upto a depth of 70 metres is muddy from south west of Alleppey to Cannanore, while it is sandy from south west of Alleppey to further south. From 70 to 90 metres depth sea bottom is hard and slightly uneven, whereas from 90 to 110 metres, the sea bottom is greatly unlevelled with rocks and coral formations. At a few places it has been observed that there are 2 to 3 rows of rocks lying side by side as long ridges each often with moderately high peaks. In few cases the ridge has a long raised or rounded top. The width of this unlevelled stretch between 90 to 110 metres depth is estimated to be approximately 1 to $1^1/_2$ nautical miles. It is on this rocky ridge of sea bottom that rock cods abound. Cod dwelling rocks is neither uniform nor extensive in length. However, occasional long unbroken stretches also abound. One such ridge was found to be 10 to 15 miles in length. In between these rocks and ridges there are small stretches of more or less even but hard bottom. From 110 to 200 metres depth the sea bottom continues to be more or less hard, but rather even. Rock-cod fishing grounds appear to form a more or less continuous stretch, throughout the length of the shelf, at a distance of about 20 to 25 nautical miles off the coast and 3 to 5 nautical miles inwards of the 200 metres line.

Rock cods generally appear in schools along the edges of rock peaks.

The catches that are consolidately called rock cods, consists of a number of species belonging to the families *Serranidae* and *Lutjanidae*. The four important species that are found in the catches in the Cochin–northwards regions are *Epinephelus aerolatus, E. chlorostigma, E. diacanthus* and *Pristipomoides typus*. In the exploratory fishing, off Travancore coasts, though *E. chlorostigma* showed the largest individual landings, there were appreciable quantities of *Lutjanus malabaricus* and *Latjanus argentimaculatus*. The large rock cod, *E. tauvina* also occured in fair numbers. In the Wadge Bank, *E. tauvina* and *E. bleekeri* formed the larger proportion in the catches. *E. chlorostigma* and *E. diacauthus* were not caught from the Wadge Bank.

Larger species of rock cod are more abundant in the southern regions than on the northern areas. Some of the rock cods attain large sizes. *E. pantherinus* and *E. tauvina* reaching lengths of 2 to 2.5 metres weights 150 to 160 kgs. The colour pattern in this group in brilliant and variegated. Rock cods are predaceous fish living on crabs, prawns, polychaete worms and other fishes like *Therapon* and *Ambassis*. Some of the species are long living. A few specimens of *E. pantherinus* have been observed to live for over 25 years in the Madras aquarium. The meat of rock cod is very firm but soft and excellent to taste when cooked.

The grounds lying between latitudes 11°N and 12°N are richer in rock cods than the areas further south. The highest catch rates was obtained during the month of February, while the lowest was obtained in June. The catch rate during the other months remained more or less the same.

Chapter 4

Coastal Fishery Resources in Countries Bordering Indian Ocean

In Indian Ocean around Sri Lanka, the provisional estimates indicate a potential yield of 250000 tonnes comprising 170000 tonnes pelagic resources and 80000 tonnes demersal resources.

In Sri Lanka, scombroids (tunas, Spanish mackerel, Indian mackerel), clupeoids (sardines, anchovies, wolf-herring), carangids (especially horse mackerel) and hair-tails are the important pelagic fishes, while snappers, croakers, groupers, elasmobranch and prawns are the important demersal resources.

In India the estimates indicate a harvestable potential of 343000 tonnes from the demersal resources. Individual groups of pelagic species or groups of species showed that the total maximum yield of these groups from the inshore waters of the east coast would be about 1,24000 tonnes. A high standing stock of fish biomass mostly pelagic is observed in the shelf region off Cape Comorin and in Gulf of Mannar (7°-9°N) during the June-October period. It ranged between 1,27,000 tonnes and 9,70,000 tonnes with an average of about 6,24,000 tonnes. Of this *Stolephorus* stock was the mainstay, its abundance ranging from 55,000 to 8,04,000 tonnes with an average of 5,09,000 tonnes forming 82 per cent of the general average standing stock of the total fish biomass.

In India, the clupeoids (sardines, anchovies and Hilsa), hair tails, carangids, and Spanish mackerel are significant from the pelagic group, while in the demersal group, elasmobranchs, catfishes, croakers, leiognathids, perches, pomfrets, prawns and crabs are the major items.

In Bangladesh, the estimate of exploitable yield, as per trawl survey confined to 1,75,000 tonnes of demersal fishes and 9000 tonnes of prawn. A brief acoustic survey beyond 10m depth indicated a provisional figure of 60000 tonnes pelagic species.

In Bangladesh the important pelagic fishes are *Hilsa*, Bombay duck, hair-tails mackerel and scads. The important demersal fishes are elasmobranchs, catfishes, threadfins, croakers, pomfrets, eels, snappers, grunts and prawns.

The latest estimate in Thailand indicates an optimum yield of 205000 tonnes for demersal resources and 61,000 tonnes for pelagic resources thus totalling 2,66,000 tonnes. The estimated maximum sustainable yield for mackerel, sardines and anchovies are 20000, 5000 and 7000 tonnes respectively.

In Thailand, mackerels, scads, carangids, sardines, anchovies, king mackerel and little tunnies dominate the pelagic group and elasmobranchs catfishes, threadfin breams, croakers, snappers and prawns are the important fisheries in the demersal sector.

In Malaysia the maximum sustainable yield estimates are 1,60,000 tonnes of demersal resources and 88,000 tonnes of pelagic resources.

In Malaysia the principal pelagic stocks exploited are those of mackerels, anchovies, sardines, mullets, and scads while the commercial food fishes from the demersal stocks include snappers, groupers, croakers, flat fishes, threadfin breams, catfishes, lizard fishes and prawns.

Among the pelagic resources, the clupeoids, scombrids and carangids are the dominant groups in all the countries. In the clupeoid group, *Sardinella*, and *Stolephorus* are important in all countries except Bangladesh. *Hilsa* replaces these species in the upper east coast of India and Bangladesh. Among the scombroids, *Rastrelliger* is important in all countries. The tunas and tuna-like fishes are exploited to a large extent in Sri Lanka and to some extent in the contiguous area of the Tamil Nadu coast of India. To a somewhat similar extent, this would apply to the Spanish mackerel also, whose importance however, extends further north to the Andhra Pradesh coast of India. Carangids are represented by many genera, the most important of which are *Decapterus, Megalaspis, Carangoides* and *Caranx*. The hair-tails and Bombay duck are more of sub-regional importance, the former in Sri Lanka, India and Bangladesh and the latter in the upper east coast of India and Bangladesh.

Among the demersal resources, the snappers and groupers, croakers, elasmobranchs and penaeid prawns are the important groups in all the countries. Catfishes form a sizeable percentage in all countries except Sri Lanka. The importance of leiognathids and pomfrets is confined to sub-regional pockets-the lower east coast of India for the former and the upper east coast of India-Bangladesh for the latter are the principal areas. Similar are the threadfin breams in Thailand and Malaysia and threadfins in the upper east coast of India and Bangladesh.

There is thus a general similarity in the important exploited fish varieties both from the pelagic and demersal stocks appear in the coastal areas of the countries bordering Indian Oceans. There is also a general similarity in the species concerned, more so in the geographically contiguous waters.

Coastal Fisheries in the Bay of Bengal Region

Marine fishery resources in the east coast of India, bordering Bay of Bengal are constituted of a large variety of fin and shell fishes, typical of tropical waters. These comprise pelagic resources, such as, lesser sardines, ribbon fish, white baits, hilsa, horse mackerel, Indian mackerel, seer fish and flying fish; demersal resources such as, silver bellies, elasmobranchs, sciaenids, catfish, threadfin breams and other perches and pomfrets; crustacean resources, such as prawns, lobsters and crabs and molluscan resources, such as mussels, oysters and clams, cuttlefish and squids.

The eastern seaboard of India has a coast line of about 3000 km and the Andaman Nicobar Islands about 1500 km. The total continental shelf covers an area of about 67000 sq. km upto 50m. depth and 1,10,000 sq. km upto 200m. In the Andaman Nicobar seas the continental shelf covers about 16000 sq. km upto 200m. The average width of the continental shelf is 43 km off the Tamil Nadu coast, 32 km off the Andhra coast and 68 km off the Orissa-West Bengal coasts.

Lesser Sardines

Among lesser sardines, the commercially important species are *Sardinella gibbosa, S. fimbriata, Sualbella* and *S. sirm.* The fishery is restricted to the inshore waters within 25 km from the shore and mainly supported by 0-year class fish. They mature and breed at the end of first year of life and their life span is about 2 years. The first three species mentioned above have more or less the same rate of growth reaching about 120-130 mm total length at the end of the first year, when they attain maturity. *S sirm* attains 170-180 mm length at the end of first year. The lesser sardines have similar food preferences, zooplankton being the most dominant food item.

Ribbon Fishes

Ribbon fishes form fishery of considerable magnitude along the east coast, particularly along Tamil Nadu and Andhra coasts. A limited fishery exists in the Hoogly-Matlah estuaries. The most dominant species caught is *Trichiurus lepturus* which grows upto 1.5m. Average commercial size is 75cm. The other species *Lepturacauthus savala Eupleurogrammus intermedius* and *E. muticus* are smaller in size. Large shoals of *T. lepturus* are caught from the inshore waters of peninsular Indian during August and October. Ribbon fishes breed more than once in a year. *L savala* is the predominant species in Hoogly-Matlah estuaries. Commercial size of this species is 15-50cm. *E. intermedius* is abundant in Palk Bay and *E. muticus* along Orissa coast. Commercial size of *E. intermedius* is 14-35 cm and the life span about 4 years. Commercial size of *E. muticus* is 35-50 cm. Ribbon fishes are carnivorous, the food consisting of crustaceans and fishes. Spawning of ribbon fishes appear to be offshore waters.

Seer Fishes

The fishery is constituted by 3 species *Scomberomorus commersoni, S. guttatus,* and *S. lineolatus.* Seer fishes are caught from the inshore waters by gill nets, drift nets and hooks and line. The fishing season extends from October to March. The spawners have been met during May to July period. The size at maturity is about 50 cm. The fishery is dominated by this size group assigned to the third year class.

White Baits

Stolephorous spp. occur mostly in areas with bottom depths between 20 and 50 m. *Stolephorous heterolobus, S. zollingeri, S. bataviensis, S. commersoni, S. baganensis S. devisi* and *S. indicus* are the common species. The fishery seasons are from May to November along Tamil Nadu and November to April along Andhra coast. Large concentrations of the fish are observed in the Gulf of Mannar during the June-September period. The fishery is contributed mainly by 0-year class fish, the mean age being about 6 months when they first spawn. Spawing is noticed throughout the year in white baits. White baits mainly feed on zooplankton. Life span of the white baits is estimated to be about 2 years.

Silver Bellies

Leiognathus, Secutor and *Gazza* spp are landed in large quantities along the Tamil Nadu and Andhra coasts. *Leiognathus splendens* is the most abundant species, which grows upto a length of 15 cm. *L. equulus* attains the largest size upto 24 cm. The catches of silver bellies comprise mostly of fish less than one year old. Their life span is considered to be less than 2 years. *L. splendens* caught in November to June along south west coast are found in spawning condition. Peak catches are obtained in Madras during October to December and in Andhra, Orissa and West Bengal from January to June.

Jew Fishes

Several species of jew fishes mainly, *Pseudosciaena, Johnius, Otolithoides* and *Otolithus* spp contribute to the fishery along the east coast. *P. diacanthus* the largest of the jew fishes found in south east coast grows upto 120 cm in length. This fish attains sexual maturity at 70-80 cm. Smaller species are caught by trawlers in large quantities in coastal waters. Along the north east coast some species enter the river systems and are fished from the estuaries. The average length of the Ganges jew fish (*Pseudosciaena coiber*) from the commercial catches is about 92 cm.

Pomfrets

There are three species of pomfrets fished along the Indian coasts, the black pomfret (*Formio niger*), the white or silver pomfret (*Pompus argenteus*) and the Chinese or grey pomfret (*Pampus chinensis*).

The black pomfret is fished off the coasts of Orissa and West Bengal mostly from June to September.

Threadfin breams

Off Andhra and Orissa coasts have good resources of threadfin bream, *Nemipterus japonicus* in the area during the January-April period.

Catfishes

Tachysurus thalassinus and *T. tenuispinis* are the major catfishes available along the east coast. *T. jella* and *T. dissumieri* are two other species found in the region. Exploratory trawling upto 100m depth along Andhra, West Bengal coast showed about 28 per cent of catfishes in the catch. *T. thalassinus* of 6-80 cm and *T. tenuispinis* of 6-60cm size ranges are landed commercially.

Mackerel

Of the total mackerel landings of India, 10 per cent only is caught from the east coast. In addition to the major species, *Rastrelliger kanagurta* two others. *R. brachysoma* and *R. faughni* occur in the east coast. Most of the landings on the east coast are along Tamil Nadu and Andhra coasts. The size of *R. kanagurta* in the commercial catches range from 12-23 cm which belong to 0 and 1 year class. In the Andamans in addition to *R. kanagurta*, *R. brachysoma* is also caught. The fish are known to breed all the year round.

Penaeus indicus

Distributed along the entire coast of India upto the 50 m depth zone, juveniles occuring in estuaries and backwaters. The female attend its first maturity at 130 mm. Fecundity ranges from 68000 to 731,000 ova in females measuring 140 mm and 200 mm respectively. At Tamil Nadu, peak spawing activity is observed from May to September. The species feeds on both vegetable and animal matters, consisting of mainly crustaceans. Juveniles grow at an average monthly rate of 10 mm in Chilka lake, 14.4 mm in Ennur estuary, 16 mm at Adyar estuary and 24 mm in Covelong back waters. The adults show a monthly growth of 5.6 mm in males and 7.0 mm in females at Madras. Males and females attain a length of 156/138 mm at the end of first year and 189/181 mm at the end of second year of life. Fishing is supported by the 0-year old in the estuaries and by 0-year (80-120 mm) and 1-year olds (95-175 mm) in the marine region. Within the size range the modal sizes vary from place to place and season to season.

Penaeus monodon

The species is commonly distributed in the north-east coast. Number of eggs the females bear varies from 3 to 7 lakhs. Breeds in the same grounds at *P. indicus*. Its food consists of large crustaceans, vegetable matter, polychaetes, molluscs and fish. Largest recorded size of the species is 337 mm. In the Chilka lake the juveniles grow at a rapid rate of 25 mm per month and at Madras it reaches 160-170 mm size in six months in brackish water. Commercial catches are formed by 0-year and 1-year class. It attains about 250 mm in one year.

Penaeus semisulcatus

The species is more common on the east coast. The size of the female at first maturity is 23 mm carapace length. Fecundity ranges from 67900 to 6,60900 eggs in different sizes. June to September and January, February are peak spawning season in Gulf of Mannar and Palk Bay. The species attain a maximum size of about 250 mm. It consumes large quantities of animal matter as well as diatoms and algal filaments. In the estuary it grows to about 150 mm forming the 0-year class. The marine fishery is contributed by sizes ranging from 120 to 230 mm consisting of both 0-year and 1-year classes.

Metapenaeus monoceros

The species is distributed in both estuarine and marine regions. It attain maturity in the sea after 120 mm size. Fecundity ranges from 1,55,000 to 3,38,000 eggs. Peak spawning take place in July-August and November-December. It grows to a maximum size of 190 mm. The species feeds mostly on small crustaceans. In Godavary estuary migration out of the estuary is mostly nocturnal and immigration mostly at dawn. The estuarine fishery is contributed by 0-year class. Marine fishery mostly contributed by 125-150 mm sizes of the 1-year class.

Metapenaeus dobsoni

The species is distributed to a depth of about 40 m, with large quantities in the brackish water areas. The size at first maturity is 64 mm. Fecundity ranges from 34000 eggs in 70 mm prawn and 1,60,000 eggs in 120 mm size. The species attain a maximum size of 130 mm. It breeds in the inshore waters inside the 25 m depth region. The species is a detritus feeder. Juveniles grow in the estuarine environment at an average monthly rate of about 10 mm. Bulk of the fishery in the back waters and the sea is supported by 3-12 month old prawns.

Metapenaeus brevicornis

The species is distributed in the northern region of the coast. It attains a maximum size of 135 mm and maturity at about 75 mm. In the Hoogly estuarine system there are two spawning seasons, in March, April and July, August. Major food items of the species are vegetable matter and crustacean remains. Growth rate varies with salinity and temperature of the environment. In the estuary the sizes range from 15 to 115 mm constituted by 0 to 2nd year groups.

There are several other species of penaeid prawns which occur in small quantities in the fishery of different areas and different seasons. The important non-penaeid species which contribute to the fishery of mostly Andhra coast and in the northern region are *Acetes indicus, Palaemon tenuips, Palaemon styliferus* and *Hippolysmata ensirostris*.

Lobsters

Spiny lobsters are distributed along the south east coast of India and forms good fishery at Tuticorin, Mandapam, areas and Tamil Nadu. The important species are *Panulirus homarus, P. ornatus* and *P. versicolor*. The first two species are equally abundant. The peak seasons are January to March and July to September. Along the Bengal coast *P. polyphagus* is the dominant species. In the south east coast the sizes of lobsters in the fishery ranges from 110-370 mm.

Crabs

Crab fishery is getting importance in several centres along the east coast. The import species which are exploited from the marine sector are *Portunus pelagicus* and *Portunus sanguinolentus*. In the brackish water environments *Seylla serrata* is dominant. The carapace width of the three species in the fishery are 30-185 mm, 20-94 mm and 30-126 mm respectively.

Orissa

In Ganjain coast of Orissa 148 species of fish belonging to 61 families have been recorded from the trawl catch. This includes 19 species of elasmobranchs 5 species of flatfish and 124 species of round fish. Silverbellis dominated the catch. In all, 13 species of marine prawns were recorded from the catches, the major ones are *Penaeus monodon* (73.23 per cent), *Penaeus indicus* (13.08 per cent), *Penaeus merguiensis* (3.99 per cent), *Penaeus semisulcatus* (0.85 per cent), *Metapenaeus monoceros* (2.46 per cent). The average percentage composition of the commercially important species of fish occuring in the trawl catches off Rushikulya estuary of Ganjam coast are silver bellis (*Leognathus* spp)–22.87 per cent, goat fish (*Upeneus* spp)–10.15 per cent, jew fish and croaker (*Johnius* spp and *Pseudosciaena* spp)–9.89 per cent, silver pomfret (*Pampus argenteus*)–9.47 per cent, cat fish (*Tachysurus* spp.)–8.83 per cent, moon fish (*Drepane* spp)–6.57 per cent, lizard fish (*Saurida* spp)–6.40 per cent, flat fish (*Cynoglossus* spp)–4.17 per cent, horse mackerel (*Caranx* spp)–3.54 per cent, ribbon fish (*Trichiurus* spp)–2.18 per cent threadfin (*Polynemus* spp)–2.11 per cent, leather jackets (*Tricanthus* spp)–1.47 per cent, prawns (*Penaeus monodon, P. indicus, Metapenaeus* spp)–3.04 per cent, elasmobranchs–1.21 per cent and miscleneous fishes–8.10 per cent.

Kerala

The maritime state of Kerala is a narrow strip situated at the south-west corner of the Indian Peninsula between 8°18' and 12°48' north and 74°52' and 77°22' east. Its borders touch the Arabian sea in the west. The continental shelf off Kerala extends over an area of about 40000 sq. km. In contrast to the north west coast of India, the shelf in this area is relatively narrow, the width has an average distance of 64 km compared to 94 km in Karnataka and 174 km in Maharashtra. Nearly 30000 sq. km. or about 75 per cent of the shelf area is within the 75 m depth zone. The northern and major part of the shelf down to a depth of 60 m is muddy, while it is sandy off the two southern districts, Quilon and Trivandrum. The sea bottom is hard and slightly uneven in the 60-90 m depth range, while it is rocky with coral formations in the 90-120 m range. The slope of the shelf between 120 m and 200 m depth appears to be hard and even, especially the area between Cochin and Cannanore. Stray occurrences of rocky grounds are noticed in this depth range between Ponnani and Alleppey.

During the southwest monsoon from May to September a southerly current is generated with a thermocline tilt along Kerala coast. This upwelling is a very shallow one, but the surface water is about 6°C, cooler here than elsewhere. Combined with relatively low salinity of the coastal waters due to the influx of freshwater from rivers and backwaters and by precipitation, the peak algae production occurs during this time. In the post-monsoon period of September-October there is in addition a true wind-induced upwelling area between Alleppey and Quilon, with positive effects on plankton production. In several locations along the coast between Quilon and Cannanore the annual phenomenon, mud-banks are formed during the south-west monsoon in June-July, which lie in the shallow sea adjacent to the coast. These mud-banks extend seawards for a few kilometers in a semicircular form and are storehouses of nutrients like phosphates and promote fish plankton production. Exceptionally high amounts of fish and prawns collect around these mud banks. The sea between the mud banks and the coast is calm inspite of the monsoon thus facilitating fishing operations by traditional craft. The yearly occurrence of upwellings and mud banks and thus rich plankton production results large assembledge of so many species of fish and crustaceans in Kerala's coastal waters.

Six groups of fish, namely, sardines, mackerels, anchoviellas, ribbon fish, prawn and perches account for 73 per cent of total marine landings. The peak fishing season lasts from July to December during which 70 per cent of the total landings are obtained. Contribution of important pelagic and demersal fish groups are:

Fish Group	*Season*	*In Percentage of Total Marine Fish Landings*
Pelagic		
Sardines	October–January	34.4
Mackerels	September–March	6.3
Ribbonfish	June–October	6.3
Anchoviellas	September–October	5.1
Demersal		
Prawns	June–September	13.0
Perches	July–October	7.5
Others		27.4

Commercially important marine fish species occur in Kerala coast are; *Sardinella longiceps, Sardinella fimbriata, Pellona indica, Kowala coval, Thrissocles mystax, Anchoviella indicus, Chirocentrus dorab, Megalops cyprinoids, Tachysurus dussumieri, Mugil* spp, *Polynemus* spp, *Serranus* spp, *Sear crumenophthalmus; Caranx* spp, *Chorinemus laysan, Rachycentron canadus Lutjanus* spp, *Sacutor* spp, *Otolithus ruber O. argenteus, Pseudosciaena diacanthus, Trichurus haumela, Trichurus savala, Rastrelliger* spp, *Katsuwonus pelamis, Scomberomorus* spp, *Parastromateus niger, Pampus argenteus, Cynoglossus* spp, *Scoliodon* spp, *Caracharinus* spp, *Sphyrna blochii, Dasyatis* spp, *Penaeus monodon, Penaeus indicus Metapenaeus dobsoni, Parapaenopsis* spp.

Sardines comprising of *Sardinella longiceps* and *Sardinella fimbriata* form 34.4 per cent of total fish landings in Kerala coast, including white sardine, *Kowala coval.* Next in importance include 13.9 per cent of prawns and other crustaceans. Important species are *Penaeus monodon, Penaeus indicus, Metapenaeus dobsoni, Parapenaeopsis.* Perches contribute 7.5 per cent of the catch represented by *Pseudosciaena diacanthus, Lutjanus* sp. Mackerel and ribbon fish each contributing 6.3 per cent by *Rastrelliger kanagurta, Trichurus haumela* and *T. savala.* Anchoviella form 5.1 per cent mainly by *Anchoviella indicus* and *Thrissocles mystax* Sciaenids, comprising of *Otolithus ruber* and *O. argenteus* from 3.3 per cent of the total landings. Elasmobranchs contribute 2.6 per cent of the catch mainly by the species like *Scoliodon* spp, *Carcharinus* spp, *Sphyrna blochii* and *Dasyatis* spp. Catfishes comprise 2.4 per cent of the catch mainly by *Tachysurus dussumieri.* Other varieties of 18.2 per cent include species like, *Pellona indica, Chirocentrus dorab Megalops cyprinoides, Mugil* spp, *Polynemus* spp, *Serranus* spp, *Caranx* spp, *Chorinemus laysan, Euthunnus pelamis, E. thratus, Scomberomorus* spp, *Parastromateus niger, Pampus argenteus, Cynoglossus* spp.

South-west Coast of India

In the shallow waters of south west coast of India, the catch composition of the major fish groups for a three year period from 1969-72 has been observed as 13.2-19.00 per cent catfishes, 6.08-22.03 per cent kilimeen, 1.71-3.00 per cent lizard fish, 0.34-2.66 per cent ribbon fish, 7.62-15.2 per cent elasmobranchs and 54.06-62.97 per cent miscellaneous and other varieties. The less dominant varieties quality fishes include pomfrets, perches, carangids, barracuda, white fish (*Lactarius* spp), sciaenids, green fish etc.

The average maximum catch of 115 kg of cat fishes per trawling hour has been obtained between the depth range of 32-40 m. and the maximum individual catch of 1000 kg per hour was obtained at 42 m. depth. As such the catfish concentrations are maximum at 32-50 m and the intensity of catch is found to decrease as depth increases and towards the shallow depths the catch intensity is found to be comparatively better than the deeper zones beyond 50 m.

The average maximum catch of 188 kg of kilimeen per trawling hour has been obtained between 42 and 50 m. depth with the individual maximum of 893 kg/hour at a depth of 42 m. The intensity of kilimeen is comparatively high between the depth range of 32-50 m. The catch intensity increases as the depth increases, but from the depth range of 52-60 m onwards a reverse effect is noticed.

The maximum average catch of elasmobranchs per hour of 83 kg was obtained from a depth range of 42-50m. with an individual maximum catch of 294 kg. per hour at a depth of 36 m. In the case of elasmobranchs also, the catch intensity is high between 32-50 m. range. A gradual depth wise increase upto 50 m is noticed from where it decreases.

A maximum average catch of lizard fish at 167 kg. per hour was obtained at a depth range of 42-50 m. with the individual maximum catch of 333 kg/hour at 42 m. depth. The intensity of lizard fish is high between the depths of 32-50 m.

The catches of ribbon fish is comparatively better in the shallow depths of 20-40 m.

Dense shoals of catfishes were not observed very often along these areas as against those found in east coast. It is found that catfish is making diurnal vertical migration. The fishes are found to descend from the mid-water during the early hours of the day and then scattered over the bottom. The depth at which they remain during the day time vary between 4-6m. from the bottom. The shoaling tendency is seldom noticed in the catfish when they are at the bottom. The downward migration of catfish during the morning hours other than light is also influenced by its feeding habits. Similarly the upward movement during the evening hours may be because the bottom living organisms also move up and spread during night. Catfish can be grouped as demersal fish with day time occurrence on or close to the bottom, migration and dispersal into the water mass above during sun set and discend to the bottom during sun rise.

Out of the 37 fishing grounds between Tanur and Cape Comorin about 11 grounds are comparatively more potential for the dominant fish groups considering the average catch per hour. Six grounds located between off Ponnani and Carangannore and five grounds located between off Alleppy and Quilon have better concentrations of catfishes. For kilimeen four grounds located off Cannanore and three grounds off Cochin, one located south west of Alleppy and five located off Quilon have yielded a maximum. Four grounds between Ponnani and Cochin and seven located between Cochin and Alleppy are better for elasmobranchs. For lizard fish, four grounds between Cochin and Trivandrum yielded the highest. Regarding ribbon fish catches above 70 kg/hour of trawling was not obtained from even a single ground.

Of five depth ranges higher catches for all the species except ribbon fish was observed to be between the depth range of 50-60 m. For ribbon fish comparatively better catches were obtained from the shallow depths of 20-40 metres.

A maximum appearance during March for catfishes, July-August for kilimeen, July for elasmobranchs, October for lizard fish and March for ribbon fish have been observed.

On the occurrence and distribution of some commercially important fishes in shallows waters (20-62m) of the south-west coast of India, between Tanur and Cape comorin (latitude 8°-11°N and longitude 75°-78°) reveals that the predominant varieties are.

1. *Cat fish*: The common species obtained under this group are (a) *Tachysurus thalassinus*, (b) *T. dussumurei*, (c) *T. maculata* and *T. sona*.
2. Kilimeen, composed of single species, namely, *Nemepterus japonicus*.
3. Elasmobranchs, composed of sharks, skates and rays belonging to the genera; (a) *Carcharhinus*, (b) *Scoliodon*, (c) *Dasyatis*, (d) *Actomylus*, (e) *Aetobatus* and (f) *Rhynchobatus*.
4. Lizard fish, composed of single species, *Saurida tumbil*.
5. Ribbon fish, consisting of dominant species, like, (a) *Trichiurus lepturus*, (b) *Lepturacanthus savala* and (c) *Euplereogrammes muticus*.

Proportionately smaller quantities of (a) pomfrets, (b) perches, (c) carangids, (d) barracuda, (e) white fish (*Lactarius* sp), (f) sciaenids and seer fish were also found in the trawl catch.

The catfish concentrations are maximum between the depth range of 32-50m and the intensity of the catch is found to decrease as depth increases and towards the shallow depths the catch intensity is found to be comparatively better than the deeper zones beyond 50 metres. They concentrate in large numbers during February-March and during October-November. Cat fishes appear in more numbers gradually towards mid day. They leave the area during late hours of the day. In the early morning the arrival of cat fishes are less.

The concentrations of *Nemipterus japonicus* is comparatively high between the depth range of 32-50m. The availability of the species increases as the depth increases, but from the depth range of 52-60m onwards a reverse effect is noticed. During the months of July and August the species appears in large numbers; while in the month of May minimum number of fishes concentrate in the region.

Nemepterus japonicus is a column water fish coming down to the bottom often in less densely packed fish shoals.

Elasmobranchs are also found to roam in large numbers at a depth rage of 32-50m. Their concentrations show a gradual increase upto 60m. depth. Maximum concentrations of elasmobranchs occur during the month of July, though only a few of them are available in the month of June.

The depth wise intensity of *Saurida tumbil* is observed to be of the same pattern as in the previous fisheries and the intensity of lizard fish is high between the depths of 32-50m. Maximum congregations of the species is observed during the months of August, September and October. No appreciable quantity of lizard fish is observed during the months January to June.

Species of ribbon fish concentrate and roam about in shallow depths, between 20-40 metres. The fishes occur in maximum numbers during the months of February and March, which decrease in numbers with the progress of summer and is observed as minimum in the month of May.

Andaman and Nocobar Islands

Andaman and Nicobar Islands are characterized by fringing coral reefs and in some cases atoll formations with their lagoons.

The islands being of a typical oceanic nature and of volcanic origin the continental shelf is either very narrow or non-existant around some of the islands. The shelf on the western side of the islands is much wider than the east coast reaching about 8 nautical miles at certain points. The length of the coast line of the Andaman and Nicobar Islands is around 1500 km. The total shelf area available for exploitation is around 16,000 sq. km. The climate of the islands is typically tropical with heavy rain receive from both the south west and north east monsoons. The temperature is equitable, the range being 23°-30°C. The average annual rainfall is around 3000mm.

Andaman and Nicobar waters are qualitatively very rich in fish fauna. Over 400 species have been reported from these areas. But only about 60 species are commercially important of which most of them are commonly represented in trawl catches. They are black shark (*Charcharius* spp), cat shark (*Chiloscyllium indicum*), balfour's shark (*Hemigaleus balfouri*), dog shark (*Scoliodon* spp), hammer head shark (*Sphyrna blochii, S. zygaena*), shovel nose (*Rhynchobatus dijiddensis*), pointed saw fish (*Pristis cuspidatus*) small toothed saw fish (*Pritis microdon*), blue spotted sting ray (*Amphotistius kuhlii*) spotted eagle ray (*Aetobatus narinari*), eagle ray (*Myliobatis nichofii*), Javanese cow ray (*Rhinoptera javanica*), devil ray, (*Mobula diabolus*), whip-tail sting ray (*Himantura uarnak*), car fish (*Tylosurus* spp), scad (*Decapterus russelli, Selar mate*), torpedo trevally (*Megalaspis cordyla*) queen fish (*Chorinemus toli*), six banded trevally (*Caranx sexfasciatus*), black tipped trevally (*Caranx melampygus*), rainbow runner (*Elegatis bipinnulatus*), black tipped sardine (*Sardinella melanura*), tembang (*Sardinella jussieu*), spotted herring (*Harengula ovalis*), toother shad (*Pellona ditchela*), spotted herring (*Herklotsichthys punctatus*), wolf herring (*Chirocentrus dorab*), rainbow sardine (*Dussumieria* spp), short jaw anchovy (*Thrissocles* spp), white bait (*Anchoviella* spp), half beak (*Hemirhamphus marginatus*), silver belly (*Leiognathus equulus*), bream (*Lethrinus* spp), white fish (*Lactarius lactarius*), cock-up (*Lates calcarifer*), snapper (*Lutjanus* spp), silver belly (*Liza* spp), grunter (*Pomadasys* spp), tassel fish (*Polynemus* spp), jew fish (*Johnius* spp), croaker (*Otolithus* spp, *Sciaena* spp), rock cod (*Epinephelus* spp,) silver whiting (*Sillago sihama*), lizard fish (*Saurida tumbil*), barracuda (*Sphyraena jello*), mackerel (*Rastrelliger kanagurta*), chub mackerel (*Rastrelliger brachysoma*), Spanish mackerel (*Scomberomorus commersoni*), spotted spanish mackerel (*Indocybium guttatum*), wahoo (*Acanthocybium solanderi*), frigate mackerel (*Auxis thazard*), little tuna (*Euthynnus affinis*), dog tooth tuna (*Gymnosarda unicolor*), skipjack (*Kutsuwonus pelamis*), yellowfin tuna (*Neothunnus macropterus*), northern blue fin (*Krishinoella tongol*), marlin (*Makaira* spp), short nosed marlin (*Tetrapturus brevirostris*), sail fish (*Histiophorus gladius*), threadfin bream (*Nemipterus japonicus*), goat fish (*Upeneoides* spp).

Elasmobranchs constituted 18.1 to 27.4 per cent of the catch followed by 1.15 to 6.8 per cent cat fish and 0.95 to 6.4 per cent perches. Small miscellaneous fishes, which constitute 71.9 per cent is composed of leiognathids–28.8 per cent, sciaenids–12.9 per cent, nemipterids–2.5 per cent, upeneoids–10.0 per cent, lizard fish–1.0 per cent and others–15.7 per cent. One peak representing high catches always occur during the months January-April after the north east monsoon. This period coincides with the pre-south west monsoon months, when the temperature of the sea water starts rising. Then there is generally low catch in the month of May followed by uncertain catch for next three to four months.

Small miscellaneous fishes formed the most important group comprising of 65 to 85 per cent of the total catch in different areas. Of this, leiognathids, sciaenids and upeneoids are quantitatively more abundant. The percentage of elasmobranchs varied from 18 to 27. Next in importance are cat fish and perches. The seasonal abundance and the variation of the catches show that the period from October to March is generally more productive. High catch rates have been obtained in July and August also. Good resources of sharks and marlines exist south of North Andamans and east of Little Andamans including the invisible bank. Schools of tuna particularly the skipjack (*Katsuwonus pelamis*) are sighted frequently. The occurrence of *Alopias vulpinus*, the thresher shark is also indicative of the presence of good school of lesser tunas. Schools of sardines and mackerels as well as skipjack and other varieties of tuna are very frequently encountered in the Invisible Bank. The most important constituents of the catches are, in the order of priority, perches, caranx, seer, mullets, mackerels, anchovies, sardines and silver belly. The best fishing season appears to be the south west monsoon and post-monsoon months of July to December. There is a distinct lowering of catches during January-April. The fishery begins to build up from May onwards. The best fishery of tunny and seer appears to occur during the months preceding south west monsoon and the fishing of these two are at the lower ebbs during the south-west monsoon and subsequent months. The marine fish production of Andaman and Nicobar Islands is mainly contributed by perches, sardines mackerels, caranx, seer, anchovies, silverbellies, tunnies and sharks and rays. In 1980, the per cent composition of the catch around the islands were hilsa-2.98 per cent, pomfrets–0.83 per cent, sardines–13.49 per cent, anchovies–5.47 per cent, barracuda–3.73 per cent, hemirhamphus and benone–2.26 per cent, caranx–8.16 per cent, seer–6.52 per cent, mugil–6.50 per cent, perches–16.73 per cent, sharks and rays–3.12 per cent, mackerel–10.13 per cent, silverbellies–5.68 per cent, tunnies–3.06 per cent, catfish–1.75 per cent, prawns–2.97 per cent, miscellaneous varieties–5.20 per cent.

The results of exploratory fishing recorded the highest catch of 124.1 kg/hour in 50-59m depth range followed by 102 kgs in 40-49m and 95.4 kgs in 30-39 m. depth range during 1980. The highest catch rate was observed in respect of silver bellies (44.7 kg/hour) followed by sciaenids (25.1kg/hour), goat fishes (5.4 kg/hour), sharks and rays (4.1kg/hour), perches (2.2 kg/hour) and other varieties (38.7 kg/hour). In tuna long lining, hooking rate was worked out to be 3 per 100 hooks.

The availability of perches in the rocky grounds of the Islands enables exploitation of the resources with suitable hand lines or perch traps. Tuna is reported to have significant abundance around the Great Nicobar, South of Car Nicobar and the southern regions. Stocks of tuna species such as skip jack, yellowfin, and big eye are estimated at 75,000 tonnes around the seas of the Islands.

Perches constitute about 15 to 18 per cent of the average annual landings representing the important generas, like, *Lethrinus, Lates, Lutjanus, Pomadasys, Epinephelus, Pristipomoides* and *Aprion*. Thirty three species have been reported in the catch and the main fishing season is from August to November. Perches are for the most part fished by hook and lines. A small percentage is also caught by gill nets. Off Little Andaman very good catches of perches were obtained on the limited hauls by the trawl net.

Next to perches, carangids (Horse mackerel and Scads) form the important group in artisinal fisheries. This group is represented by six generas, namely, *Decapterus, Selar, Megalapsis, Chorinemus, Caranx* and *Elegatis* and July to November is the best fishing period for them. Most of the catches are with gill nets and boat seines. An appreciable quantity is also taken through hooks and lines.

Carangids represent 28.0 per cent of the catch obtained by trolling. The areas east and south of Port Blair are very rich in carangids, some of them weighing more than 20 kg each.

Seers constitute about 6 to 8 per cent of the total annual catch. They are caught in the Islands mostly through gill nets. In trolling line however, 13.7 per cent of the catch is represented by the seer fish. Though the group is constituted of 2 genera and 3 species, namely, *Scomberomorus guttatus, S. commersonii* and *Acanthocybium solandri*, the commercial fishery is mostly made up of former two species. The later is occassionally caught near the reefs and do not constitute a significant fishery. The best season for the fishery is from March to August synchronising with that of the southern portion of the west coast of Malaysia.

The mullets contribute to almost as much as seer fish in the total landing of the island. The main genera are *Mugil* and *Liza*. Though it occurs almost throughout the year, the important fishing season appears to be from July to December, synchronising with the low salinity and high phytoplankton production period. They are mostly caught in gill nets and boat seines.

Schools of Mackerel are very frequently encountered around the Islands and are represented by two species, namely *Rastrelliger kanagurta* and *R. brachysoma* of which the former is a commercially important one. Mackerel constitute about 8 per cent of the total annual landings. They are caught in gill nets and boat seines. The fishery shows a bimodal fluctuation, one peak being March to June and another from September to December.

The anchovies constitute an important fishery representing 6 per cent of the total landings in the inshore areas of Andaman Islands, of which *Thrissina baclena* is the most important. The other genera which form the complex group of anchovies are *Anchoviella*. The best fishing period is June to November and they are caught by shore seines and boat seines.

Sardines contribute to a significant portion about 15 per cent of the annual landing. The genera contribute to this group are *Sardinella, Harengula, Pellona, Dussumeria* and *Herklotisichthys*. Of these *H. punctatus* is the most common and most important species constituting as much as 75 to 80 per cent of the sardine group in the fishery. Schools of sardine occurs in the inshore waters throughout the year, but the peak season is from July to December. The commercial fishing for sardine is restricted mostly to the eastern side of the South Andamans. A very little fishing efforts for sardines has been put in Nicobar and West of Andamans due to coral and barrier reef, which are seemed to be the abode of tertiary resources of *Sardinella* of considerable quantities.

Silver bellies, represented by two genera, namely, *Leiognathus* and *Gazza* account for about 5 to 6 per cent of the total annual catch. In the trawl catch they are represented about 28 to 29 per cent. Boat seines and shore seines account for the majors portion of the catch and the main fishing season is from June to December. More than 90 per cent of the fishery is composed of *Leiognathus*.

Tuna is one of the most important pelagic fish species occuring in the Andaman and Nicobar seas. The important tuna species found around the islands are skipjack (*Katsuwonus pelamis*), yellow fin, (*Thunnus neothunnus albacores macropterus*) and big eye (*Thunnus parathunnus abesus mebachi*). Around the Andamans, especially in the eastern region of upwelling, the hooking rates are reported to be the highest (more than 2 per cent for yellowfin and more than 1 per cent for Big eye), but then the period is limited to that of upwelling in December-January. A stock of about 25,000 tonnes of yellowfin and big eye has been estimated for Andaman and Nicobar region based on the hooking rates and yield. The region is also very rich and important for skipjack resources. Very little of this is caught in the gill nets and boat seines. Trolling was found particularly successful for schooling species like skipjack. Tuna constitute 16.5 per cent of the troll catch of the exploratory fishing. Marlins with recorded hooking rates of 1.5 per cent has been observed in the month of April to July.

Two species of lobster, namely, *Panilurus ornatus* and *P. polyphagus* are most important in the region commercially. Specimens of *Penaeid* prawn have often appeared in the fish trawl catch.

The regions around the entire Andaman group of islands have a very rich potential of sharks. The percentage of elasmobranchs varied from 18 to 27 indicating that shark fishing on a commercial way has much scope in the region. The elasmobranch fishery in the area is represented by six species of sharks, namely, *Charcharias spp*, *Chilloscyllium indicum*, *Hemigaleus balfouri*, *Scoliodon* spp, *Sphyrna blochii*, *S. zygaena*, three species of skates, namely, *Rhynchobatus dijiddensis*, *Pristis cuspidatus*, *P microdon* and six species of rays, namely, *Amphotistius kuhlii, Aetobatus narinari, Myliobatis nichofii, Rhinoptera javanica, Mobula diabolus* and *Himantura uarnak*. The composition of elasmobranchs in the trawl catch varied from 18.1 to 27.4 per cent. In long lining the overall hooking rates of 2.94 to 4.59 per cent were mainly due to the high percentage of sharks, *Charcharias* spp is the most common variety of sharks caught in the area. Thresher sharks *(Alopias vulpinus)* are also caught very frequently. Sharks constitute 10.8 per cent of the catch obtained by trolling line.

Small miscellaneous fishes, namely, sciaenids, nemepterids, upeneoides, lizard fish and others constitute 36 to 56 per cent of the trawl catch in different areas of the islands. Of this sciaenids and upeneoides are quantitatively more abundant. The seasonal abundance and the variation of the catches show that the period from October to March is generally more productive than the other periods, but high catch rates have been obtained in July and August also.

The Andaman and Nicobar Islands have a marine coast covering a chain of 348 islands. The topography of the Islands suggest the existence of two sperate continental shelves, one for the main Andaman and Middle Nicobar group of islands and other for Nicobar group. The continental shelf is fairly steep encountering great depths (more than 100 fathoms) within a short distance. The northern most island in the group is the Land Fall Island which is 900 km from the mouth of Hoogly river. The principal islands from north to south, the North Andaman, Middle Andaman, South Andaman, Baratang, Rutland and Little Andaman as well as the Richie's Archipelago east of South Andaman. The Andaman group of Islands form a part of the broken chain of islands extending from Mayanmar to almost the Indonesian group of Islands. The other group is the Nicobars which are seperated from the Andaman group by the 10 degree channel. The southernmost tip of the Nicobar Group *i.e.* Pygmalion Point is only 150 km from Sumatra. The Islands being of typical oceanic nature have a total shelf area of 16000 sq. km. available for fisheries exploitation. High catch rate of demersal fish by bottom trawling show high catch rate (99-241 kg/hour) in the bay off Neil, Havelock and South Andaman. The bumper catches in individual hauls in certain occassions have also been obtained from these areas ranging 462 to 1111kgs/hour. Highest concentration of fish appears to be in the depth range of 30-50 m.

The catch from tuna long lining operated by exploratory fishing indicate that east off South Andaman are most productive with hooking rate of 2.9 to 4.15 and a catch of 121.3 to 150.6 kg per 100 hooks for all types of fish. The highest catch rate per day was 504 to 690 kg per 100 hooks with hooking rates of 2.6 to 8.8 per cent in the same area. When trolling lines are operated the area is found to be more productive. Tunas and barracudas are the major constituents of the catch.

Hut Bay and Dugong Creek off Little Andaman are found to be rich in demersal fish resources. A catch rate of 304 to 620 kg of fish is recorded in the area at a depth of 40 metres, the main constituent of the catch being perches. Scombroid and shark catch of the area are significant in long line and troll line. Of the scombroid the marlins and sail fish are the most common. Rangat Bay and Henry Lawrence Island are found to be rich in seer fish, perches and sharks and a good catch is obtained in trolling lines. Penaeid prawns are also predominent in the area and are caught in shore seine and boat seine during post-monsoon period.

The Invisible Bank which is situated about 96 km south west of Port Blair and north east of Little Andaman is known to be a promising area for fishery resources. The bank has an area of about 250 sq miles. This is a submerged bank over which the depth of water varies from 20 to 200 metres. Trolling line operation in the area yield a catch rate of 25 to 40 kg per hour comprising of tuna, carangids, barracuda, shark, seer fish, perches and croakers.

Campell Bay has assumed certain strategic importance considering the concentration of tuna around 5° to 6° latitude on either side of the equator. The most significant tuna grounds are south of Çar Nicobar, especially around the Great Nicobar and southern region. A majority of fish stock is concentrated south of Car Nicobar because of the higher salinity and temperature in the region.

Bangladesh

Coastal area of Bangladesh cover 37000 square km between Lat. 20.4° N-22.0°N and long. 89°E is not deeper than 50m. The standing stock of the Bangladesh continental shelf has been estimated at 552,500 tonnes consisting of demersal fish (318500 tonnes) pelagic fish (200000 tonnes), crustaceans (9000 tonnes) and 25000 tonnes of others. Annual yield from the coastal waters has been estimated at 100 000 tonnes comprising of 475 species of fin fish and 25 species of shrimps. Of the total yield 95 per cent comes from small scale fishing and rest from trawl catch. Dominent species of the catch by groups are cat fish, Indian salmon, Bombay duck, sharks and skates, jew-fish and eels.

At a depth of 6-64 m, the following group of fishes appear the Bangladesh coastal waters. The highest densities of demersal fishes are observed off–Cox's Bazar at 20-30 m depths. The dominant species in the area of Cox's Bazar are cat fish and Bombay duck, *Harpodon nehereus*. In the depth zone of 10-24 m, the catches are dominated by croakers, scianidae sp. catfish and Bombay duck. Catfish and croakers are also among the important species at 25-49 m depth zone. At 50-74 m depth. catfish contribute about 75 per cent of the catch. At 75-99m depth the contribution of catfish to the total catches decrease sharply and the most important species appear to be threadfin bream, *Nemipterus japonicus* followed by bigeye, *Priecanthas hamrur* and lizard fish, *Saurida* spp. At depths deeper than 100 m, these species also dominated the catches.

The highest concentrations of pelagic fishes are found in N 20°15', E 91°20' which consist of Indian mackerel, *Rastrelliger, kanagurta.* At depths of 10-24 m, most of the pelagic fish catches consisted of members of the clupeidae and Engraulidae families with *Sardinella fimbriata* and *Setipinna taty* as dominant species. The most abundant species down to 75m. depth is *Carangoides malabaricus*, while round scad *Decapterus maruadsi* are dominant at 75-100m. At depths deeper than 25m *Ilisha megaloptera* dominate catches of the clupeidae family. The most abundant species of Scombridae and Leiognathidae families are *Rastrelliger kangurta* and *Leiognathus bindus* respectively. Lantern fishes of Myctophidae dominate the pelagic trawl catches over bottom depths of more than 150m.

The species composition of trawl catch operating at 6-64m depth are, catfish (30 per cent), sharks, skates (16 per cent) jew fish (15 per cent) salmon (14 per cent), ribbon fish (7 per cent), Bombay duck (6 per cent), eel (4 per cent), croakers (4 per cent), pomfrets (2 per cent) and misc. fishes (2 per cent).

Hilsa ilisha

It is a medium sized (upto 60 cm), planktonivore, essentially an estuarine fish but is abundantly caught in the Meghna, the Padma and some other rivers in Bangladesh. It shows a typical anadromous migratory pattern ascending major rivers for spawning. It migrates to fresh water in two runs, in monsoon and in winter. The area from Chandpur to Hatia and Chandpur to Goalando are the major hilsa fishing ground. Mature fishes are mostly caught in the Meghna and Padma rivers from middle of May to middle of October. The fingerlings, *Jatka* are caught in the rivers Buriganga, Sitalakhya and Meghna from January to June. The growth rate of the species is more rapid during the hot season than in winter. Fry increases in length 2.5 cm per month. It generally attains maturity at the age of two years. It feeds on zooplankton and phytoplankton. From the rivers Meghna and Padma, hilsa are being caught by mainly drift gill net and shangla jal, kona and bundh. Each individual hilsa carry 2,34170 to 15,77600 eggs.

Tenualosa ilisha (Ham), a diadromous fish, is the largest single fishery in both the inland and marine waters of Bangladesh. About 16 per cent of the country's total production of 1.3 million tonnes, is generated by this fishery. Hilsa fishery provides livelihood directly or indirectly to about 2.5 million people *i.e.*, 2 per cent of the total population. The reduction in depth and discharge of rivers due to construction of dams and barrages has affected the spawning, feeding and migration of this species. Despite increasing fishing effort in the upstream rivers, landing of hilsa from the inland waters has sharply declined. However, total production has remained stable due to an increased harvest in the marine sector. Until 1972, the upstream rivers were characterized by high abundance of hilsa. Subsequently, abundance in the upstream rivers started declining gradually, though simultaneously it increased in the downstream, the estuaries and the sea.

T. ilisha are mainly caught by gill nets from Bangladesh waters and the size of the catch ranged from 3.0 to 57.0 cm. The species is recruited in the fishery throughout the year with one peak. The peak pulse produced 19.81 per cent of the recruits on an average. Maximum number of hilsa are caught between 31 and 45 cm, 53 and 55 cm with a fishing mortality rate exceeding 2.29 per year in the mid length of 43 cm. The highest peak of fishing mortality occurs in the length range between 31 to 45 cm, which is due to the excessive use of monofilament nets of mesh size 6 to 10 cm. Another smaller peak in the catch occurs in the length range between 53 to 55 cm which is due to use of *Chandi jal* (gill net, mesh size 12 to 16 cm). Individuals of *T. ilisha* of 3.0 to 57.0 cm in total length correspond to 0.44 to 2450 g in weight.

The value of annual catch, total annual stock, standing stock and maximum sustainable yield recorded were 2,14519 t, 3,35,185 t, 86152 t and 1,62,396t respectively. The value of maximum sustainable yield is below the annual catch indicating high fishing pressure on the hilsa stock in Bangladesh. The present fishing pressure of 2.49 per year needs to be reduced near to 1.88 per year to obtain maximum sustainable yield from the stock.

Hilsa kanagurta

The species almost resembles *H. ilisha* but it inhabits only in the marine environment with its spawning and feeding ground in the sea. It grows to a maximum size of 52 cm. The fish is caught in large commercial quantities from the coastal waters of. Bangladesh, particularly off shore waters of Cox's Bazar. Gill nets and local purse-seine nets are used to catch this species.

Harpodon nehereus

Bombay duck is small to medium-sized (40 cm maximum length), a predator–cum–scavenger which ascends to the upper part of the estuarine zone. It is one of the important fishery species of the upper Bay of Bengal. The fish is caught in good quantities from the shallow waters of the estuaries, mainly in the river months by trawl nets.

Tachysurus sp.

This species is purely a marine form and found in large quantities in the estuaries and adjacent off shore waters. *Tachysurus thalassinus*, also called giant catfish grows to a maximum of 150 cm. One of the important commercial species of Bay of Bengal, caught by trawl nets.

Polynemus indicus

One of the most abundant commercial marine species which are caught in large quantities from marine habitat by trawl nets and from shallow estuarine waters by fixed gill nets. The species attain an average size of 70-80 cm and the maximum size recorded was 142 cm.

Lates calcarifer

Large predatory fish, grows to 180 cm in length and may weigh 90 kg. It frequents the mouths of rivers and occassionally ascends them for fair distances. It is usually taken by gill nets in good quantities from the near-shore shallow estuaries and river mouths at depths upto 8-10 m. The species spawns in the high saline zone of the estuary from May to October and feeds on fishes, shrimps, snails and worms.

Rastrelliger kanagurta

Indian mackerel is one of the most abundant commercial species of the Bay of Bengal. It is a column feeding planktonivore. Maximum size of the species is recorded at 35 cm length, but the most common size is 20-25 cm. The species is caught mainly by gill nets and purse seine nets.

Palaemon styliforus

It inhabits shallow coastal waters and brackish waters of the estuaries. The maximum size of the species recorded is 10 cm. The breeding period extends from October to July. Hatching of the species occurs in the more saline zone of the estuary or inshore waters.

Penaeus indicus

The adults inhabit the sea, where as fry and juveniles are carried into the estuary and inshore waters by the tide and current. It is the most important species of shrimps on account of its great abundance both in the estuarine and open waters. It is a bottom feeder and omnivorous. The species is reported to have a deep water spawing habits.

Penaeus monodon

Tiger prawn is a large 300 mm in length and one of the most important commercial shrimps of Bangladesh. Fry are reported to be available the year round from the estuarine area of Bangladesh. Adults and mature specimens are abundantly available in Bay of Bengal off Cox's Bazar. The species is omnivarous and a bottom feeder. The breeding period of the species extends from January to

April. The spawning ground of the species seems to be the off shore of Cox's Bazar. Chakria–Sundarban and Khulna-Sundarban estuaries are known as the main seed collecting centres in Bangladesh.

Metapenaeus monoceros

The species inhabits sea and brackish water zones of the estuary. Juveniles are found in estuaries and back waters of reduced salinity, adults occur in the sea. The species is reported to breed throughout the year with two peaks, one in July and August and the second in November and December. The adults are generally available in the sea in slightly deeper waters than the other species of *Metapenaeus*. It is recorded to grow upto 155 mm from Bangladesh waters.

Sri Lanka

Sri Lanka is situated in the Indian Ocean, south east of India, between lat. 6-10°N and long. 80-82°E. The coast line is 1760 km in length. Fishing takes place all round the coast. It is concentrated primarily within the continental edge, which rarely extend beyond 40 km, and average 23 km in width. In this area there are good resources of pelagic and demersal fish species with an annual sustainable yield of 250000 tonnes.

The mackerel shark, *Lemna litropis* appears in the catches from grounds south of 30°S, the lancet fish, *Alepisaurus borealis* appears in the long-line catches from all latitudinal ranges but probably more frequently in the latitudes of the south. In the central part of the north-equatorial region *Lemna falciformis* has the highest density of distribution for any shark species in the tuna ground of the Indian Ocean.

Skipjack tunas contribute about 10-12 per cent out of the estimated annual production of 100000 tonnes of fish.

Fishes available around Sri Lanka are, seer fish (3 per cent), carangids (6.8 per cent), skip jack (5.6 per cent), yellowfin tuna (4.14 per cent), sharks (5.33 per cent), skates (3.3 per cent), rock fish (12.1 per cent).

Malaysia

Marine fishes which appear in the coastal waters of the east and west coasts of Peninsular Malaysia bordering Indian Ocean are quite distinct. The coastal fishery resources found within 80-96 km of the coast line have been grouped into demersal and pelagic varieties. The demersal fish resource contains a wide variety of demersal and semi-pelagic species ranging from large fish of commercial importance, such as, snappers and groupers to small ones which are generally classified as trash fish for animal feed and fish meal. The pelagic fish resource includes important pelagic schooling species like the chub mackerels, anchovies, tunas, sardines, and herrings as well as round scads and yellow banded scads. The prawn resources, on the other hand, contains only the penaeid prawns which include high priced species like, *Penaeus monodon* and *Penaeus merguiensis*.

The bulk of the demersal fish stocks are exploited by the otter trawl. Other fishing gear employed include bag nets, hand lines and portable traps. Demersal fish landed by the trawlers accounted for some 84 per cent of the demersal fish landings. The demersal fish landing includes a very large number of species, none of which is individually very abundant. Some of the more common families landed are the Sciaenidae, Cynoglossidae, Nemipteridae, Mullidae, Tachysuridae and Synodontidae. In addition of these, the higher priced demersal fish species landed include those from the families Serranidae and Lutjanidae.

The major groups of pelagic fishes that are currently being exploited are the mackerels, anchovies, clupeids, carangids, and tunas consisting of *Rastrelliger* sp. *Stolephorus* sp., *Decapterus* sp, *Megalaspis cordyla, Euthynnus* sp, *Kishinoella tonggol, Scomberomerus* sp, *Chirocentrus dorab, Caranx* sp, *Selar* sp, *Selaroides leptolepis, Clupea* sp and *Parastromateus niger*.

In *Rastrelliger*, like in some other scombrid fishes, the eggs in the ovary ripen asynchronously. Only a small portion of the eggs are fully mature at any one time. This results in the release of different batches of eggs over an extended spawning period. At least two batches of eggs could be spawned. In *R. kanagurta* the spawing season extends from October/November to April. In Penang, the spawning season of the species extend from May to January. The main spawning season of *R. neglectus* extends from October to December annually.

The size at first maturity of *R. kanagurta* is reported to be 18.75 cm total length and 16.6 cm standard length respectively. *R. neglectus* was reported to have attained first maturity at 18.5 cm total length.

Only a small portion of eggs in the ovary of *Rastrelliger* spp seem to be spawned at one spawning season. It was estimated that a female would release only about one-tenth of the total number of eggs in an ovary, 20000 to 30000 at a spawing season.

The main food items consumed in order of percentage occurrence in the stomachs of *R. kanagurta* are phytoplankton, copepods, decapods, ostracods, bivalve larvae and dinoflagellates.

The mainstay of the *Rastrelliger* fishery on the west coast of Peninsular Malaysia are medium sized *R. kanagurta* between 16.75 cm and 20.26 cm.

As for *R. neglectus*, a recruitment population sizes of 10 to 14 cm appeared in the landings of January to March while adult fish are caught from May till September.

Stolephorus fishery is important in the straits of Malacca *Stolephorus heterolobus* spawns throughout the year at an intervals of one to two months. Some of these spawning periods may be more intense than others. The spawning grounds of *Stolephorus* may be located further out at sea.

The size at first maturity of *S. heterolobus*, *S. bataviaensis* and *S. indicus* are 50 mm, 50 mm and 116-125 mm in standard length respectively.

The number of matured eggs in a single female of *S. heterolobus* of 55 mm length was found to be 1000. A fish of 62 mm fish reported to have 2500 eggs. *S. commersoni* and *S. bataviaensis* reported to have 5000 to 10000 eggs, while *S. indicus* has 9000 to 14000 eggs per fish.

S. heterolobus mainly feeds on calanoid and harpacticoid copepods, ostracods *Leptochela*, brachyuran zoea, other decapod larvae, polychaete larvae and phytoplankton. The main food item of *S. bataviaensis* is *Leptochela*. But *S. indicus* take calanoid copepods, mysids, *Leptochela*, brachyuran megalopa, squilla larvae and fish larvae as food.

Fish between 30 to 60 mm are commonly encountered in the catch throughout the year. In the first half of the year small fish appear along the coast. Larger fish are more commonly caught during the last 6 months of the year. *S. heterolobus* enter the commercial fishery at a standard length of 15 mm.

Thailand

The west coast or Indian Ocean coast of Thailand extends westward to Andaman and Nicobar Islands, northward to Mayanmar waters and southward to the west coast of Peninsular Malaysia and Sumatra Island. The coast line which starts from Ranong to Satul Province, is about 740 km long. The fishing area of the shelf from the coast to the 100 m. depth contour is about 44000 sq. km. In the north off Ranong Province or the Mayanmar/Thailand border the continental shelf to 100m depth is some 96 km wide, narrowed to about 32 km off Phuket Island and then widens to 256 km off Satul Province or the Thailand/Malaysian border.

In comparison with the Gulf of Thailand, the grounds on Indian Ocean side are much rougher, consisting of rocks and sea mounts, and the depth increases abruptly outward from the shore. There are vast area of mangrove forests along the coast and numerous islands with hard coral in offshore areas. Nevertheless, along this coast is the good area for nursery and living grounds of marine animals. The Andaman sea has been the traditional fishing ground for the fishermen who live along the coastal area from Ranong to Satul Province. All the fishing activities however are carried on near shore at depths less than 90 m.

The mackerel, *Rastrelliger* spp, are the most important pelagic species group in the west coast of Thailand, comprising more than 45 per cent by weight of the total pelagic catches.

Sardines in the west coast of Thailand include several species of which *Sardinella gibbosa* are the most abundant.

Anchovies caught along the west coast of Thailand are of several species of the genus *Stolephorus*, the most important species is *Stolephorus heterolobus*. Anchovies are mostly caught by small meshed anchovy purse seines, push nets and tidal traps. Some species of anchovies are also caught by trawlers. especially, *Stolephorus indicus* and *S. buccaneeri*.

Rastrelliger brachysoma is available in Ranong, Phang Nga, Phuket, Karbi and Satul. The species attain first maturity at 17.5 cm and spawn during February-March and July-August. The sex ratio of the species has been calculated at 1:1.7. Fishes of 9.5-12.5 cm size appear along the coast during April, August and December, The species feed on phytoplankton, zooplankton, diatom, copepods, dinoflagellates, poriferans, ciliophorans. The mean length in the catch is 17.5 cm.

Rastrelliger kangurta feeds on phytoplanktons, zooplanktons, crustaceans and fish larvae. The species is abundant in Ranong-Satul and very abundant at Surin Islands, Ranong Province, Kantang Bay and Adang Islands. The species attain first maturity at 19 cm and breed in December-February. Their sex ratio being 1 : 1.2. 13 to 14 cm fish enter the fishery during May and December with a mean length of 19.2 cm.

Carangoides malabaricus appear in Ranong, Phang Nga, Phuket, Krabi and Satul during January, March and July with a size range of 10-21 cm., the mean length of the catch being 14.7 cm. The species attain its first maturity at 19.5 cm and spawn throughout the year with the peak in February-May, March-April and August-September. The sex ratio in the population were found to be 1 : 1.08.

Megalaspis cordyla attain first maturity at 28.0cm and spawns during March-April. 15.2 cm long fishes enter the fishery during September–December.

Euthynnus affinis spawn in January, while *Auxis thazard* spawn during February-March.

Demersal fishes are distributed throughout the coral and rock lines along Thailand coast and around islands from Ranong to Satul province.

Among them *Sillago sihama* attain its first maturity at 11.6 cm and breed during January-March and May-June. Their sex ratio is 1 : 0.72. They enter the fishery during November with 14.8 cm size fishes. *Sillago ciliata* attaining maturity at 11.8 cm breed in November-December and June-July with their sex ratio being 1 : 0.95.

Nemipterus delagoae feeds on Metapenaeus, crabs, squids, annellids, Attain its sexual maturity at 16.0 cm and breed during January-February and April-May. The species enter the fishery during December, April and July when they attain 9.5-12 cm. *N. tolu* attain maturity at 17.0 cm and spawn during December-January. February-March and June-July. Their sex ratio being 1 : 1.06. At 13.6 cm they are recruited in the fishery.

Scolopsis taeniopterus attain maturity at 16 cm and breed on October-November, January-February, June-July and December-January. Their sex ratio being 1 : 0.125. At 15 cm length the species enter the fishery.

Priacanthus tayenus attain maturity at 17 cm and enter the fishery at a mean length of 13.5 cm

Epinophelus sexfasciatus attain sexual maturity at 14.7 cm. *Penaeus monodon*, abundant in sandy bottom at a depth of 35 m attain maturity at 20 cm and breed in June-August and February. *P. merguiensis* along the coast attain maturity at 13 cm and breed during July–August.

Chapter 5
Dominant Fish Groups Occurring in Indian Coasts

Elasmobranch

Sharks, skates and rays together in Tamil Nadu coast form an average annual catch of 167 tonnes, mainly by gill net (16.7 per cent) followed by hooks and line (6.6 per cent) and trawl net (2.2 per cent). The appearance of sharks and rays along the coast does not fluctuate very much throughout the year barring the low values during November and December. Skates on the other hand occur in large numbers along the coast from November to March.

An adult oviparous female zebra shark, *Stegostoma faciatum* was caught off Pamban in the Gulf of Mannar. The shark measuring 207 cm in length and 49 kg by weight was caught while trawling at 20 m depth. Although landing of juveniles of *S. faciatum* are not uncommon along the Indian coasts a large adult specimen was first recorded from the south east coast of India.

An unusually large Devil ray measuring 386 cm in total length and 447 cm in breadth, caught from the inshore waters of south west coast of Karwar at 40 m depth by a mechanized gill net boat, was landed at Karwar on December, 1987. This female species, identified as *Manta birostris*, weighting 800 kg had a large liver weighting 40 kg.

Elasmobranchs constitute one of the important group of fishes to form commercial fisheries of Mandapam and Rameswarm region of the east coast of India. On an average 2933 tonnes of elasmobranchs are exploited from this region comprising 13.3 per cent of total fish production. They are mainly caught by trawl net (70.5 per cent) followed by gill net (28.48 per cent) and hooks and line (1.01 per cent). The important species among the group occur in the region are *Dasyatis uarnak, D. bleekeri, D sephen, Amphotistius imbricatus, A. kuhli, Gymnura poecilura, Aetomylaeus maculatus, Rhinobatus grannulatus, Aetobatus narinari, A. flagellum Rhinoptera javanica* and *Mobula diabolus.*

Dasyatis bleekeri are mainly caught in trawl net (28.3 per cent) from the region between March to December from 10 to 20 m depth. The species of 145-1060 mm length form the commercial fishery. The species attain its first maturity at 670 mm length and spawn during April and September.

Dasyatis uarnak also caught mainly in the trawl net (27.1 per cent) between August and March at a depth of 10-20 m. The species attain its first maturity at 760 mm length and breed during December and July. Fishes of 215-1200 mm length form the commercial fishery in the region.

The other species of this genus, *Dasyatis sephen* are caught in small numbers in the trawl net (10.12 per cent) during April-June and October-December from the same depth of 10-20 m. The fish attain its first maturity at 660 mm and breed during December-January. Fishes of 180-1460 mm length form the commercial fishery in the region.

The eagle ray, *Rhinoptera javanica* occur mostly during August to November at a depth of 10-20 m and are caught by trawl net (3.33 per cent). Fishes of 480-1500 mm length form the commercial fishery in the region and are matured when they attain a length of 540 mm. They breed during January-February.

Amphotistius umbricatus attain its first maturity at 180 mm and breed through out the year. Fishes of 140-249 mm form the commercial fishery in the region with peak period of occurrence during February to July. They are mainly caught by trawl net (14.14 per cent) from a depth of 10-20 m.

The other species of the genera, *Amphotistius kuhli* appear in small numbers in the region and are caught by trawl net (2.55 per cent) mostly during August to October from a depth of 10-20 mm. Fishes of 120-339 mm length form the commercial fishery. They mature at 240 mm size for the first time and breed during January to April.

Aetobatus narinari appear in the region twice a year. from April to May and from August to November in large numbers at a depth of 10-20 m and are caught by the trawl net (3.33 per cent). In the commercial fishery, fishes of 520 to 1060 mm size are caught. They get matured when they attain 740 mm in length and breed during November.

Aetobatus flagellum of 420 to 1300 mm size form the commercial fishery in the region during November-December at a depth of 10-20m, from where they are caught by trawl net (2.81 per cent). The species attain its first maturity at 860mm length and breed during March and April.

Gymnura poecilura occur in large numbers in the region, during July and October at a depth of 10-20 m from where they are caught by trawl net (2.24 per cent). The fishes attain its first maturity at 570 mm and breed during March and September. The commercial fishery of the species is dominated by 240-960 mm size groups.

The shark, *Scoliodon palasorrah* in the region are mainly caught by gill net (10 per cent) followed by hooks and line (4 per cent) and trawl net (0.5 per cent) from a depth of 10-40 mm during April to October. The species attain its first maturity at 1285 mm size and breed during April. Fishes of 220-1500 mm length range form the commercial fishery in the area.

Small pelagic sharks belonging to the family *Galeidae* constitute 0.33 per cent of total catch in and around Minicoy island and are mostly caught by troll line (95 per cent) and pole and line (5 per cent).

In and around Kakinada, Andhra coast sharks under the family Carcharhinidae forms 2 per cent in total fish catch with a peak period of occurence in February to May and October to November. They are caught from a depth of 5 to 50 m by gill net (13.2 per cent) and trawl net (0.3 per cent).

Skates under Rhinobatidae form 0.4 per cent in total catch in Kakinada coast mainly during December to March at a depth of 3 to 80 m. The species is caught in trawl net (0.4 per cent).

Rays under the families Trygonidae and Myliobatidae in Kakinada coast form 2.4 per cent in total catch are caught by trawl net (1.9 per cent). The species appear at a depth of 5 to 50 m with a peak during November to April.

Sharks, represented by a single species, *Carcharhinus melanopterus* form only 0.04 per cent in total catch along Karwar coast. The peak period of occurrence of the species is between August to May, when fishes of 200 to 1000mm length form the commercial fishery at a depth of 5 to 50 m. They are mainly caught by purse seine, trawl net and gill net.

Rays under the family Dasyatidae at Karwar coast form a negligible catch of 0.01 per cent and are caught by trawl and gill net.

Sharks, skates and rays occur in Calicut coast of Kerala and are caught by drift hook and lines, hook and lines and drift net. On an average 25.69 tonnes of elasmobranch fishes are caught from this coast forming 0.42 per cent, of the total catch.

A female whale shark, *Thiniodon typus* of 5.06 metre length along with other fishes namely *Scomberomorus* spp, and *Euthynnus affinis* got entangled in gill net at about 30 km north of Chennai at 70 m depth. The shark weighed about 1250 kg.

Sharks and rays at veraval coast, Gujrat form 3.3 per cent in total catch by gill and trawl net. Sting ray, *Himantura bleekeri* of 1000 to 1400 mm size group are mainly caught between January to June by trawl net. Eagle ray, *Aetobatus narinari* of 2000 to 3000 mm long are caught between April to May in trawl net.

Hammer-head sharks comprising of *Sphyrna mokarran* and *Sphyrna lewini* of 2000 to 2500 mm and 700 to 2000 mm respectively occur in large numbers during April-May at a depth of 80 m when they are mainly caught by gill net.

Sharks comprising of species like, *Carcharhinus limbatus, C. melanopterus, Scoliodon laticaudus, Rhizoprionodon acutus* and *Rhizoprionodon oligolinx* form the commercial fishery in the region between October to May at a depth of 30 to 80 m. The dominant size group in commercial fishery is 500-1500 mm; for the species of genus *Carcharhinus* and 300-850 mm for the fishes of rest of the two genus.

Sharks and rays together contribute 378 tonnes at Cochin coast of Kerala in almost equal proportions, the major season being April to December for sharks and November to March for rays. The average annual catch of sharks is 163 tonnes, comprising 0.86 per cent in total catch, mainly caught by drift net (76 per cent) and trawl net (24 per cent). The major species contributing shark fishery at Cochin coast are *Carcharhinus melanopterus, C. brevepinna, C. plumbeus, Rhizoprionodon acutus, R. oligolinx, C. limbatus* and *Sphyrna zygaena. R. acutus* of 485-780 mm length occur in large numbers during February to May at 50 m depth. *R. oligolinx* of 370-800 mm appear in large numbers during January and July-August at 50 m. depth *C. limbatus* of 620-1080 mm length form the commercial fishery at 50 m. depth during October-November. *C. melanopterus* of 530-2120 mm length appear in large numbers during April to July at 50 m. depth. *Scoliodon laticaudus* of 290-565 mm form the commercial fishery at 40 m. depth. *Sphyrna zygaena* at Cochin contribute 0.05 per cent in total catch are exclusively caught by drift net. Fishes of 700-1640 mm length form the commercial fishery at 50m. depth during June to September.

In Vizhinjam coast sharks, rays and skates form 1.3 per cent in total catch and are mainly caught by hooks and line (0.37 per cent), drift net (0.64 per cent) and boat seine (0.19 per cent).

Rays, sharks and skates in Visakhapatnam coast form an average annual catch of 60.8 tonnes (Rays) and 55.7 tonnes (sharks and skates) and are caught by hooks and line and trawl net.

Sharks, skates and rays in Bombay coast form 17.2 per cent of total catch and are mainly caught in trawl net. Sharks assemble in the coast almost throughout the year with peak in October to December. A small quantity of rays appear throughout the year with peak at January-March; with the abundance of skates in the coast showing two peaks, one during October and the other at December.

Stromateidae

The pomfrets forming 1.1 per cent of the total landings at Calicut coast are exploited by drift net, boat seine and trawl net. *Parastromateus niger* and *Pampus argenteus* are the chief species contributing to the pomfret fisheries in the region. *Parastromateus niger* of 230-350 mm size group form the commercial fishery at a depth of 15-40 m during September to December and are caught by drift net (6.06 per cent) and boat seine (0.51 per cent). The species attain maturity at 300 mm length and spawn during July to October.

Pampus argenteus of 160 to 250 mm size group form the commercial fishery at a depth of 10-40 m during November to January and are caught by drift net (2.92 per cent) and trawl net (0.85 per cent). The species attain first maturity at 220 mm length and spawn during May to September.

Pomfrets in Cochin form 1.01 per cent in total catch, which are mainly caught by purse seine (56.9 per cent), drift net (38.6 per cent) and trawl net (4.5 per cent). *Parastromateus niger* of 180-520 mm size group congregate during August to November in the region at 30-40 m. depth. Matured species (280 mm length) spawn in September to December and in February. *Pampus argenteus* on the other hand assemble during August to October at 25-40 m depth and fishes of 100-350 mm length form the commercial fishery.

Pomfrets at Vizhinjam coast form 0.23 per cent in total catch and are caught by boat seine (0.09 per cent) and drift net (0.14 per cent).

Pomfrets in Tamil Nadu coast occur throughout the year with an average annual catch of 19.5 tonnes in trawl net (0.3 per cent) and gill net (1.3 per cent)

Pomfrets, chiefly, represented by *P. argenteus* and *P. niger*, form 6.6 per cent in total catch in Bombay coast, mainly by trawl net, gill net and dol net. *P. argenteus* of 40-320 mm size form the commercial fishery at 10-90 m. depth during August-May. The species attain maturity at 220-240 mm length and spawn during January to March. *P. niger* of 150-500 mm length form commercial fishery along Bombay coast during August-May at 30-90 m depth of water.

Heavy catches of silver pomfret *Pampus argenteus* were obtained at Satpati, about 100 km north of Bombay within a span of four days from 7.9.87 to 10.9.87 mainly by surface gill nets operating for 12 hours in each trip. The silver pomfrets were in the range of 170 to 320 mm length. As regards to range in weight they form 47.96 per cent for fishes above 600gm, 25.93 per cent for fishes of 350-590 gram, 11.01 per cent for fishes of 250-340 gram, 2.82 per cent for fishes of 200-240 grams and 3.3 per cent for fishes of less than 190

Pampus argenteus

gram. The catches of promfrets by the bottom gill nets during this period was very poor.

Pomfrets in Kakinada coast, represented by *Pampus argenteus* and *Parastromateus niger* form only 0.3 per cent in total catch and are caught by gill net (0.7 per cent) and trawl net (0.2 per cent)

Parastromateus niger have been found to occur during February to May and August and November at a depth of 5 to 50 m. At the

Parastromateus niger

Pampus chinensis

same depth *Pampus argenteus* appear during April to July and during December.

Pomfrets in Karwar coast form 1.84 per cent in total catch and are represented by *Parestromateus niger, Pampus argenteus* and *Pampus chiensis.* They are mainly caught by gill net, purse seine and trawl net. *Parastromateus niger* of 50 to 450 mm size group form the commercial fishery during September-October at 5-50m depth. At the same time *Pampus argenteus* of 40 to 290 mm size group and *Pampus chinensis* of 50 to 250 mm size group form the commercial fishery at 5 to 50 depth. Pomfrets like *P. niger* and *P. argenteus* form 1.5 per cent in total catch by gill net and trawl net. *P. niger* of 100-400 mm and *P. argenteus* of 100-300mm size groups form commercial fishery in April-June at a depth of 20-40m. *P. chiensis* of 250-350 mm size form commercial fishery at 20-40 m depth of water.

Isphyraenidae

The peak period of occurrence of barracuda along Tamil Nadu coast is September with an average annual catch of 43 tonnes mainly by trawl net (0.7 per cent). In Minicoy, *Sphyraena* sp. form 1.8 per cent of total troll line catches. Barracuda form 0.08 per cent of the total catch exclusively by troll line in Minicoy.

Barracudas form 0.95 per cent in total catch along Vizhinjam coast and are caught by drift net (0.43 per cent), hooks and line (0.09 per cent), boat seine (0.33 per cent) and shore seine (0.01 per cent).

Barracuda at Kakinada coast constitute 0.9 per cent in total catch and are mainly caught by gill net (1.0 per cent), trawl net (0.6 per cent) and other gears (0.6 per cent). The peak period of their occurrence is December to June at a depth upto 80 m.

Barracuda form 0.01 per cent in total catch at Karwar and are mainly caught by purse seine and trawl net. Barracuda in Veraval coast contribute 0.01 per cent in total catch by trawl net.

Scombridae

Though mackerel occur abundantly in the month of July along Tamilnadu coast, but the seer fish occur throughout the year forming an average annual catch of 111.6 tonnes mostly by trawl net (19.4 per cent), gill net (52.8 per cent), hooks and line (4.6 per cent) and bag net (17.5 per cent).

Indian mackerel form a very minor fishery in Mandapam region. They are mostly caught by drift gill net (23 per cent) and shore seine (20 per cent). *Rastrelliger kanagurta* occur in large numbers during September to March at a depth of 5-18m forming a commercial fishery with a size group of 220-240 mm. Their spawning season lasts from June-July and January to March.

Scombroid fishes of Minicoy constitutes 98.96 per cent of the total fish catch forming a very important fishery of the region. Species like *Katsuwonus pelamis* (about 90 per cent of the total marine fish catch), *Thunnus albacares, Euthynnus affinis* and *Auxis thazard* contribute to the fishery among the group. They are mainly caught by pole and line (96.9 per cent) and troll line (3.1 per cent).

Katsuwonus pelamis form 88.6 per cent of the group occur in large numbers during December to April. Fishes of 300-720 mm size form the commercial fishery. The species attain its first maturity at a length of 430 mm and spawn during October to March and May to October.

Euthynnus affinis

Katsuwonus pelamis

Next in importance is *Thunnus albacares* which form 10.79 per cent of the group with a peak period of occurrence in December to May at the surface water from where 320-1020 mm sized fishes are caught with pole and line and troll line.

Euthnnus affinis, a minor contributor (0.13 per cent) of the group appear in surface water during December to April, from where fishes of 460 to 580 mm size groups are caught in pole and line.

Thunnus albacares

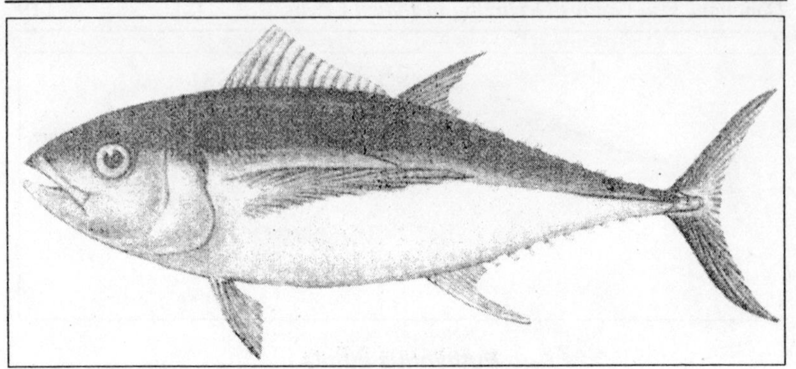

Thunnus albacares

Auxis thazard, the smallest contributor of the group (0.04 per cent) are caught by pole and line and by troll line during December to April, when fishes of 350 to 370 mm size group from the commercial fishery.

Rastrelliger kanagurta under Scombroid fishes form 6.5 per cent in total catch from the Kakinada coast and are mainly caught by gill net (48.0 per cent), trawl net (0.6 per cent) and other gears 5.7 per cent. Fishes of 35 to 235 mm size group forms the commercial fishery during January to April and August to December at a depth upto 80 m.

Fishes under the family Scombridae, represented by *Scomberomorus commerson*, *S. lineolatus* and *S.* guttatus form 1.3 per cent in total catch in and around Tuticorin off Tamil Nadu coast. The king seer *S. commerson* forms not less than 90 per cent of total seer catch and the rest by the streaked seer, *S. lineolatus* and spotted seer, *S. guttatus*. Though seer fish occur throughout the year, the peak period of abundance coincides with the tuna season during July to September. They are mainly caught in drift gill net (0.7 per cent), trawl net (0.4 per cent) and hooks and line (0.2 per cent).

Scomberomorus commerson occur in large numbers during June to October and January to March at a depth of 20-60 m from where they are caught by drift net, hooks and line and trawl net. Fishes of 100-1400 mm size group form the commercial fishery in the region.

Auxis thazard, *Euthynnus affinis*, *Sarda orientalis* and *Thunnus tonggol*, *Thunnus albacares* and *Katsuwonus pelamis* support the tuna

The Indian Mackerel

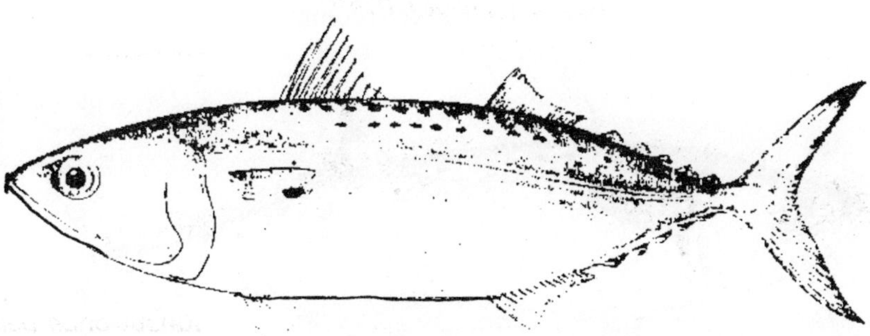

Rastrelliger kanagurta

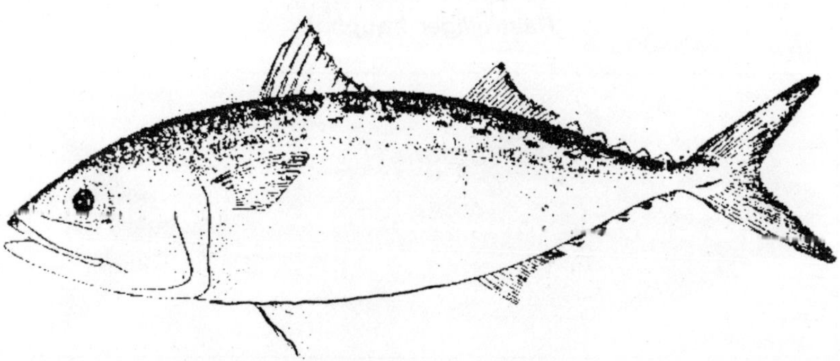

Rastrelliger brachysoma

fishery in Tuticorin region forming 1.7 per cent in total catch. They are mainly caught by drift gill net (1.7 per cent).

Euthynnus affinis form 70.7 per cent in the catch of the group. Their peak period of occurrence is June to August at a depth of 30 to

Rastrelliger brachysoma

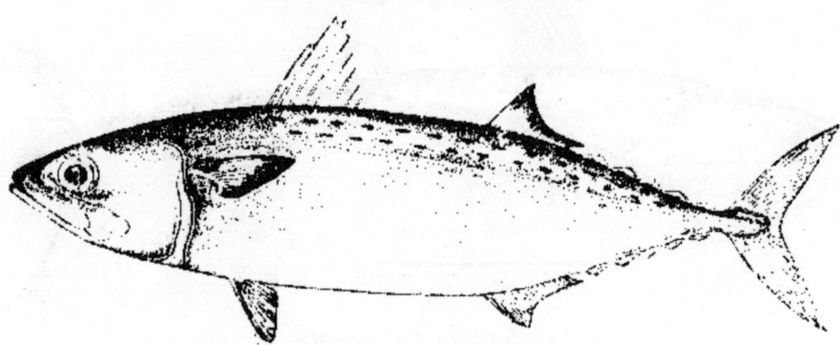

Rastrelliger fraughni

Indian Mackerel *Rastrelliger kanagurta*

40m. Fishes of 440-600 mm form the commercial fishery. The species attains it's first maturity at 480 mm length and spawn during September to November.

Auxis thazard forming 14.6 per cent in the catch of the group occur in large numbers during June to August at a depth of 30-40 m when they are caught by drift gill net. Fishes of 300-420 mm size group form the commercial fishery in Tuticorin region. The species attain its first maturity at 300 mm and spawn between September and March.

Thunnus albacares form only 7 per cent in the catch of the group and are caught by drift gill net from a depth of 30-40 m.

Katsuwonus pelamis are also caught from 30-40 m depth by drift gill net which form 1.1 per cent in the catch of the group.

From the same depth (30-40m) *Thunnus tonggol* are caught in drift gill net. The species form 0.7 per cent in the catch of the group.

Mackerel, tuna and seer fish together constitute 17.02 per cent in total catch along Karwar coast and are mainly caught by purse seine and gill net. Of these tuna, consisting of species like *Auxis thazard, Auxis rochei,* and *Euthynnus affinis* form 1.97 per cent in total catch along the Karwar coast. Both *Auxis thazard* of 200 to 350 mm length group and *Auxis rochei* of 200 to 300 mm size group appear in large numbers along the coast at a depth of 10 to 75 m when they are caught by purse seine and gill net. *Euthynnus affinis* of 250 to 680 mm size group form the commercial fishery along the coast during August to May at a depth of 5 to 50 m. from where they are caught by purse seine and gill net.

Scomberomorus commerson and *Scomberomorus guttatus* among seer fish form 0.09 per cent in total catch along Karwar coast. *Scomberomorus commerson* of 410 to 770 mm size group and *Scomberomorus guttatus* of 400 to 610 mm size group form the commercial fishery in the region at 10 to 75 m depth during August to May. They are mainly caught by purse seine and gill net.

The only species of mackerel, *Rastrelliger kanagurta,* mainly caught by purse seine and gill net form the major share of 14.97 per cent in total catch in the region. Fishes of 70 to 270 mm size range form the commercial fishery at 2 to 50 m depth during August to May. The species attain its first maturity at 220 mm length and spawn during May to December.

Scombroid fishes form 6.0 per cent in the total catch along Calicut coast are exploited mostly by drift net (45.3 per cent) boat seines (34.4 per cent) and gill net (19.9 per cent). The Indian mackerel

Rastrelliger kanagurta forming 209 tonnes on an average is exploited by gill net, boat seines and drift net. The species shows wide fluctuation in the yield from 709 tonnes to 385 tonnes. The peak fishing season for mackerel which occur from 5 to 40 m depth is from August to October. The species attain its first maturity at 200 mm length and spawn during May to August. Fishes of 180 to 220mm form the commercial fishery in the region.

Seer fishes *Scomberomorus commerson* and *S. gultatus* contribute on an average 89.6 tonnes annually representing 22.3 per cent in total catch in the region. They are mainly landed by drift net. The peak period of abundance is from October to December, but they spawn during April to May. *S. guttatus* attain its first maturity at 410 mm length and fishes of 300-600 mm size group form the commercial fishery at a depth of 20 to 50 m. *S. commerson* of 500 to 800 mm size group form the commercial fishery at the same depth of 20 to 50m. and fishes of 750 mm size attain its first maturity.

The average annual catch of tunas is 64.2 tonnes which are exclusively landed by drift net. The dominant species is the little tunny, *Euthynnus affinis* and their peak abundance extends from March to May. Fairly off shore waters from 30 to 50 m depth is the region of abundance of tuna. The species attain its first maturity when they are 430 mm long and spawn during September to October. Fishes of 300 to 500 mm size group form the commercial fishery in the region.

Seer fishes, like, *Scomberomorus guttatus, Scomberomorus commerson, Scomberomorus lineolatus,* tuna, like *Euthynnus affinis, Auxis thazard, Katsuwonus pelamis, Thunnus albacares* and Indian mackerel, like *Rastrelliger kanagurta* under the Scombroid group constitute 2.1 per cent in total catch along Veraval coast of Gujarat and are caught by gill net and trawl net mainly.

S. guttatus of 200-500 mm size group occur in maximum numbers during July to March at 40-80m depth. At the same time and depth *S. commerson* of 300-900 mm size group form the commercial fishery. Along with the two former species 400-700 mm long *S. lineolatus* form the commercial fishery at the same depth and same period of the year.

E. affinis of 400-600 mm size group occur in large numbers during October to March at a depth of 40m. At the same depth of

Scomberomorus lineolatus

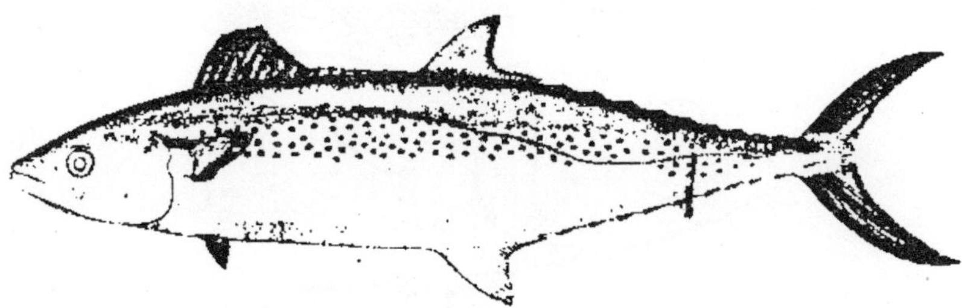

Scomberomorus guttatus

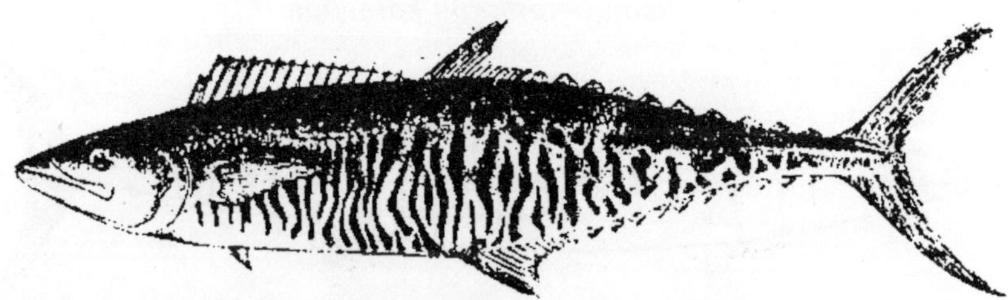

Scomberomorus commerson

Scomberomorus koreanus

water and at the same time *A. thazard* of 300-400 mm size group, *K. pelamis* of 400-800 mm size group, *T. albacares* of 1000-1500 mm size group form the commercial fishery at Veraval coast of Gujarat. *R. kanagurta* however occur at 40-60 m depth and form the commercial fishery.

Auxillary branches of lateral line

Scomberomorus koreanus

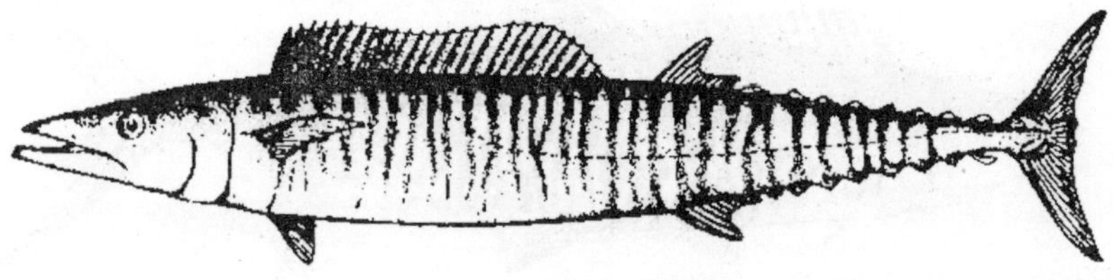

Acanthocybium solandri

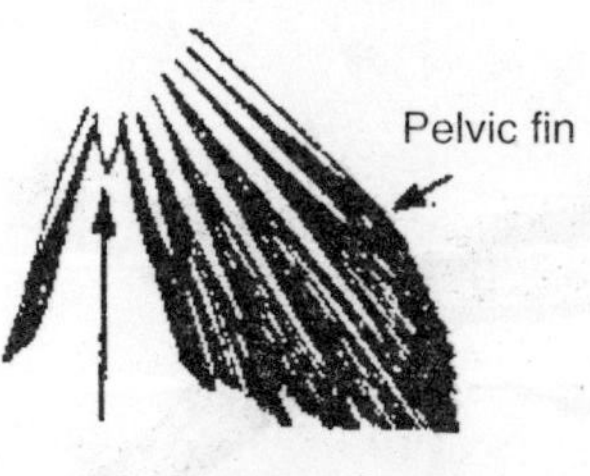

(a) Interpelvic process single

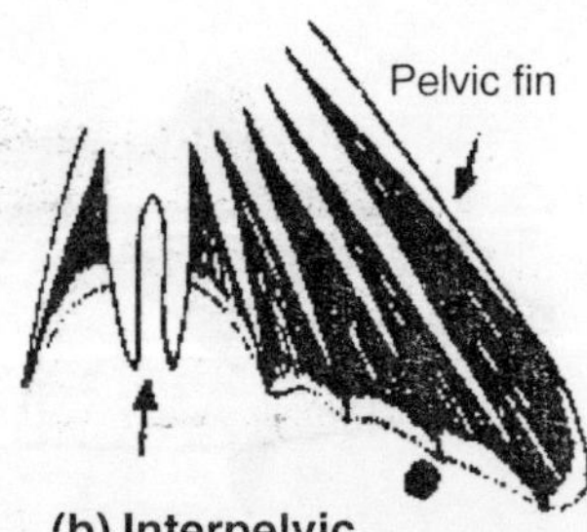

(b) Interpelvic process double

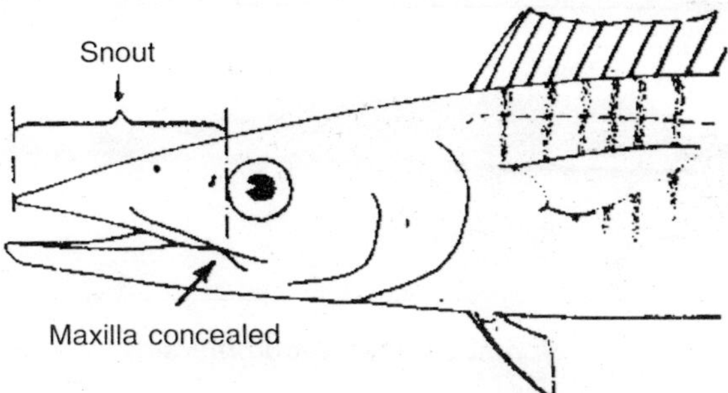

Acanthocybium solandri

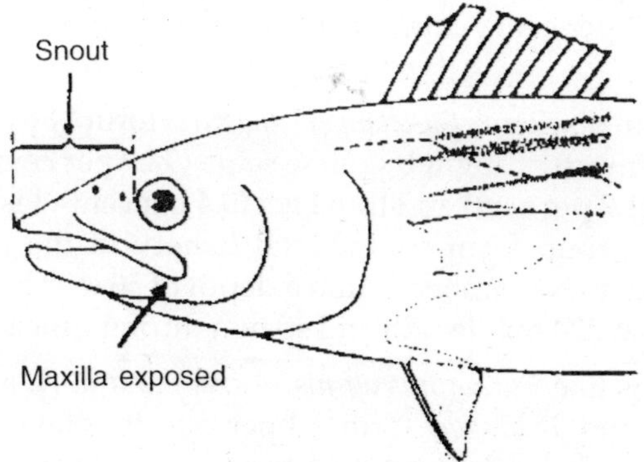

Scomber

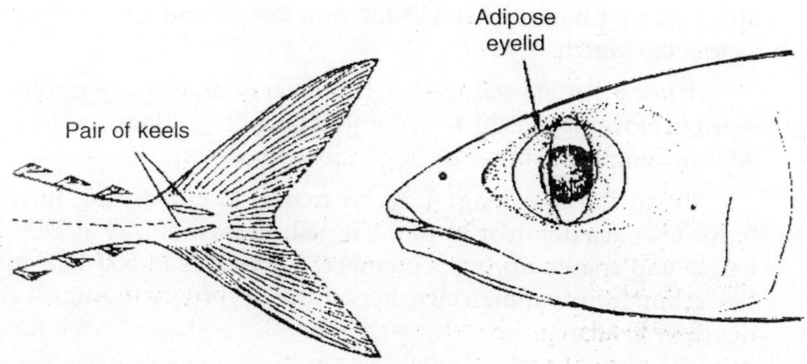

Scomber

Scomberomorus commerson

Scombrid fishes, like mackerel, tunas and seer fishes together form 16.2 per cent in total catch and are mainly caught by purse seine (58.2 per cent), drift net (41.0 per cent) and trawl net (0.8 per cent)

Indian mackerel, *Rastrelliger kanagurta* form 11.6 per cent in total catch are mostly caught by purse seine (97.9 per cent) followed by drift net (1.7 per cent) and trawl net (0.4 per cent). Fishes of 160-240 mm size group form commercial fishery in the region during September to November within a depth of 50 m. The species attain maturity at 220 mm length and spawn during June to August.

Tunas like *Euthynnus affinis, Auxis thazard, A. rochei, Thunnus albacares,* and *T. tonggol* form 3.3 per cent in total catch in Cochin region, mostly by drift net (97.6 per cent) and purse seine (2.4 per cent). *E. affinis* of 220-720 mm length group form the commercial fishery at a depth of 20-40m during April to November. The species attain its first maturity at 420-430 mm length and spawn during October to March.

A. thazard of 200-480 mm length appear in large numbers during April to November at 20-40 m depth. The species attain maturity at 300 mm length and breed during October to December.

200 to 320 mm long *A. rochei* mainly occur during July to November at a depth of 30-40m. The fish attain maturity at 240mm length and spawn during September. *T. albacares* of 500-1200 mm size group from commercial fishery during February to August at a depth of 30-40 m.

Seer fishes, eg, *Scomberomorus commerson* and *S*. guttatus form 2 per cent in total catch at Cochin coast, and are mainly caught by drift net (99.8 per cent) and trawl net (0.2 per cent). 350-1250mm long *S. commerson* form commercial fishery at a depth of 30-40 m during August-September in the region. The species attain sexual maturity at 750 mm length and spawn during January to September, while *S. guttatus* of 300-500 mm length congregate at a depth of 30-40m during September to November in the region.

Tuna, seer fishes and mackerel together form 18.4 per cent in total catch of Vizhinjam coast and are mainly caught by hooks and line (9.7 per cent), drift net (8.4 per cent), and shore seine (0.1 per cent). Tunas form the major share 12.58 per cent in total catch. *Katsuwonus pelamis* appear in the coast during January at a depth of 40-60m. *Thunnus albacares* of 400-800 mm size group form the commercial fishery at a depth of 40-60m during October to February. *Euthunnus affinis* of 280-720 mm length appear at 40-60 m depth during September to June. *Auxis thazard* of 260-400 mm size group visit the coast in February–June and September–December at 40-60 m depth to form commercial fishery. *Auxis rochei* of 180-280 mm length group form commercial fishery during July to December at 30-60 m. depth.

Seer fishes at Vizhinjam coast form 2.27 per cent in total catch and are mainly caught in drift net (2.20 per cent) and hooks and line (0.01 per cent).

Indian mackerel, *Rastrelliger kanagurta* in Vizhinjam coast form 3.54 per cent in total catch, 60-300 mm sized fishes form the commercial fishery within a depth of 60m. during November–May. The species spawn during March–June and September–October.

In Visakhapatnam coast *Euthunnus affinis, Thunnus albacares* and *Katsuwonus pelamis* among tuna are predominant. *E. affinis*, 79.9 per cent among the group, form commercial fishery at 30-60 m. depth during June to August. *T. albacares*, 15.2 per cent among the group, form commercial fishery at 30-60 m depth during June–July and November *K. pelamis*, 2.2 per cent of the group, congregate at 30-60 m depth during October-November.

Among seer fishes, *Scomberomorus guttatus* and *S. commerson* are prevalent along Visakhapatnam coast and are caught by hooks and line (55 per cent) and bottom set gill net (98 per cent) *S. guttatus* of 300-800 mm length form the commercial fishery at 20-100m depth during May–July, September–October and January. The species attain maturity at 300 mm length and spawn during February–March and July. *S. commerson* of 400-1000 mm length form commercial fishery at 50-100m depth during February–April and July.

Mackerel, comprising of single species *Rastrelliger kanagurta* of 40-270 mm length from the commercial fishery at Visakhapatnam coast at 1-60 m depth during January–May. The species attain maturity at 185 mm length and spawn during October–April.

Seer fish and tuna in Bombay coast comprise 4.2 per cent in total catch and are caught by gill net. *Scomberomorus commerson* of 550-1200 mm size form commercial fishery at 30-60 m depth during October–December. *S. guttatus* of 350–550 mm length form commercial fishery at 30-60m depth during October–December and found to spawn during April–June.

Among tuna, *Euthunnus affinis* of 300–700 mm size group form the commercial fishery at 30-60 m depth during October–December. The species spawn during April to September. *Thunnus tonggol* of 350–750 mm length group appear in large numbers along the coast during October–December at a depth of 30–60 m. to form commercial fishery.

Leiognathidae

Silver bellies, along Tamilnadu coast, represented by *Leiognathus dussumieri L. splendens, L. bindus, L. berbis, L. fasciatus, L. equulus, L. smithursti, L. leusiscus, L. daura, L. lineolatus, L. elongatus, Secutor insidiator, S. ruconius, Gazza minuta* and *G. achlamys* occur throughout the year. Silver bellies form 9.9 per cent in total catch along the coast are mainly caught by trawl net (10 per cent). Among the group *Leiognathus bindus* represent 30.9 per cent and form a commercial fishery with 60-90 mm size group. The peak period of occurrence of *L. bindus* is February to May at a depth of 15-40m. The species attain its first maturity at 87 mm and spawn during December to April and June to September.

The next in importance among the group (27.8 per cent) is *Secutor insidiator* which appear in large numbers during September at a depth of 15-40m. The species of 60-90 mm length form the commercial fishery and breed during December to April and June to September.

Silver bellies in Mandapam-Rameswaram region accounts 51 per cent of the total trawler catch, though the annual yield show a great fluctuation from 7474 to 14800 tonnes. They are mostly caught by trawl net (70 per cent) and comprise of 13 major species, namely, *Leiognathus jonesi, L brevirostris, L dussumieri, L berbis, L lineolatus, L leuciscus, L bindus, L duara, L equulus, Gazza achiemys, G minuta, Secutor insidiator and S. ruconius.*

Leiognathus jonesi occurs in large numbers throughout the year at a depth of 5-20 m and are caught by trawl net and shore seine. The species of 35-95 mm. size range form the commercial fishery. Males and females of this species attain its first maturity at 70 and 65 mm respectively and they spawn throughout the year.

Leiognathus brevirostris also occur in the region throughout the year at depth of 12-15 m and are caught by trawl net and shore seine. The dominant size in the commercial fishery ranges from 45-95 mm. Males of 68mm and females of 63 mm length attain its first maturity and they spawn twice during May–June and October–November.

In addition to trawl net and shore seine, *Leiognathus dussumieri* are also caught in gill net throughout the year from a depth of 20-40 m, mainly by a size group of 40-90mm. Males of 78 mm and females of 83 mm length attain its first maturity and breed during April–May and November–December.

Leiognathus berbis also occur throughout the year at a 3-8 m depth and are caught by trawl net and shore seine. The dominant size group in commercial fishery being 60–90 mm.

Leiognathus equulus spawn in January to March and also in May and occur in the region throughout the year at a depth of 10-25 m from where they are caught in trawl and gill net. Fishes of 125–200 mm length range forms the commercial fishery.

Gazza minuta occur throughout the year at a depth of 7–20 m in the Gulf of Mannar from where they are caught by trawl net and shore seine. Fishes of 45-115 mm length range form the commercial fishery. The species spawn in two spells, from January to April and in August to December.

Secutor reconius of 40 to 50 mm length form the commercial fishery in the region throughout the year, at a depth of 5–20 m from where they are caught in trawl net and shore seine. The species attain its first maturity at 45 mm and spawn during October to December.

Secutor insidiator of 15-85 mm length forms the commercial fishery throughout the year, in this region at a depth of 7-20 m from where they are caught in trawl and gill net. The species matures at 70 mm length and spawn during October to February.

Silver bellies comprising of *Leiognathus bindus, Leiognathus brevirostris, Leiognathus lineolatus, Secutor insidiator* and *Gazza minuta* from 10.5 per cent in total catch along Kakinada coast and are caught by trawl net (7.9 per cent) and other gears (6.7 per cent)

Leiognathus bindus form 45.2 per cent among the group are mainly caught from a depth of 5 to 50 m during January to May and October to December. Fishes of 17 to 129 mm size group form the commercial fishery. The species attain its first maturity at 80 mm length and spawn during December to February.

Leiognathus brevirostris forming 4.4 per cent among the group occur in large numbers during February, June, July and November at 5 to 50 m depth of water. The species of 60 to 105 mm size group form the commercial fishery in Kakinada coast.

Secutor insidiator forming 30 per cent of the group occur in large numbers during June to September at 5 to 50 m depth of water. The species form a commercial fishery with the size group of 20 to 117 mm. Ninty millimeter long fishes attain its first maturity and breed during January to March.

Gazza minuta constitute 4.7 per cent among the group and occur along Kakinada coast during May, August and December at a depth of 5 to 50 m. Fishes of 62 to 137 mm size groups form the commercial fishery in the region.

Silver bellies represented by *Leiognathus bindus* and *Leiognathus splendens* in Karwar coast form 3.58 per cent in total catch and are mainly caught by trawl net and purse seine. *Leiognathus bindus* of 50-100 mm length group form commercial fishery during September to April at a depth of 5 to 25 m. *Leiognathus splendens* of same size group (50-100 mm) appear in large numbers in the same time of the year and at the same depth as that of *L. bindus*.

Leiognathus splendens in Veraval coast form 0.1 per cent in total catch by trawl net. Fishes of 75-100 mm length group from the commercial fishery in the region during September to May at a depth of 20–80 m. Silver bellies in Vizhinjam coast form 3.95 per cent in total catch mainly by boat seine (3.39 per cent), shore seine (0.12 per cent) and gill net (0.37 per cent).

Eight species of silver bellies, namely, *Leiognathus dussumieri, L. bindus L. eqqulus, L. leuciscus, Gazza minuta, Secutor insidiator, S. rucanius, S. splendens* appear along Visakhapatnam coast of which the major one, *L. bindus* form 69 per cent in the catch of the group appear in large numbers during May–September at 10-50 m depth and form commercial fishery with a size group of 20-129 mm fishes. The species attain maturity at 80 mm length and spawn during December–February.

Engraulidae

Fishes of this group, representing by *Thryssa dussumieri, T. malabaricus, T. mystax, T. setirostris, Stolephorus commersonii, S. devisi, S. indicus* and *S. bataviensis* from 7.6 per cent in total marine catch along Tamilnadu coast. They are caught by trawl net (6.1 per cent) and gill net (2.6 per cent). *Stolephorus bataviensis* and *Thryssa dussumieri* constitute the major species of the group. The occurrence of *Thryssa* does not fluctuate very much throughout the year except during November and December, when they occur in small numbers. *Stolephorus bataviensis* form 44 per cent in the catch of the group and their peak period of occurrence being July and October at water depth upto 30m. *S. bataviensis* of 70-90 mm form the commercial fishery in the region.

Anchovies of the family Engraulidae comprising of *Stolephorus devisi, Stolephorus bataviensis, Stolephorus macrops, Stolephorus commersonii, Stolephorus indicus, Stolephorus andhraensis* form 8.6 per cent in total catch along Kakinada coast, which are caught by trawl net (6.1 per cent), gill net (2.0 per cent) and other gears (12.5 per cent)

Stolephorus bataviensis from 41.1 per cent among the group and are caught from a depth upto 50 m during January to May and July to September. Fishes of 40 to 109 mm size group form the commercial fishery in the region.

Next in importance *Stolephorus commersonii* constitute 23.4 per cent within the group and form the commercial fishery with a size group of 70 to 139 mm. They are caught from a depth upto 50 m in large numbers during January to May and July to September.

Stolephorus devisi, 15.9 per cent among the group, occur in Kakinada coast in large number during January to May and July to September. 55-94 mm size group form the commercial fishery in the region and are caught from a depth upto 50 m.

Stolephorus macrops of 50-100 mm size group form the commercial fishery along Kakinada coast at a depth upto 50 m and form 15 per cent in the catch of the group. Its peak period of occurrence are January to May and July to September.

Three major species of the group, namely, *Stolephorus devisi, Stolephorus bataviensis* and *Thryssa mystax* form 4.7 per cent in total catch along Karwar coast. They are mainly caught by purse seine and trawl net. *Stolephorus devisi* of 60-115 mm length range form the commercial fishery at a depth of 11 to 19 m during October to May. *Stolephorus bataviensis* of 60-115 mm size group also form the commercial fishery at a same depth (11-19m) during the same period (October to May). While *Thryssa mystax* of 80-190 mm length group form the commercial fishery during September to April at the same depth (11-19m)

Anchovies in Calicut coast form 6.3 per cent in the total landings are exploited by boat seines (83.6 per cent) and trawl net (16.4 per cent). The peak period of abundance of anchovies is from October to January along the coast.

Anchovies represented by a single species *Coilia dussumieri* form 8.1 per cent in total catch at Veraval coast which are mostly caught by trawl and dol net. Fishes of 60 to 160 mm size group form the commercial fishery during October to March in the region.

White baits in Cochin, represented by *Stolephorus bataviensis, S. devisi, S. buccaneeri, S. commersonii, S. macrops* form 1.4 per cent in total catch and are mostly caught by purse seine (46.9 per cent) and trawl net (53.1 per cent). *S. bataviensis* of 45-105 mm length contribute to commercial fishery during January–February within a depth of 50 m. Male and female fishes attain first maturity when they are 57 and 61 mm long. They spawn during December to March.

S. devisi appear in Cochin coast during October and November in large numbers within a depth of 50m. and fishes of 40-100mm length group form the commercial fishery. Male fishes of 57 mm length and female fishes of 61 mm attain their first maturity and they spawn during December to March.

S. buccaneeri also congregate at the same depth of 50 m during April to June and fishes of 50-90 mm length group form the commercial fishery. They spawns during December to March.

S. commersonii of 75-125 mm size group form the commercial fishery within 50 m depth during April–May. The species attain its first maturity when they are 110 mm long. They breed during December to March.

S. macrops of 45–85 mm length form the commercial fishery within a depth upto 50m. The species attain its first maturity at 67 mm length and spawn during December to March.

Anchovy along Vizhinjam coast form 11.41 per cent in total catch. *Stolephorus indicus* of 70-149 mm length group form commercial fishery during August to October at 5-15m depth. The species attain its first maturity at 120mm and spawn in October to June *S. devisi* of 30-140mm length group form the commercial fishery along Vizhinjam coast during April–November at 5-16m depth. The fish is sexually matured at 60 mm length and is an extended spawner.

S. bataviensis of 45-109 mm length appear during May–October within 15m depth along the coast. This species is also an extended spawner and attain maturity at 75mm length.

S. bucceneeri of 30-99 mm length congregate along the coast during July-August at 5-15m depth. The species attain maturity at 45 mm length and spawn during October-June.

Species like *Thryssa mystax* and *Thryssa setirostris* also appear along Vizhinjam coast during May, August and December within a depth of 30 m. *T. setirostris* of 85-170mm length group form commercial fishery in the region. The species spawn in November–December.

Six species of *Stolephorus* form commercial fishery along Visakhapatnam coast and are caught by shore seine (60 per cent), boat seine (3 per cent) and trawl net (1.9 per cent). *S. devisi* of 30-95 mm length appear during August–May at 10–20 m and 40–60 m depth to form commercial fishry. The species attain maturity at 60-64 mm length and spawn during June–August and October–December *S. bataviensis* of 35-133 mm length appear at the same depth as that of *S. devisi* during September–January. The species attain maturity at 80-84 mm and spawn in February-March, June-

July and December. *S. indicus* of 45-160mm form commercial fishery during April–September at 10-20 m. depth.

Thryssa mystax comprise of 46 per cent in the catch of the group occur in Visakhapatnam coast during June–October at 40-60m depth and form commercial fishery with fishes of 65-220 mm size group. *T. dussumieri* forming 33 per cent in the catch of the group occur in June–August at 40-60 m depth to form commercial fishery with fishes of 70-150mm size group, while *T. setirostris* comprising 21 per cent in the catch of the group forms commercial fishery with fishes of 80-195 mm length group during July-September at 40-60m. depth.

Gerridae

Fishes of this group appear along the Tamilnadu coast mainly in July and August and form a minor fishery (Annual average catch of 120 tonnes) in the region. They are caught by trawl net and their appearance are insignificant at other times of the year.

Istiophoridae

Sail fish, *Istiophorus platypterus* mainly caught by gill net (6.5 per cent) along Tamilnadu coast only form a minor fishery (6.5 tonnes annually) and mostly occur during April to November and again in January to March.

Istiophorus platypterus form 5.4 per cent of the troll line catches in Minicoy. They are exclusively caught in troll line from the sea surface during July to November, when fishes of 700-1250 mm size group form the commercial fishery. The species represent 0.23 per cent of the total catch in the region.

Istiophorus platypterus mainly caught in drift gill net in Tuticorin region from a depth of 30-40m. The species form 4.1 per cent in the catch of the group.

Sail fish along Vizhinjam coast form 0.48 per cent in total catch and are caught by hooks and line (0.42 per cent) and drift net (0.06 per cent)

Istiophorus platypterus in Visakhapatnam appear during January-April at 30-60m. depth and are caught by hooks and lines. Fishes of 1500-1700 mm form the commercial fishery in the region.

Sciaenidae

Sciaenids in Tamilnadu coast, represented by *Johnius carutta, J. sina, J. vogleri, J. dussumieri, J. belengari, J. macrorhynus, Pennahia aneus, Otolithus argenteus, Kathalg axillaris, Nibea maculata, Otolithus ruber, Otolithus cuvieri, Atrobucca nibe, Chrysochir aureus* and *Pseudosciaena diacanthus* form 5.7 per cent in total catch mainly by trawl net (5.8 per cent). Sciaenid concentration in the region is higher from November to March. Among the group, *Johnius carutta* form 25.5 per cent. The peak period of occurrence of the species is January-February at 15-40m depth of water. During the period fishes of 130-150mm size group form the commercial fishery in the region. The species attains its first maturity at 140mm length and spawn during June and July.

Jew fishes in Kakinada coast, comprising of *Johnius carutta, Johnius dussumieri, Johnius vogleri, Pennahia macrophthalmus, Atrobucca nibe, Nibea maculata* and *Otolittes ruber* form 8.1 per cent in total fish catch. They are caught by trawl net (6.3 per cent), gill net (1.4 per cent) and other gears (2.1)

Atrobucca nibe, 20.1 per cent in the catch of the group occur in large numbers between January to May at a depth of 5 to 60m. Fishes of 90-235 mm form the commercial fishery in Kakinada coast. At 147 mm length the fish attains its first maturity and spawn in October to June.

Johnius carutta form 11.3 per cent in the catch of the group. The peak period of occurrence of the species is between September to November at a depth of 5-50m and a size group of 75-225 mm form commercial fishery in the region. The species attain its first maturity at 155 mm length and spawn during January to June.

Johnius dussumieri forming 7.4 per cent catch among the group appear in large numbers during April to July at a depth of 5-50m. Fishes of 81 to 168 mm form the commercial fishery in the region and attain its first maturity at 110mm length. The species breed during March to August.

Johnius vogleri attain its first maturity at 190mm length and spawn during November to June. Fishes of 90-238 mm size group form the commercial fishery between January to June at 5-50m depth forming 10.7 per cent in the catch of the group.

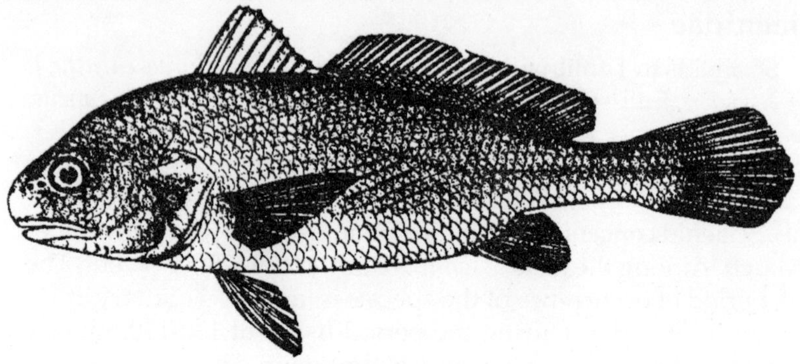

Johnius sina

Johnius dussumieri

Pterotolithus maculatus

Otolithus ruber comprising 9.6 per cent in the catch of the group, occur in large numbers during April to November at 5 to 50m depth. Fishes of 30 to 250mm size group form the commercial fishery in the region.

Nibea maculata appear in Kakinada coast in large numbers during July and August at 5 to 50m depth forming 7.9 per cent in the catch of the group.

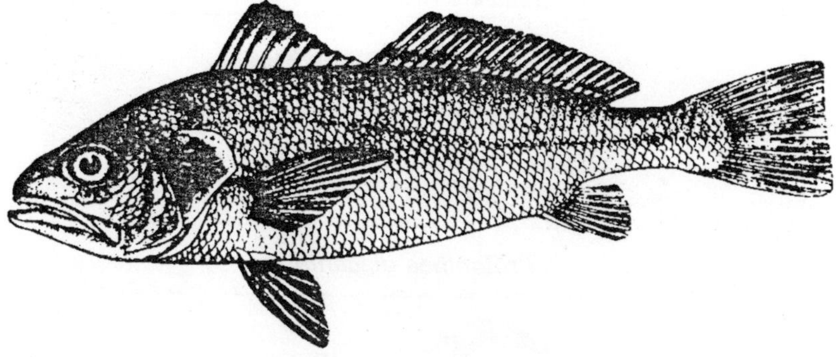

Pennahia macrophthalmus

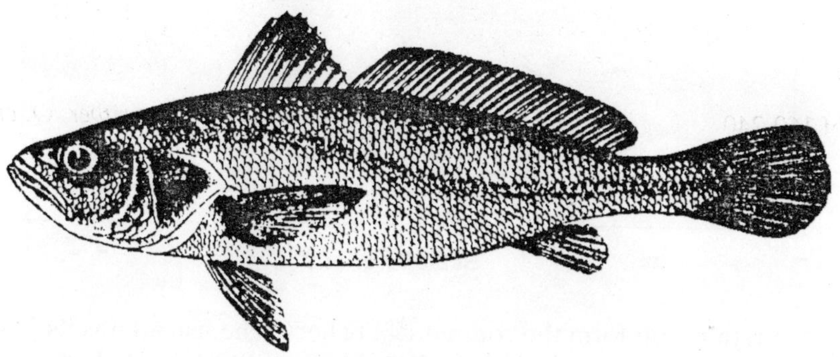

Otolithus cuvieri

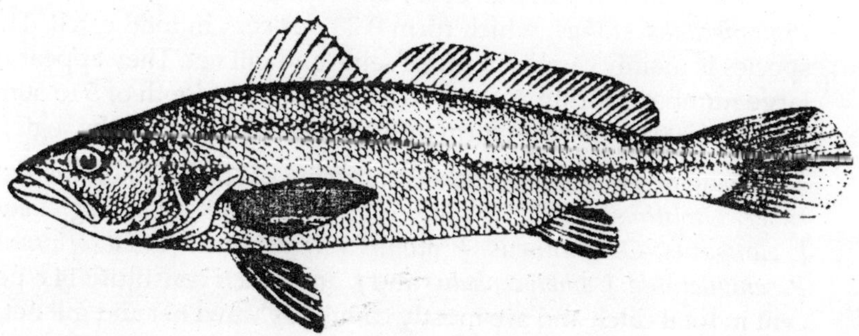

Otolithus ruber

Pennahia macrophthalmus form 6.3 per cent catch in the group at Kakinada coast. The species occur in large numbers during January to March and in December at 5 to 70m depth. Fishes of 40-260mm

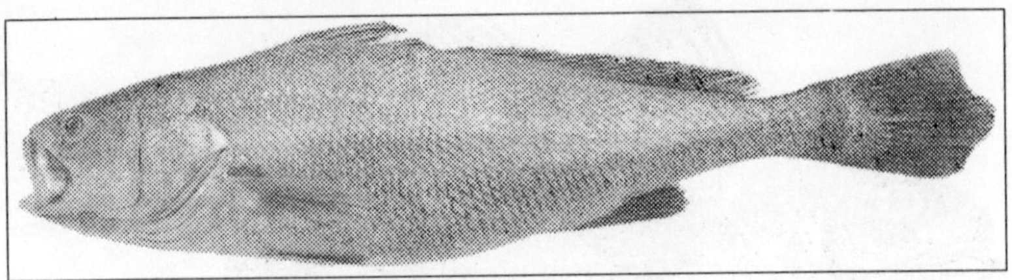

Protonibea diacanthus

Sciaena aneus

length group form the commercial fishery. The fish attains its first maturity at 147 mm length and spawn during October to June.

Jew fishes in Karwar coast are represented by the species, *Protonibea diacanthus,* which form 0.23 per cent in total catch. The species is mainly caught in purse seine and gill net. They appear in large numbers during August to December at a depth of 5 to 50m. Fishes of 200 to 1500 mm form the commercial fishery in the region.

Sciaenids in Veraval coast comprising of species like, *Otolithus ruber, Otolithus cuvieri, Johnius vogleri, Johnieops sina, J. glaucus, J. elongatus, O. brunneus, Protonibea diacanthus, J. macrorhynus, P. semiluctuosa, J. amblycephalus* and *J. dussumieri* constitute 14.8 per cent in total catch and are mostly caught by trawl net and gill net.

J. sina of 150-200 mm size group form commercial fishery during October to April at 60m. depth and are caught by trawl net.

P. diacanthus of 200-400mm size group appear in large numbers during December to January at 30-40 m depth and are caught by trawl net (8.2 per cent) and gill net (9.6 per cent).

J. vogleri of 150-200 mm size group occur in large numbers during October to March at a depth of 10-40m and are caught by trawl net (10.2 per cent) and gill net (8.2 per cent). The species attain sexual maturity on attaining a length of 170mm and spawn during June-July and October-November.

O. ruber in Veraval coast are caught in trawl net (6.2 per cent) and gill net (18.1 per cent). Fishes of 150-240mm length occur in large numbers during October to March at 10-40m depth.

O. cuvieri in Gujrat coast are caught in trawl net (35.5 per cent) and gill net (30.6 per cent). Large number of fishes, mostly of 150-250mm length appear in the coast during October to January and March–May at a depth of 10-40 m. The species attain maturity at 150-180mm length and spawn during November to April.

Scianids represented by species, like *Johnieops sina, J. dissumieri, Kathala axillaris, Otolithus ruber, O. cuvieri* and *Johnius carutta* from 2.1 per cent in total in Cochin coast and are exclusively caught by trawl net.

J. sina of 90-170mm length group form commercial fishery within a depth of 40m during January to May. The species attain maturity at 125 mm length and spawn during February to May. *J. dussumieri* appear in Cochin coast in large numbers during June and September–December within a depth of 40m and fishes of 85-170mm length group form commercial fishery. The species get matured at 125 mm length and breed during November to February. *K. axillaris* of 95-175mm size group form the commercial fishery within 40m depth during November to February and the species breed in June to September. *O. ruber* of 85-250 mm length group appear at a depth of 40m during January to April and spawn during June-October. While *O. cuvieri* of 80-240mm size group form commercial fishery within 40m depth during April–March and spawn during June to October. *J. carutta* of 90-170mm size group appear in the same depth as that of other sciaenids and form a commercial fishery during November–December and February.

Croakers at Vizhinjam coast form 1.51 per cent in totals catch and are caught by boat seine (1.07 per cent) shore seine (0.04 per cent), hooks and line (0.07 per cent) and gill net (0.33 per cent).

Eleven species of the group form commercial fishery along Visakhapatnam coast which are mainly caught by trawl net. 40-240 mm *Johnius carutta* form the commercial fishery along the coast at a depth of 20-70m during October-February. The species, which form 26 per cent in the catch of the group attain first maturity at 170mm and spawn during January-March. *Kathala axillaris*, 25.3 per cent in the catch of the group, appear in large numbers at 30-70m depth during July-August and October. The species of 50-150mm length form the commercial fishery and attain maturity at 155mm length. *Nibea maculata* comprising 12.6 per cent in the catch of the group, congregate along the coastal water at 30-70m depth during May-September and January-February to form commercial fishery with a size group of 70-240mm in length. The species attain maturity at 160mm length. *Pennahia macrophthalmus*, comprising 5.7 per cent in the catch of the group, appear at Visakhapatnam coast at 20-70m depth during November-February and form commercial fishery with the size group of 100-240mm length. The species is sexually matured at 172 mm length and spawn during December-March. *Otolithus ruber*, constituting 5.6 per cent in the catch of the group, appear the coast during September at a depth of 30-70m and fishes of 150-300mm form the commercial fishery. *Johnieops vogleri* constituting 16 per cent in the catch of the group, occur along Visakhapatnam coast during March–June and December at 30-70m. depth and form commercial fishery with a length group of 100-235mm. The species attain maturity at 152 mm and spawn in February to April. *Johnieops dussumieri*, 4.3 per cent in the catch of the group occur in large numbers during February at 30-70m. depth and form commercial fishery with a size group of 100-180mm. *Protonibea diacanthus* (Ghol), comprising 1 per cent in the catch of the group appear along the coastal waters during September at a depth of 30-70m and form commercial fishery with a size groups of 150-750mm.

Eleven species of croakers, jew fish and drummers form 14.9 per cent in total catch along Bombay coast and are mainly caught by trawl net, gill net and dol net. *Protonibea diacanthus* forming 8.7 per cent of trawl net catch form commercial fishery with a size group of 200-1200mm during August-May at 10-90m depth. The species attain maturity at 850mm and spawn during June-September. *Otolithus ruber* forming 6.0 per cent of trawl catch appear along the coast at 60-90m. depth. They spawn in July-October. *Otolithus cuvieri* forming 26.2 per cent in trawl net catch occur during October-December at a depth of 60-90m and form commercial fishery with 140-310mm size group. The species attain maturity at 170mm length and spawn during July-September. *Johnius dussumieri* forming 6.2 per cent of the trawl net catch appear in large numbers at 60-90m depth during March-April to form commercial fishery.

Johnieops sina forming 4.2 per cent in the catch of the group occur in large numbers during January-May at a depth of 60-90m. *Johnieops vogleri* forming 23.8 per cent in the catch of the group appear in large numbers during January-March at a depth of 60-90m to form commercial fishery with 120-290mm size group. The species attain first maturity at 150mm and spawn during July-September. Other species of the group appear in Bombay coast are *P. anneus, N. semiluctuosa, J. macrorhynus, U. sinuta, J. elongata* and *D. russeli.*

Tachysuridae

Cat fish under the family Tachysuridae at Kakinada coast form 3.6 per cent in total catch mainly from trawl net (2.6 per cent) and gill net (1.7 per cent). Their peak period of occurrence is January to June at a depth of 5 to 80 m.

Species like *Tachysurus thalassinus* and *Tachysurus tenuispinis* under the family Tachysuridae form 2.79 per cent in total catch along Karwar coast are mostly caught by purse seine, gill net and trawl net. The former species of 50 to 585 mm length group form the commercial fishery during August to May at a depth of 5 to 25 m. At the same depth the later species of 250 to 400 mm size group predominates during August to December.

Cat fishes forming 4.3 per cent in the total landings of Calicut coast is an important component in the demersal fishery resources and is exploited by hooks and line (73.08 per cent), drift net (23.14 per cent) and trawl net (3.78 per cent). *Tachysurus dussumieri, T. thalassinus T. tenuispinis* and *T. serratus* are the most important species forming the commercial fishery in the region. The peak period of the fishery generally coincides with the peak breeding season, often with mass destructions of gesting males leading to mass mortality of eggs and embryos.

Tachysurus dussumieri of 500 to 1000mm size group form the commercial fishery at 15 to 40m depth during March to May and are mainly caught by hooks and line (14.89 per cent) and drift net (5.91 per cent). The species mature at 500 mm length and spawn during April to August.

Tachysurus thalassinus of 170 to 600 mm length group form the commercial fishery at 25 to 60 m depth during August to October and are caught by hooks and line (11.11 per cent) and drift net (0.27 per cent). The species attain its first maturity at 280mm length and spawn during April to August.

Tachysurus tenuispinis of 230 to 400 mm size group form the commercial fishery at 35 to 60 m depth during September to March and are caught by hooks and line (27.46 per cent), drift net (7.15 per cent) and trawl net (1.22 per cent). The species attain maturity at 280mm length and spawn during May to September.

Tachysurus serratus of 600 to 1000mm length group form the commercial fishery at 25 to 40m. depth during September to December and are caught by hooks and line (3.50 per cent) and drift net (1.82 per cent). The species attain its first maturity at 600mm length and spawn during April to July.

Osteogeneiosus militaris, Tachysurus dussumieri, T. thalassinus, T. tenuispinis and *T. caelatus* in Veraval coast form 1.6 per cent in total catch caught by trawl net (1.1 per cent) and gill net (0.5 per cent). *O. militaris* of 240-490mm length form the commercial fishery during January to May at 20-60m depth. *T. sona* of 600-700mm length occur in large numbers at a depth of 30-60m. *T. tenuiospinis* of 260-590mm length appear in large numbers during January to April and June at 20-60m depth. The species spawn during December to April. *T. thalassinus* of 260-550mm length appear in large numbers during May to December at 20-30m depth. *T. dussumieri* of 300-750mm length form commercial fishery during January to May at 20-60m depth. *T. thalassinus* spawn during October to March.

Cat fishes in Cochin coast consisting of *Tachysurus dussumieri, T. thalassinus, T. serratus, T. tenuispinis* form 0.03 per cent in total catch are mainly caught by drift net (55.8 per cent), trawl net (41.4 per cent) purse seine (2.8 per cent) *T. dussumieri* of 500–800mm form the commercial fishery during July-February within 70m depth. The species attain first maturity at 520mm and spawn during September–January. *T. thalassinus* appear in the coast in large numbers during May to October within a depth of 70m. The species is matured at 370mm length and spawn during August to October. Fishes of 300–600 mm form the commercial fishery. *T. serratus* of 600–900 mm length form the commercial fishery during June to October within a depth of 70m. The species attain maturity at 600mm length and breed during September–October. *T. tenuispinis* of 300-450mm congregate in the region during May to October within 80m depth to form commercial fishery. The species spawn during April–September after attaining maturity at 320mm. length.

Cat fishes at Vizhinjam form 1.34 per cent in total catch and are caught by drift net (0.26 per cent), hooks and line (0.74 per cent), boat seine (0.30 per cent).

Tachysurus thalassinus and *T. tenuispinis* are found in Visakhapatnam coast of which *T. thalassinus* form the major share and are caught by hooks and line and trawl net. *T. thalassinus* of 80-660mm length form the commercial fishery at 10-80m depth during March to December. At 360mm length the species attain maturity and spawn during May–July.

Species under the family *Tachysuridae* form 12 per cent in total catch in Bombay coast and are caught by trawl net.

Bumper catch of *Tachysurus dussumieri* consisting of 2000 fishes were obtained in Tamil Nadu coast in shore seine net. Their total length ranged from 611-710mm. of which 68.7 per cent were male and 33.3 per cent female. Large shoals of cat fish in west coast and east coast especially at Pondicherry and Mandapam coast have been reported. Spawning migration of *T. thalassinus* and *T. tenuispinis* towards shallow waters of less than 10m. depth has been reported during south-west monsoon (May–July) from the west coast. *T. dussumieri* migrate to shallow waters during north-east monsoon (October-December) along the east coast.

Muraenidae and Congridae

Eels of both the group, Muraenidae and Congridae form 1.9 per cent in total catch in trawl net mostly during July to November at a depth of 5 to 50m.

Muraenesox talabonoides form 1.6 per cent in total catch in trawl net along Veraval coast. 500 to 2500 mm long eel form the commercial fishery at 60m depth during October to June.

Eel in Bombay coast form 0.4 per cent in total catch and are caught by trawl and dol net. *Muraenesox talabonoides* appear in the coast in large numbers during January-May and October–December at 10-60 m dpeth and fishes of 400-2000mm size range form the commercial fishery. The species attain sexual maturity at 1200mm length and spawn during April–May and September–October.

Trichiuridae

Ribbon fish at Kakinada coast, represented by *Lepturacanthus gangeticus, Eupleurogrammus muticus, Eupleurogrammus glossodon, Lepturacanthus savala Trichiurus lepturus*, and *Trichiurus russelli* form 10.9 per cent in total catch are mainly caught by trawl net (7.1 per cent), gill net (2.4 per cent) and other gears (49.3 per cent).

Trichiurus lepturus, 73 per cent in the catch of the group occur in large numbers during October to December at a depth of 5-50m. Fishes of 165 to 1150mm size group form the commercial fishery in the region. The species attain its first maturity at 525 mm length and spawn during February to June.

Lepturacanthus gangeticus form 12.9 per cent in the catch of the group. Their peak period of occurrence is January to December at a depth of 5 to 50m and form the commercial fishery with a size range of 195 to 585 mm. The species breed during May to July.

Lepturacanthus savala comprising 4.6 per cent catch among the group appear in large numbers during July and August at a depth of 5-50m. Fishes of 200 to 700mm size group form the commercial fishery in the region.

Eupleurogrammus muticus forming 4.2 per cent in the catch of the group appear in large numbers during July and August at a depth of 5-50 m. Fishes of 160 to 460 mm form the commercial fishery in the region and breed during May to November.

Trichiurus lepturus and *Lepturacanthus savala* constitute 24.8 per cent in total catch along Veraval coast, Gujrat. They are mainly caught by trawl and dol net *T. lepturus* of 300-1200mm length appear in large numbers during December to February at 40-80m depth. While *L. savala* of 300–900mm length form the commercial fishery during December to May at 40–80 m depth.

Ribbon fish in Bombay coast, form 8.5 per cent in total catch and are caught by trawl net and dol net. *T. lepturus* of 410–950 mm appear in Bombay coast during October–December at 60–90 m depth to form commercial fishery. The species spawn in June and July. *Eupleurogrammus muticus* of 250–400 mm length form commercial fishery at 10-30 m. depth in Bombay coast during July–September. The species spawn in February to May.

The peak period of abundance of ribbon fish is November every year in Tamilnadu coast, where they form an average annual catch of 221 tonnes mainly in trawl net (2.0 per cent)

Ribbon fish, comprising of single species *Trichiurus lepturus* in Karwar coast form 0.21 per cent in total catch are mainly caught in purse seine and trawl net. Fishes of 150 to 1500 mm form the commercial fishery at 5 to 50 m depth during August to May.

Ribbon fishes at Vizhinjam, comprising of single species *Trichiurus lepturus* form 26.65 per cent in total catch, which are mainly caught by boat seine (25.80 per cent), hooks and line (0.43 per cent) and drift net (0.43 per cent). 640-1130mm long *T. lepturus* congregate during June to September in the area at 15-40m depth and form commercial fishery. The species spawn in July to September.

Ribbon fishes comprising of species like *T. lepturus, T. russelli, E. muticus* and *L. savala* are found in Visakhapatnam coast of which *T. lepturus* form 96.4 per cent of the group appear during July–November at 10-80m depth. Fishes of 150-1000mm length form the commercial fishery. *T. lepturus* mature at 525mm length and spawn during February–June.

Dussumieridae

Rainbow sardine, *Dussumieria acuta* form 0.1 per cent in total catch at Kakinada coast from a depth upto 50m mainly by gill nets (1.1 per cent) and other gears (0.9 per cent). The peak period of occurrence of the species is October to December.

Rainbow sardine, along Karwar coast form 1.04 per cent in total catch and are mainly caught by purse seine.

Rainbow sardine, *Dussumieria hasseltii* form 2.33 per cent in total catch along Vizhinjam coast. Fishes of 80-224 mm length group form commercial fishery during May–September within 40m. depth. The species gets sexually matured at 170mm length.

Harpodontidae

Bombay duck, *Harpodon nehereus* in Kakinada coast form 1.2 per cent in total catch and are caught by trawl net (0.9 per cent), gill net (0.4 per cent) and other gears (0.3 per cent). The species of 170-

300mm size group form the commercial fishery during May to October. The fishes occur at a depth upto 60m.

Bombay duck, *Harpodon nehereus* at Veraval coast Gujrat mostly occur during March, May and November and are mainly caught by dol net. Fishes of 30–330 mm size group form the commercial fishery. Bombay duck in Bombay coast form 3.5 per cent in total catch and are caught in dol net. *Harpodon nehereus* of 40-350 mm size group appear along the coast during June to October at 10-30m. depth to form the commercial fishery. The species mature at 210 mm length and spawn during November–December.

Lactaridae

White fish, *Lactarius lactarius* form 0.7 per cent in total catch along Kakinada coast and are mostly caught in trawl net during July to September in depth between 5-50m.

Also known as big-jawed jumper, *Lactarius lactarius* in Karwar coast are mainly caught by purse seine, trawl net gill net, and cast net, when the species occur in large quantities at 2 to 10m depth during August to May. Fishes at 60 to 180mm length range form the commercial fishery in the region.

The species *Lactarius lactarius* form 5.7 per cent in total catch at Veraval coast. They are exclusively caught by trawl net. Fishes of 60 to 120mm length group form the commercial fishery at 60-80m depth in the region during October to June.

White fish in Vizhinjam coast form 0.20 per cent in total catch and are mainly caught by boat seine.

Lactarius lactarius in Visakhapatnam coast appear during July-December at 10-60m. depth and form commercial fishery with fishes of 45-250mm length group and are exclusively caught by trawl net. The species attain maturity at 150mm length and spawn during March-April.

Polynemidae

Thread fins, *Polynemus* spp constitute 0.6 per cent in total catch along Kakinada coast are chiefly caught by trawl net (0.4 per cent) during December to February from a depth of 5 to 50m.

Negligible quantities of threadfins occur along Karwar coast which are mostly caught by purse seine, gill net and trawl net.

The largest *Polynemus indicus* (female) measuring 170 cm in total length and weighing 50.08 kg was caught at 28m depth from 80 km north of Bombay during 1987. Day had recorded a maximum length of the species at 126.6 cm weighing 9 kg. Mohamed in 1955 observed 142.3 cm in length with the weight of 27 kg for the species. The fish reported to attain a maximum length 142 cm on 1984. The weight of the ovary of 170cm length fish was 1.5 kg.

Thread fins represented by *Polynemus sextarius* and *Polynemus indicus* form 0.5 per cent in total catch at Veraval coast are exploited mainly by gill net and trawl net. 50–200 mm long *P. sextarius* form the commercial fishery at 60 m depth during January. While *P. indicus* of 50-100 mm form the commercial fishery at 40 m depth during January–February.

Polynemus sextarius, the only species under threadfin groups form commercial fishery along Visakhapatnam coast at 10-60m. depth during August–December by fishes of 100-210 mm size group. The species is matured at 170 mm length and spawn during January–May.

Belonidae and Hemiramphidae

Gar fish in Karwar coast form 0.03 per cent in total catch, which are mainly caught by purse seine, trawl net and gill net. The single species of the group *Strongylura strongylura* appear in large number in the northern most tip of the Karnataka coast during August to December at a depth of 5 to 25m. Fishes of 100 to 1000 mm length form the commercial fishery in the region.

Needle fish in Vizhinjam area comprising of 0.31 per cent in total catch are caught by hooks and line (0.17 per cent) and drift net (0.14 per cent).

Half beak in Vizhinjam coast form 0.04 per cent in total catch.

Mugilidae

Insignificant quantities of grey mullets (14.6 tonnes annually) are caught in bag net (14.6 per cent) along Tamilnadu coast.

Grey mullets are also caught in purse seine, trawl net and gill net in Karwar coast.

Mullets along Vizhinjam coast form 0.38 per cent in total catch and are exclusively caught by boat seine.

Mullidae

Goat fish occur in Tamil Nadu coast almost throughout the year and are mainly caught in trawl net (2.4 per cent) with an average annual catch of 145 tonnes.

The species form 1.6 per cent in total catch from Kakinada coast and are mainly caught by trawl net (1.2 per cent) during July to September and November to December from a depth of 5 to 50m.

Upeneus sulphureus among goat fish form 0.3 per cent in total catch at Veraval coast. Fishes of 100 to 150 mm length form the commercial fishery in the coast during October and December at 80m depth and are caught by trawl net.

Goat fishes in Vizhinjam coast form 0.09 per cent in total catch and are caught by boat seine (0.05 per cent) and gill net (0.04 per cent)

Upeneus sulphureus in Visakhapatnam coast form 52 per cent in the catch of the group and form commercial fishery with 65-235 mm size group during August–March at 10-60 m. depth. The species attain maturity at 130 mm length and spawn during February–March. *U. vittatus* on the other hand form 33 per cent in the catch of the group and form commercial fishery with a size group of 100–180 mm length during April–July at 10–60 m. depth. The species attain maturity at 125 mm length and spawn during February–March. Goat fish at Bombay coast form 2.1 per cent in total catch and are caught by trawl net.

Nomeidae

Drift fishes (butter fish) mainly occur in August and March along Tamilnadu coast with an average annual catch of 48 tonnes mainly caught by trawl net (0.8 per cent).

Drift fish represented by *Psenes indicus* in Kakinada coast constitute 6.6 per cent in total catch. The species are caught by trawl net (5 per cent). Fishes of 55-217 mm size group forms the commercial fishery at a depth of 5-70m and occur in large numbers during October to January.

Psenes indicus of 60-275 mm length group form commercial fishery along Visakhapatnam coast during February–July at 10-60 m. depth. They are mainly caught by trawl net. The fish attain maturity at 160mm length and spawn during February–June.

Pomadasyidae

Higher catch of grunter have been observed in the month of March along Tamilnadu coast followed by September, when 39 tonnes of this species are caught annually by the trawl net (0.6 per cent)

Grunters in Vizhinjam coast form 0.08 per cent in total catch and are caught by hooks and line (0.01 per cent) boat seine (0.01 per cent) and drift net (0.06 per cent)

Synodontidae

Lizard fish exhibits a higher catch from April to August along Tamilnadu coast landing 331 tonnes annually, mainly by the trawl net (5.4 per cent)

Lizard fish represented by two species, namely *Saurida tumbil* and *Saurida undosquamis* form 2.8 per cent in total catch along Kakinada coast and are caught by trawl net (2.1 per cent) and gill net (0.1 per cent)

Saurida tumbil comprising 40 per cent in the catch among the group occur at a depth of 5 to 70 m during January, July and October to December and form a commercial fishery with size group of 170-350mm. At the same depth of 5 to 70m, *Saurida undosquamis*, comprising of 20 per cent catch of the group appear during the same time as that of *Saurida tumbil*

Saurida tumbil form 0.3 per cent in total catch at Veraval coast mainly by trawl net. Fishes of 300-400 mm size group form the commercial fishery in the region during November to January at a depth of 40-80m.

Lizard fishes consisting of *Saurida tumbil* and *S. undosquamis* in Cochin coast form 1.8 per cent in total catch exclusively by trawl net. *S. tumbil* of 160-420mm length occur during January to September within 70m depth and form commercial fishery with a size group of 160-420 mm. They spawn from March to September. While the other species, *S. undosquamis* appear the coast during April–May, November and January and February within 70m depth. Fishes of 90-200 mm size group form the commercial fishery and the species spawn in November and January–February.

Saurida tumbil and *S. undosquamis* together form 0.65 per cent in total catch along Vizhinjam coast. 130-349mm long *S. tumbil* congregate at 20-40 m depth during July–August to form commercial fishery. The species spawn during August–September. *S. undosquamis* of 50-329mm length group form commercial fishery at 20-40 m depth during February–March, May–June and September–October.

S. tumbil and *S. undosquamis* appear in Visakhapatnam coast are caught by trawl net. 40–320 mm long *S. tumbil* form commercial fishery at 30–70 m depth during July–September and spawn during October–March. *S. undosquamis* of 50-210 mm length group form the commercial fishery in the region at 30-70m depth during July–September.

Lizard fish, represented by *Saurida tumbil* in Bombay coast form 5.3 per cent in total catch by trawl net. Fishes of 210-540 mm length form commercial fishery during October-November at 60-90 m. depth.

Nemipteridae

Thread fin bream fishery in Madras coast is constituted by five major species namely, *Nemipterus japonicus, N. mesoprion, N. tolu, N luteus* and *N. delagoae* form 12.5 per cent of the total marine landings of the region. 12.7 per cent of them are caught by trawl net. *Nemipterus japonicus,* the principal of the group (53.0 per cent) are caught at a depth range of 15-40m. and their peak period of occurrence is July and August. The commercial fishery is sustained by 130-150mm length group. The species attain its first maturity at 145 mm length and they spawn during December to March.

Fishes under this group (Threadfin bream) form 6.3 per cent of total marine fish catch in Kakinada. They are mainly caught by trawl nets and comprise of species like *Nemipterus mesoprion, N. japonicus, N. delagoae* and *N. luteus.* Among the group, *N. mesoprion* contribute 48.3 per cent followed by *N. japonicus* (44.4 per cent), *N. delagoae* (2.0 per cent) and *N. luteus* (0.6 per cent). They are mainly caught from a depth range of 10-80m. *N. mesoprion* occur in maximum numbers in the region during December–March and form a commercial fishery with 30–215 mm size groups. The species attain its first maturity at a length of 100 mm. and they spawn between December and April. The peak period of abundance of *N. japonicus,* however, is November–March and form a commercial fishery with a size range of 35-305mm. The fish attain its first maturity at 125 mm length and breed between August and April.

N. delagoae of 90-250 mm size group form the commercial fishery during September to December in the region, while *N. luteus* of 90-216 mm size group appear in large numbers twice at January–April and July–December in the region.

Represented by a single species *Nemipterus japonicus* form commercial fishery with a size groups of 90–180 mm in Karwar coast and occur in large numbers during September to April. They are caught form a depth of 5-50 m by trawl net and purse seine.

Threadfin bream fishery around Visakhapatnam coast comprise of *Nemipterus japonicus, N. mesoprion, N. delagoae, N. luteus* and N. *tolu* contribute 13.1 per cent of trawl net catch. *N. japonicus* 38 per cent of the group, occur in maximum numbers during August–January and May–June and form a commercial fishery with a size group between 50 and 280mm. They mostly occur at 30-70m depth. The fish attain its first maturity when they are 148mm in length. Their breeding season is from December–February and June–July. The *N. mesoprion*, largest among the group (51.8 per cent) appear in large numbers during December–April and July–August at a depth of 30–70 m and form a commercial fishery with 80–210 mm size group and are caught by trawl net. The species attain its first sexual maturity at 147 mm length and spawn during January-April. Comparatively negligible (6 per cent) among the group, *N. delagoae* form a commercial fishery with 100-200 mm size group at 30-70m depth during May–June and October–November. At the time they are caught by trawl net.

Contributed by three species, namely, *Nemipterus japonicus, N mesoprion* and *N. delagoae* the thread fin bream form 3.5 per cent of the total catch at Gujrat coast, which are exploited by trawl net from a depth of 80m. Largest among the group (81.2 per cent) *N. japonicus* of 140-250mm size group form a commercial fishery during April–March along the coast. The fish is found to attain its first maturity at 157mm length and they breed in October and April. Next in abundance of the group (17.2 per cent). *N. mesoprion* of 120-200 mm size group form a commercial fishery during November–December. They attain first maturity at 135 mm length and breed on October and April. Lowest occuring group (1.6 per cent). *N. delagoae* at 50-500mm size group form the commercial fishery in the region during January.

Threadfin bream fishery in Cochin, mainly contributed by *Nempterus japonicus, N. mesoprion* and *N. delagoae,* occupy 37.6 per cent of the total marine fish catch are exclusively caught by trawl net *N. japonicus* of 70-300 mm size occur in July–September at a depth of 75m. and form the commercial fishery. The fish of 180mm attain first maturity and spawn during July–November. The commercial fishery of *N. mesoprion* is formed with 70-260mm size group during June to October and occur at a depth of 75m. At 155 mm length the fish attain first maturity and spawn in July–November the *N. delagoae,* however, form the commercial fishery in the month of December.

Three species of threadfin bream, namely, *Nemipterus japonicus, N. mesoprion* and *N. delagoe* comprising 65.2 per cent, 32.2 per cent and 2.7 per cent of the group is exclusively caught by trawl net from 60-90m depth along the Bombay coast. *N. japonicus* of 130-290 mm size form commercial fishery, while *N. mesoprion* of 140-320 mm size occur in abundance in November–January (both the species). *N. delagoe* form the commercial fishery in the month of December.

Thread fin bream in Vizhinjam coast form 2.63 per cent in total catch and are caught by hooks and line (2.55 per cent), boat seine (0.05 per cent) and gill net (0.04 per cent).

Theraponidae

Thorn fish (crescent perch) in Vizhinjam coast form 0.40 per cent in total catch and are caught by hooks and line (0.19 per cent), drift net (0.02 per cent) and boat seine (0.03 per cent). Crescent perch under the family Theraponidae, represented by a single species, *Therapon jarbua* form a very small fishery along Karwar coast by contributing 0.07 per cent in total catch which are mostly caught by purse seine and trawl net.

Sillaginidae

Whiting in Vizhinjam coast form 0.18 per cent in total catch and are caught by boat seine only.

Menidae

Moon fishes in Vizhinjam coast form 0.03 per cent in total catch and are exclusively caught by boat seine.

Holocentridae

Squirrel fish in Vizhinjam coast appear in small numbers and are caught by hooks and line and drift net.

Platycephalidae

Flat heads in Vizhinjam coast form 0.44 per cent in total catch and are caught by gill nets (0.39 per cent), hooks and line (0.02 per cent), boat seine (0.01 per cent) and drift net (0.03 per cent). Represented by a single species, *Platycephalus scaber*, appear in Karwar coast mainly during August to May at a depth of 5 to 50m. Fishes of 100-220 mm size group form the commercial fishery in the region. *Platycephalus maculipinna* in Cochin coast form 0.18 per cent in total catch and are exclusively caught by trawl net. Fishes of 160-290 mm size group form the commercial fishery in the region within a depth of 40 m during February–March and July–August.

Ambassidae

Glassy perchlets at Vizhinjam coast form 0.09 per cent in total catch and are caught by boat seine (0.08 per cent) and shore seine (0.01 per cent)

Atherinidae

Hardy heads at Vizhinjam coast form 0.04 per cent in total catch mainly by boat seine and shore seine

Balistidae

File fish in Vizhinjam coast form 0.67 per cent in total catch and are caught by hooks and line (0.67 per cent)

Xiphidae

Sword fish, in Bombay coast form 0.9 per cent in total catch and are caught in gill net. The species are found almost throughout the year except in the month of July.

Coryphaenidae

Dolphin fish, *Coryphaena hippurus* form 0.03 per cent in total catch in and around Minicoy Island. The species is exclusively caught in pole and line and negligible quantity (0.4 per cent) by troll line. The species is mainly caught from surface water during

November and fishes of 550-580 mm size group form the commercial fishery in the region.

Represented by a single species *Coryphaena hippurus* constitute 0.19 per cent in total catch along Karwar coast. They are mainly caught by purse seine and gill net. Fishes of 210 to 460 mm form the commercial fishery at a depth of 5 to 50 m during September to April.

Dolphin fish at Vizhinjam coast form 0.39 per cent in total catch and are caught by hooks and line (0.37 per cent) and drift net (0.02 per cent).

Haemulidae

Spotted grunter under the family Haemulidae form 0.2 per cent in total catch at Veraval coast and are mainly caught by trawl net.

Carangidae

In the Tamil Nadu coast carangids trevally and scad, have a peak period of abundance in the month of March every year, with an annual average catch of 247 tonnes. Trawl net gill net and hook and lines contribute 37.0 per cent, 4.7 per cent and 13.1 per cent respectively.

Carangid fishes, namely horse mackerel, scads and leather jackets form an important fishery in Gulf of Mannar and Palk Bay. The group is represented by the species like, *Carangoides malabaricus, Atule mate, Selar leptolepis C. sexfasciatus* and *C. ignobilis* mainly. The group forms 42 per cent of the total marine catch in the region. The seasonal abundance of carangids in the inshore waters are April to July and again in November when they are caught by shore seine (40 per cent). The trawl net catch comprise of 2 per cent only, mainly in the month of January–February.

In the sea around Minicoy, the rainbow runner (*Elagatis bipinnulatus*) form 74 per cent catch in pole and line, while *Caranx* sp (3.9 per cent) and *Elagatis bipinnalatus* (2.3 per cent) form 26 per cent catch of troll line fishery.

In the Tuticorin coast, carangid fishes form 2.5 per cent of total marine fish catch of which 2.5 per cent are caught in trawl net and 0.03 per cent in drift net. The dominant species under the group are *Caranx carangus, Caranx leptolepis, Chorinemus* spp mainly. They are found to occur during August to February at a depth of 20-50 m.

Carangids along Kakinada coast of Andhra Pradesh are represented by scad, horse mackerel and trevallies. They together form 13.8 per cent of total marine fish landings around the region, of which 10.2 per cent are caught in trawl net and 3.5 per cent and 5.5 per cent in gill nets and other fishing gears respectively. February–March are the peak period of catch in the trawl net.

Decapterus russelli form 83 per cent catch of the group in trawl net. Fishes of 52–217 mm size group form the commercial fishery at a depth range of 10-80 m and their peak period of occurrence is January to April. The species attain its first maturity at 150mm length and breed during December–January.

Carangid fishes form 8.1 per cent of total marine fish landings at Karwar coast. They are caught by purse seine, trawl net and gill nets.

Caranx kalla of 50-200mm length forms the commercial fishery in the region with peak period of occurrence during August–May at a depth of 2–10 m.

Decapterus russelli, mainly caught in purse seine, appear during September–April in large numbers at a depth of 5–50 m and form a commercial fishery with 70–200 mm size group. *Megalaspis cordyla* of 100-300mm size group occur abundantly during September–April at 5-50m depth. *Chorinemus tala* of 125–250mm appear at 5–50m depth in large numbers during September–April to form commercial fishery. At the same time and at the same depth another species, namely, *Chorinemus tol* of 200-350 mm size form the commercial fishery.

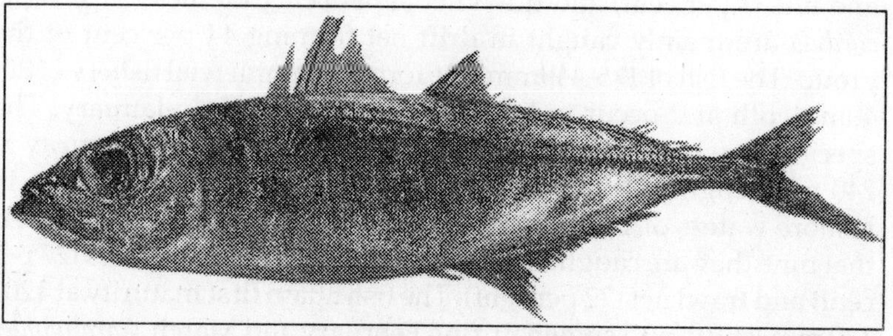

Decapterus russelli

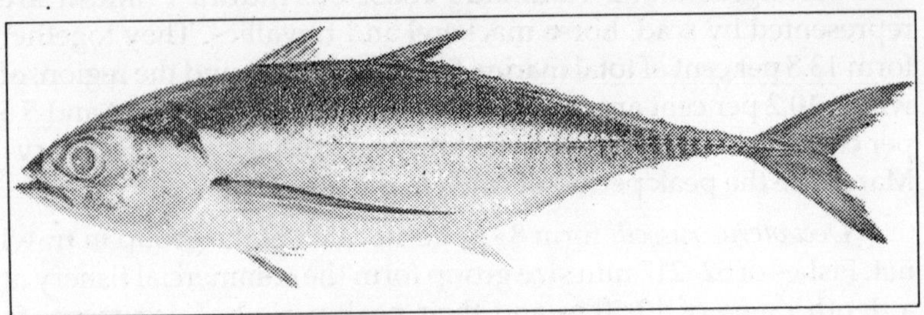

Megalaspis cordyla

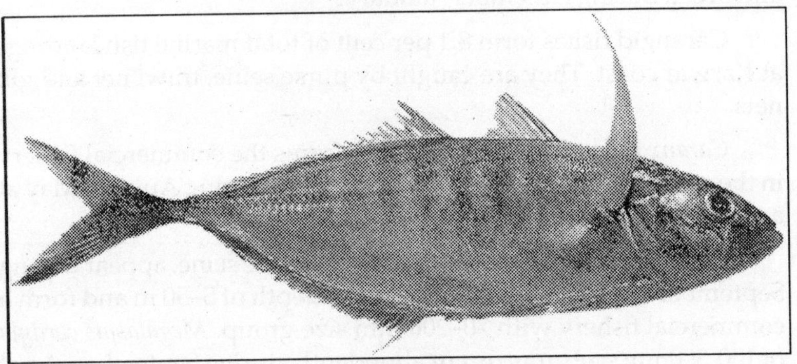

Alepes mate

Jacks, trevallies, scads and horse mackerels together form 7 per cent of marine fish landings along Visakhapatnam coast. They are mainly caught in drift gill net (14 per cent), shore seine and hooks and line (7 per cent) and trawl net (4 per cent). Of these *Megalaspis cordyla* are mainly caught in drift net forming 44 per cent of the group. The fish of 175-440mm size forms a commercial fishery of 20-40m depth and occur in large numbers during July-January. The species spawn during March–April after attaining first maturity at 240 mm length. *Decapterus dayi* of 30-234 mm length appears in inshore waters of 1-60m depth profusely during January–May. At that time they are caught in shore seines and hooks and lines (27 per cent) and trawl net (72 per cent). The fish attain first maturity at 120–130 mm length and spawn during February and March. *Carangoides malabaricus* are mostly caught in the trawl net (20 per cent of the group) from a depth of 10-60m during June–January. 110-270 mm size group form the commercial fishery. The species spawn in March and attain first maturity at 150mm length.

In Veraval coast of Gujrat trevally and scads form 2.1 per cent of total marine catch in trawl and gill net. *Decapterus russelli* of 50-150mm length congregate at 20-40m depth during December to March to form commercial fishery. *Megalaspis cordyla* of the same size group as that of *D. russelli* occur at the same depth and time of *D. russelli* while *Atropus atropus* of 100–300 mm size groups forms commercial fishery during November and December, and *Alepes djedaba* of 150-300 mm length form the commercial fishery and appear in maximum numbers in September–November at 40–80m. depth.

In seas around Cochin, carangid fishes constitute 1 per cent of total marine fish catch. The fishes are mainly caught in purse seine (66.5 per cent), drift net (22.4 per cent) and trawl net (11.1 per cent). The group is mainly represented by species like, *Megalaspis cordyla, Decapterus russelli* and *Chorinemus* spp. *D. russelli* of 70-220mm size form the commercial fishery during January–February and July–September, at a depth of 20-50m. The species attain its first maturity at 130mm and spawn during May–July. *M. cordyla* occur in large numbers during November–December at 20–80 m depth in the region and form commercial fishery by 150-390 mm size groups. The species attain first maturity at 270 mm length and spawn during May–July. Carangid fishes (Trevallies, horse mackerels) in Vizhinjam coast, represented mainly by *Decapterus dayi, Atule mate, Selar crumenophthalms, Megalaspis cordyla, Carangoides malabaricus, Chorinemus* spp form 15.47 per cent of total marine catch. They are mainly caught by drift net (2.49 per cent), hooks and line (9.92 per cent), boat seine (2.40 per cent) and shore seine (0.11 per cent). The peak period of occurrence of the group is August–October. *M. cordyla* occur in large numbers during October–January and fishes of 130–374 mm size group form the commercial fishery at 40-60m depth. The species is found to spawn in November–December. *Decapterus dayi* of 60-229 mm length form the commercial fishery and occur in large numbers during May–October. The fish attain sexual maturity at 130mm length and breed on March–May and August to October.

Atule mate occur in large numbers in March–July and September–November. The species form commercial fishery with 100–314 mm size groups at 30–60 m depth. The species breed in April to July.

Epinephelus diacanthus

Epinephelus undulosus

Lutjanus sanguineus

C. *malabaricus* of 110-215 mm size contribute to the commercial fishery at a depth of 30-60m with their peak period of occurrence in April–May and October.

Lutjanus malabaricus

A single species, *Megalaspis cordyla* of carangid group form 0.4 per cent of total marine fish catch along Bombay coast. They are mostly caught by gill net. Their peak period of occurrence is August to December at a depth of 30–60 m. Fishes of 110–350 mm form the commercial fishery and the species spawn during December to February.

Serranidae, Siganidae, Scaridae and Lethrinidae

Perches in Mandamapam, Rameswaram area form only 0.46 per cent of total catch and are mainly represented by *Lethrinus nebulosus, Siganus canaliculatus, Psamoperca weigiensis Callyodon ghobbon, Epinephalus tauvina, Lutjanus fulviflama.* Fishing methods are mainly hooks and line (80.9 per cent) and trap (19.1 per cent)

Epinephelus tauvina are also caught in traps (3.9 per cent) and hooks and line (16 per cent). The species occur in the region during October to March at a depth of 10–30m and form a commercial fishery with 180-790 mm size groups.

Siganus canaliculatus occur in the region during October to May to 2-5 m depth of water, mostly with 150-300 mm size group to form the commercial fishery. At that time they are caught by traps (19.9 per cent) and hooks and line (3.0 per cent).

Lethrinus nebulosus of 70 to 320 mm size group form the commercial fishery in the region during October to March at a depth of 10-30 m, when they are caught by traps (45.7 per cent), gill net (16.0 per cent) and hooks and line (28.0 per cent).

Callyodon ghobban of 150 to 300 mm size range form the commercial fishery in the area during November to May at a depth of 2-5 m, when they are caught by traps (8.9 per cent) and occasionally by hooks and line.

Snappers, reef cods and grunters together constitute 3.3 per cent in total catch at Kakinada coast where they are caught by trawl net (2.5 per cent) during February, May and July from a depth of 5 to 50m.

Pig face bream under the family Lethrinidae form 0.1 per cent in total catch along Veraval coast and are caught by trawl net.

Red snapper form 0.2 per cent in total catch at Veraval coast which are caught by trawl net.

Pig face breams form 0.59 per cent in total catch at Vizhinjam coast and are caught by hooks and line (0.34 per cent), drift net (0.21 per cent) and boat seine (0.03 per cent).

Snapper job fishes at Vizhinjam coast form 0.52 per cent in total catch and are caught by hooks and line (0.31 per cent), drift net (0.14 per cent) and boat seine (0.02 per cent).

Epinephelus malabaricus, E. diacanthus and *E. tauvina* form 0.3 per cent in total fish catch along Veraval coast, which are caught by gill and trawl net. *E. malabaricus* of 1000–1500 mm, *E. diacanthus* of 150–350mm and *E. tauvina* of 400–650 mm size groups form commercial fishery in Veraval coast during November–December at a depth of 20–80 m.

Reef cod and groupers consisting of *Epinephelus areolatus, E. diacanthus E. tauvina, E. chlrostigma, Pristipomoides typus* and *Lutjanus* sp. in Cochin coast form 0.1 per cent in total catch and are exclusively caught by hooks and line.

E. areolatus of 290-650 mm size group form commercial fishery in the region at 50 m depth during January–February. At the same time and same depth the other three species namely, *E. chlorostigma* of 270–580 mm length, *E. diacanthus* of 190–620 mm length and *E. tauvina* of 300–970 mm length form the commercial fishery. The bass, *P. typus* of 210–590 mm length form 0.03 per cent in total catch and form commercial fishery at 50 m depth during January–February.

Rock cods in Vizhinjam coast form 0.23 per cent in total catch and are caught in drift net (0.05 per cent), hooks and line (0.13 per cent) and boat seine (0.05 per cent).

Chirocentridae

Wolf herring, *Chirocentrus dorab* in Kakinada coast form 0.4 per cent in total catch and are mainly caught by gill net (3.6 per cent) from a depth of 3-70 m. The peak period of occurrence of wolf herring in Kakinada coast is July to September.

Represented by a single species, *Chirocentrus dorab* of 200 to 1000 mm length form the commercial fishery during August to May along the Karwar coast. They are mainly caught by purse seine, trawl net and gill net.

Single species, *Chirocentrus nudus* form wolf herring fishery at Veraval along Gujrat coast constitute 1.6 per cent in total catch and are mainly caught in trawl and gill net. Fishes of 300 to 600 mm size group form the commercial fishery during October to March in the region.

Chirocentrus dorab and *Chirocentrus nudas* together in Vizhinjam coast form 0.35 per cent in total catch and are mainly caught by drift net (0.18 per cent), hooks and line (0.04 per cent) Both the species appear in the coast during August–May and September–November at a depth of 30–60 m to form the commercial fishery.

Chirocentrus dorab in Bombay coast are caught by trawl net (38.9 per cent), gill net (46.7 per cent) and dol net (22.7 per cent) Fishes of 420–860 mm length group occur in large numbers during August–April at 10–90 m. depth. The species breed in July–August.

Priacanthidae

Represented by a single species, *Priacanthus hamrur*, Bull's eye form 0.1 per cent in total catch along Veraval coast of Gujarat. Fishes of 200–250 mm size group form the commercial fishery during November–December at a depth of 80 m from where they are caught by trawl net.

Bull's eye at Vizhinjam coast form 0.47 per cent in total catch and are caught by drift net (0.20 per cent) and hooks and line (0.27 per cent)

Priacanthus macrocanthus comprising 90 per cent in the catch among the group are found to occur along Visakhapatnam coast during August–March at 30–70 m depth to form commercial fishery with 100–230 mm size group. The species attain its first maturity at 170 mm length and spawn during November–May.

Clupeidae

Lesser sardines, comprising of species like *Sardinella gibbosa, S. dayi, S. sirm* and *S. fimbriata* are caught from Tamil Nadu coast mainly by gill net (28.0 per cent) and bag net (8.8 per cent) of which *S. gibbosa* constitute the commercial fishery. *S. gibbosa* are mainly caught in trawl net and form 80.9 per cent in the catch of the group. They occur in the coastal waters upto 20 m in depth almost throughout the year, with peak period of occurrence from October to January. Length range of 120-130 mm group form the commercial fishery and the species spawn during January to April.

Sardines form 13.2 per cent of total fish catch landed in Tuticorin coast and comprise of species like, *Sardinella dayi, S. clupeoides, S. sirm, S. albella, S. gibbosa* and *S. longiceps.* The goup together form 13.2 per cent of the gill net catches.

Sardinella gibbosa, mainly caught by gill net, form 52.1 per cent catch among the group, from a depth of 5-10 m with peak period of occurrence during May to July. The species attain its first maturity at 120mm length, spawn in the month of June and form commercial fishery with a size group of 100 to 180 mm length.

Next in commercial importance, *Sardinella sirm* constitute 27.8 per cent of the catch in the group are mainly caught in gill net from 5-10 m depth in large qualities during November–December. The species attain its first maturity at 170 mm length and spawn during August. Fishes of 145-220 mm length form the commercial fishery in the region.

Sardinella albella form 10.8 per cent in the catch of the group is also caught in gill net, mostly during April from inshore waters of 5-10 m depth. Fishes of 100-165 mm length forms the commercial fishery in the region, though the species attain its maturity at 110 mm length and spawn during June.

Sardinella dayi of 100–165 mm size form the commercial fishery in the region and occur in large numbers during April–May when they are caught in gill net from a depth of 5-10 m and form 7.3 per cent catch of the group. The species attain its first maturity at 120 mm length and breed in the month of February.

Sardinella longiceps forms only 0.9 per cent catch of the group from 5-10 m depth. Fishes of 120-165 mm size groups form the

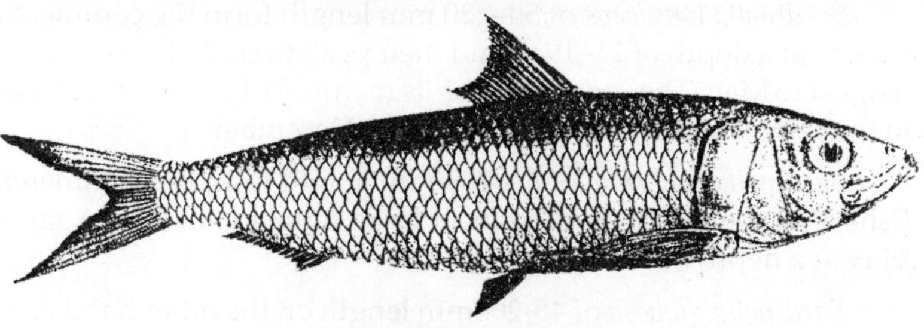

Sardinella longiceps

Sardinella longiceps

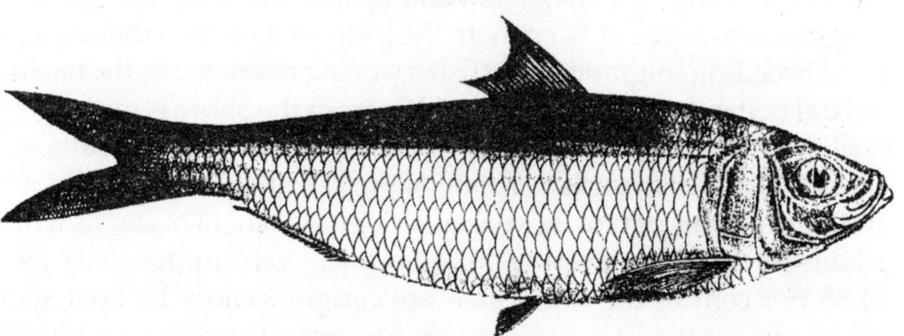

Clupea fimbriata

commercial fishery during November and December when they appear in maximum numbers and caught in gill net. The species attain its first maturity at 140 mm and spawn during March.

The clupeid fishes at Karwar coast represented by *Sardinella longiceps, S. fimbriata S. gibbosa* and *Escualosa thoracata* form 60 per cent of the total catch along the coast. The fishing methods that contribute the clupeid catch are purse seine, cast net, shore seine and gill net.

Sardinella longiceps of 50–220 mm length form the commercial fishery at a depth of 11–19 m and their peak period of occurrence is August to May. The species attains its maturity of 140-145 mm length in the region and spawn during June to December.

Sardinella fimbriata of 60–205 mm length form the commercial fishery in the region and occur in large numbers during August–May at a depth of 11–19 m.

Sardinella gibbosa of 45-200mm length on the other hand occur in coastal waters of 11-19m depth in large numbers during August-May to form the commercial fishery in the region.

The white sardine (*E. thoracata*) of 60–175 mm length form the commercial fishery at a depth of 11–19 m and their peak period of occurrence is October to May.

The annual yield of oil sardine, *Sardinella longiceps* which is the most important species among clupeids having a wide fluctuations in Calicut coast from year to year with an average of 3288 tonnes, has been showing a declining trend in recent years. Fishes of 0–year class contribute to a major portion of the catch. As the maturity progresses to a size of 14 cm with the onset of monsoon there is a sea ward breeding migration. The fishery being restricted to the narrow coastal belt extending to about 10 km from the shore is exclusively exploited by the artisanal fishermen using indigenous crafts and gears. The wide spread operation of purse seiners along both south and north of this part of the coast restricts movement of shoals which adversely affects spawning. *Sardinella longiceps* in the coast form 54.88 per cent of the total catch are caught mainly be boat seine (74.85 per cent) and gill net (39.43 per cent). The fishes of 130-210 mm size group occur in large numbers during September–February at 5–15 m depth to form the commercial fishery. The species at 140 mm length attains maturity and spawn during June to August.

Clupeids represented by *Sardinella fimbriata, S. longiceps* and *S. gibbosa* are met in the commercial fish landings off Visakhapatnam. They are mainly caught by gill net (90 per cent) and boat seine (47 per cent). *Sardinella fimbriata* of 40-195 mm length from the commercial fishery in the area with peak period of occurrence during October–March at 10–20 m depth and form 66 per cent catch in gill net and 46 per cent in boat seine among the group. Their spawning season is July–August. *S. gibbosa* of 35–180 mm length form the commercial

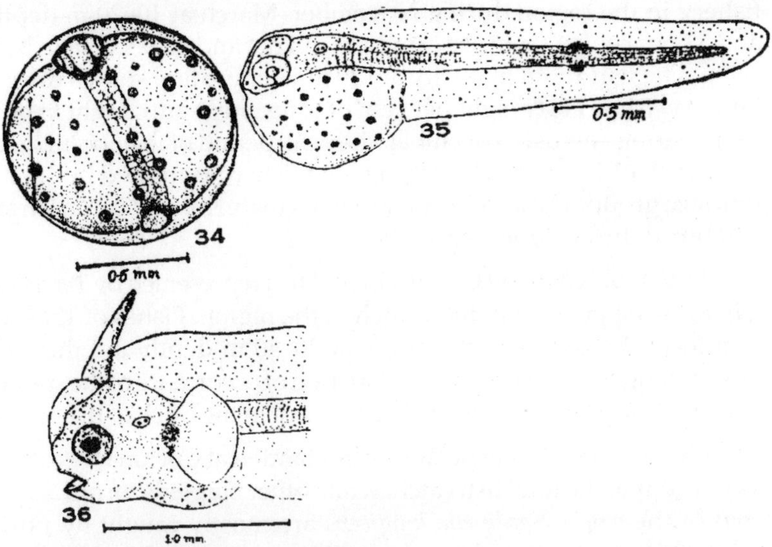

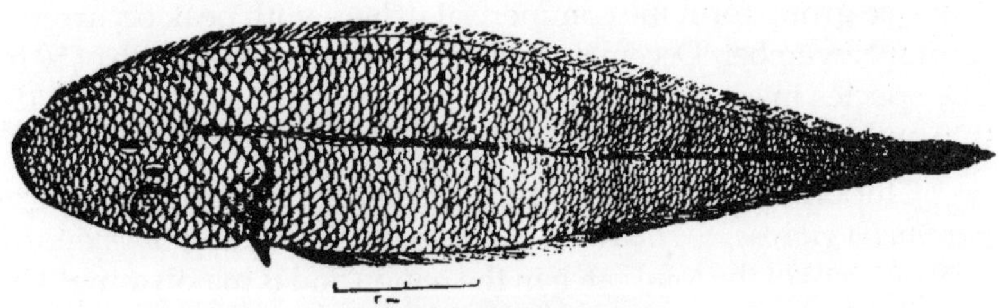

Paraplagusia blochii

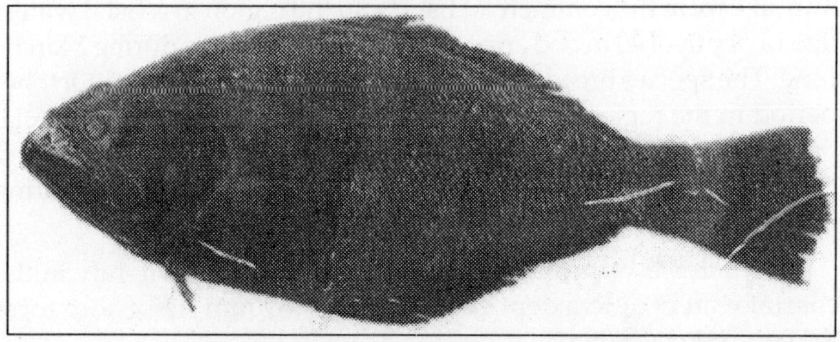

Indian Halibut (*Psettodes erumei*)

fishery in the region during November–March at 10–20 m depth, and 33 per cent of them are caught in gill net and 46 per cent in boat seine. The fish attain maturity at 120 mm length and spawn during January–May. Forty eight per cent of *S. longiceps* are caught in boat seine during August–November, the peak period of their occurrence at a depth of 10–20 m. The commercial fishery is formed by 40-205 mm size group. The species attain sexual maturity at 150 mm length and breed during June–September.

In Veraval coast, off Gujrat, clupeid fish represented by *Tenualosa toli*, form 1.4 per cent of total catch in the region. Fishes of 150–200 mm form the commercial fishery at 20–40 m depth and they are mostly caught by gill net. The peak period of occurrence of the species is October to March and form 92.7 per cent in the catch of the group.

Oil sardines among pelagic fishes landed at Cochin form about 44 per cent of the total fish catch, while other clupeids form 1.72 per cent of the catch. *Sardinella longiceps* are mostly caught by purse seine (97.8 per cent) and trawl net (2.2 per cent). Fishes of 145-185 mm size group form the commercial fishery with peak occurrence during November–December in the coastal sea upto a depth of 50 m. The species in sexually matured when they attain a length of 150 mm and spawn during June to July.

Clupeids in Vizhinjam, represented by *Sardinella longiceps, Sardinella gibbosa, S. sindensis, S. fimbriata, S. sirm* and *S. dayi* form 4.89 per cent of the total catch in the region and is mainly caught by gill net (4.14 per cent), shore seine (0.15 per cent), hooks and line (0.07 per cent) and boat seine (0.53 per cent). *S. longiceps* of 88–224 mm size form the commercial fishery in the region in coastal waters upto a depth of 40 m and appear in the large numbers during March–June. The species breed during April–May and September–October period in the region. *S. gibbosa* of 84–194 mm form the commercial fishery in the region and their peak period of occurrence being January–June upto a depth of 20 m region. The species spawn during April–August.

S. fimbriata appear in large numbers during April–July in the coastal waters upto a depth of 30 m. 120–189 mm size group form the commercial fishery.

Clupeids in Bombay coast, represented by *Tenualosa toli, Coilia dussumieri, Ilisha filigera, Sardinella longiceps, Sardinella fimbriata,*

Covala coval form 4.83 per cent of the marine catch. They are mainly caught by trawl net, gill net and dol net. *Escualosa thoracata*, 37.9 per cent of the group are caught in dol net. Fishes of 60–80 mm size group appear in large groups during May–July at 10–15 m depth and form commercial fishery. *Tenualosa toli* of 320–400 mm form commercial fishery at a depth of 30–60 m during May–January and are caught by trawl net (3.4 per cent) and dol net (30.3 per cent). *Coilia dussumieri* in Bombay coast are mainly caught by trawl net (33.6 per cent) and dol net (31.8 per cent) The peak periods of its occurrence are June–August and December–February at 10–30 m and 60–90 m depth respectively. The species of 80–180 mm length group forms the commercial fishery and attain its first maturity at 140 mm length. They breed during September to March.

Five–spotted herring *Hilsa kelee* occurs in small quantities in shore seines, drift net and in bottom set gill net at Mandapam–Rameswaram region of Tamil nadu. In this area, the fishing season for the species extends from April to October in the Palk Bay and from November to March in the Gulf of Mannar. In the latter region, the catch rate ranges from 5 to 30 kg per drift net unit. At Dhanushkodi, where relatively more intense fishing is observed, a catch rate of 380 kg per unit is obtained by the shore seine. In the bottom set gill net operated from this region, the catch rate of *Hilsa kelee* ranges from 1 to 5 kg per unit.

A bumper catch of *Hilsa kelee* was observed in the shore seines operated at the Palk Bay side at Dhanushkodi and in the Rameswaram Island in April 1987. The total catch at Rameswaram Island was 51.9 tonnes of *Hilsa kelee* of 15–20.9 cm dominated by 18–18.9 cm size group. In Dhanushkodi 14.45 tonnes of *H. kelee* of 15–19.9 cm size was landed of which dominant group was 16–18.9 cm length. Sardines (*Sardinella longiceps*) and lesser sardines together form 3.6 per cent in total catch in Kakinada coast and are mostly caught by gill net (11.5 per cent), trawl net (1.6 per cent) and other gears (4 per cent) from a depth of 1 to 60 m. Their peak period of occurrence being January to November.

Cynoglossidae (Flat Fishes)

Sole and flat fishes in Tamil Nadu coast occur throughout the year with a spurt in June and January. They are mostly caught by trawl net and contribute 0.6 per cent of total marine fish landings along the coast.

Flat fishes in Kakinada coast form 2.5 per cent in total catch and are mainly caught by trawl net, where they form 1.9 per cent of the trawl catch. The peak period and depth of their occurrence being December–May and 5–50 m respectively. The only species, *Psettodes erumei* form the fishery with 100 per cent catch among the group. The peak period of their occurrence and depth of occurrence being December–February and 5–50 m.

Tongue sole, *Cynoglossus bilineatus* in Karwar coast are mostly caught in trawl net. Fishes of 90–150 mm size group form the commercial fishery during August to May at a depth of 5–50 m.

Soles form on an average 4.8 per cent in the total landings at Calicut and is exploited by trawl net, boat seine and gill net and the most important commercial species is Malabar sole, *Cynoglossus macrostomus*. Fishes of 40-160 mm size group form the commercial fishery along Calicut coast at a depth upto 10 m. The peak period of occurrence of the species is from November to February. The fish attain its sexual maturity at a length of 120 mm and spawn in October to January.

Tongue soles form 1.4 per cent in total catch along Veraval coast and are caught mainly by trawl net. The chief species, *Cynoglossus macrostomus* of 100–150 mm size group form the commercial fishery in the region and occur at a depth of 25 m and appear in large numbers during March to May in the region.

Soles in Cochin, represented by *Psettodes erumei* and *Cynoglossus macrostomus* constitute 4.6 per cent of total marine fish landings are exclusively caught by trawl net. *C. macrostomus* of 87–150 mm group form commercial fishery in the area and they appear in large numbers during June to August at a depth upto 40 m.

Flat fishes in Vizhinjam form only 0.09 per cent in total catch and are mainly landed by boat seine (0.08 per cent) and gill net (0.01 per cent).

Indian halibut, *Psettodes erumei* of 140-620 mm size group form commercial fishery in Bombay coast. Their peak period of occurrence being December–January at a depth of 30–60 m. The species spawn during September–October. The species form only 0.2 per cent in total catch at Veraval coast of Gujrat. Fishes of 200–400 mm size group form the commercial fishery during October to January at 80 m depth from where they are caught by trawl net. Soles, *C. macrostomus* appear in Bombay coast during August–October at 30–60 m depth and form commercial fishery with fishes of 120–340 mm length group.

Chapter 6

Fishing Ground and Landing Centres along Indian Coasts

Sankarpur

In the Sand heads area of coastal Bay of Bengal the following species of marine fishes are caught abundantly in trawler fishing during October and November.

Sciaenidae (Croakers or jew fishes)

Leiognathidae (Silverbellies/pony fish)

Nemipteridae (Threadfin bream)

Clupeidae (Sardines and shads)

Trichiuridae (Ribbon fish)

Carangidae (Jacks, trevallys)

Mullidae (Goat fish)

Harpodontidae (Bombay duck)

Ariidae (Cat fish)

Stromateidae (Pomfret)

Lutjanidae (Snapper)

Polynemidae (Threadfins)

Chirocentridae (Wolf herring)

Muraenesocidae (Eel)

Synodonitidae (Lizard fish)

Scombridae (Seer fish, Mackerel)

Chandipur

Fishing zone of Balasore coast, Orissa is limited from North Dhamra to South Digha (West Bengal) between lat. 20°-50'–21°–25' and long. 87°–0'–87°–30' with a water depth of 45-52 m. Continental shelf covers an area of approximately 10,000 sq. km and extends upto 40 km in southern region. Marine fishery resources in the coast include pelagic and semipelagic fishes like hilsa, seer and pomfrets. Demersal fishes like sciaenids, cat fishes and shrimps are also available in the coast. Drift gill net and trawls are mainly used for catching these fishes. Fish landing data from 1986-87 to 1990-91 reveals the percentage composition of the catch of important group of sea fishes, as;

Hilsa–8.7–15.0 per cent

Pomfret 4.20–7.0 per cent

Cat fishes–13.0–16.35 per cent

Ribbon fish–1.02–5.0 per cent

Seer fish–1.0–2.0 per cent

Other clupeoids–3.25–9.0 per cent

Sciaenids–10.0–22.4 per cent

Prawn–1.7–4.7 per cent

Miscellaneous varieties–28.83–45.5 per cent

Paradeep

The catch of bottom trawling at Paradeep (Lat. 20°15' N. Long. 86°45' E), off Astarang (Lat. 20° N, Long. 86°23′E) and off Damodarpur (Lat. 19°25' N. Long. 85°5' E) situated on Mahanadi, Devi and Ruskhikulya estuaries at Orissa coast consists of pomfrets, chupeiform fishes, polynemids, sciaenids, ribbon fishes, silver bellies, glass fish and cat fishes,. The broad continental shelf in the northern region of Orissa coast extends from 25 to 35 km and is a rich ground for pelagic fishes. The continental shelf narrows towards south and has a favourable ground for semi-pelagic and demersal fishes. The average catch per trawling hour during 1962-1978 was 68.56 kg for the different groups, namely, miscellaneous fishes, sciaenids, prawns, pomfrets, ribbon fishes, clupeids, polynemids, silverbellies, glass fishes and cat fishes in order of abundance. The percentage composition of these groups to total catch also followed a similar trend.

Per cent Composition to the Total Catch

	Paradeep	*Astarang*	*Damodarpur*
Prawn	9.72	18.18	3.02
Pomfret	6.60	8.38	9.22
Clupeiform fishes	4.63	1.48	3.08
Ploynemids	1.14	0.07	0.69
Sciaenids	30.92	41.76	29.89
Ribbon fishes	5.49	3.96	5.21
Silver bellies	1.11	2.85	1.97
Glass fishes			
Cat fishes	0.97	6.70	1.36
Eel .	–	7.08	–
Carangids	–	2.21	–
Soles	–	1.50	–
Elasmobranchs	–	1.68	–
Miscellaneous	39.4	4.15	45.56

The average monthly variation of catch increased from 16.92 to 40.91 kg/hour (November to March). The catch rate of prawn, pomfret, ribbon fish and cat fish declined from March, their peak landing being December to February. In other groups the catch increased still further in March, but to a lesser degree, compared to the percentage of their increase in previous months. The catch rate of total landings and the majority of fish groups were more in Astarang than at Damodarpur but slightly less when compared to Paradeep. Miscellaneous fishes were significantly fewer and quality fishes dominated the catches at Astarang compared to Paradeep. The catch rate from October onwards increased progressively reaching the maximum in December and thereafter gradually declined till March both at Paradeep and Damodarpur. But the highest catch of prawns was in December at Paradeep followed by catfish in January. Ribbon fishes, clupeiform fishes, polynemids, sciaenids and silver bellies concentrated in the area during February and March respectively. The highest catch rate of miscellaneous fishes and total stock was noticed in March. At Astarang during the same period, prawns, clupeiform fishes, soles, elasmobranchs and miscellaneous fishes showed a peak in November followed by pomfrets, ribbon fishes and silver bellies in December. In January polynemids, sciaenids, catfishes, eel and carangids dominated the catch. The demersal fish stock was maximum from November to January. The fishes appear to migrate to inshore waters from north to south, first to off Mahanadi estuary in October and then off Devi river month in November.

Kakinada

Fishes like scad, horse mackerel and trevallies forming 13.8 per cent of the total catch, sharks (2 per cent of the total catch), wolf herring (0.4 per cent of total catch), flat fishes (2.5 per cent), white bait and anchovies (8.6 per cent); Bombay duck (1.2 per cent), white fish (0.7 per cent), silver bellies and pony fish (10.5 per cent); snappers, reef cods and grunts (3.3 per cent), eel (1.9 per cent); threadfin bream (6.3 per cent); drift fish (6.6 per cent); threadfins (0.6 per cent); sciaenids (8.1 per cent); scombroid fishes (6.5 per cent); barracuda (0.9 per cent); pomfrets (0.3 per cent); lizard fish (2.8 per cent); skate and rays (2.8 per cent); and ribbon fish (10.9 per cent) are the main fish groups occurring in Andhra coast of Bay of Bengal.

They occur in the depth range of 10-80 m, 3-70m, 5-50m, upto 50m, and upto 60m.

Visakhapatnam

Forty two species of fishes belonging to nineteen groups and sixteen families, besides sharks, skates and rays, have been found to occur off Visakhapatnam coast in Bay of Bengal. They occur in area extending upto 50km north–south and 15 km off the coast of Visakhapatnam all the year round. Their intensity increases during February–April followed by June–July, the highest being February–August. The concentration of bottom dwelling fishes are maximum during July–September.

The important groups of pelagic fishes met with are seer fishes 18 per cent, tunas and bill fishes 13 per cent, mackerel 13 per cent, sardines 11 per cent, cat fishes 9 per cent, jacks 7 per cent, white bait anchovy 5 per cent, sharks 5 per cent, ribbon fishes 4 per cent, rainbow sardine 2 per cent, silver bellies 2 per cent and bulls eye 1 per cent

Eight groups of demersal fishes form the bulk fish stock in the area. They are threadfin breams 13.5 per cent, prawns 13 per cent, lizard fishes 9.7 per cent, croakers 7.7 per cent, ribbon fishes 7.4 per cent, silver bellies 5.9 per cent, crabs 5.5 per cent and goat fishes 4.9 per cent, Six other groups account for 17 per cent of the total population. They are jacks 3.3 per cent, bulls eye 3.4 per cent, squids and cuttle fishes 3.1 per cent, white bait and thryss anchovies 3.0 per cent.

The fisheries activities off Visakhapatnam is a combination of artisanal fishing carried out within 8-10 km from shore by a variety of indigenous gears from country crafts and of mechanised fishing by shrimp trawls carried out in deeper waters by 200 small mechanized boats and by about 30 mini and 100 large trawlers. The range of operation of the small mechanized boats extends 15 km off the Visakhapatnam coast making fishing trips lasting for 1-3 days, while the mini and large trawlers fish mostly in the Gopalpur–Sandheads area making longer voyages lasting for 10-30 days. The country crafts and the small mechanized boats land almost all of their catches, whereas the mini and large trawlers land mostly shrimp, discarding the fish at sea.

Though year–round fishing activity is noticed in the area, fishing effort by the artisanal sector is intense during February–April followed by June–July. It is quite low during November–January. Bulk of the annual fish catch is obtained during February–August. The average catch per unit of effort for hook and lines is 10 kg, bottom set gill net 20 kg, gill net 40 kg and boat seine 50 kg. Hook and lines are operated from December to August, bottom–set gill nets from May to September, gill nets from November to July and boat seines from August to February.

Mechanized fishing by small boats is also carried out throughout the year. All the boats go for daily day-fishing trips during March–October. During November–February 75-80 per cent of these boats goes for either daily night or stay fishing. During the course of a single trip a boat expends about six hours of actual fishing effort in daily day fishing, nine hours during daily night fishing and about 24 hours during stay fishing with an average catch of 23 kg/hour. Both total fish catches as well as catch rates are better during July–September and this is considered as the best period for trawl fishery off Visakhapatnam.

Chennai

Among demersal fishes, threadfin bream ranks first forming 12.5 per cent of the total catch followed by silver belly, anchovy croaker and lizard fish. There is no significant variations in the concentrations of threadfin bream throughout the year except during the monsoon period of November and December when their occurrence is low. In fact the concentrations of many fishes are low during November and December.

Basing on the month wise appearance of demersal fishes in this region, the different fish groups may be categorised as (a) the appearance of those groups, which does not fluctuate very much throughout the year barring the low values during November and December, namely, sole and flat fish, *Thryssa*, seer fish, pomfret, goat fish, silver belly and shark and ray; (b) the presence of those fishes which are comparatively higher in certain months, namely, carangid (September), anchovy (April), grunters (March) skate (November-March), threadfin bream (August and September), sciaenid (November-March) and lizard fish (April–August) and (c) fishes which have a definite peak period of abundance every year, namely, barracuda (September) mackerel (July), drift fishes (August and March), carangids (March), Gerridae (July and August) and ribbon fish (November). They are insignificant at other times of the year.

The landings of mechanized gill nets are constituted mainly by seer fish (44.2 per cent), and shark (16.7 per cent) and those of indigenous gears by lesser sardine (28 per cent), mackerel and *Hilsa* (17.5 per cent) each.

The occurrence of shark (May), tuna (November), seer fish (February and March), sail fish (June and February) and carangid (August and January) is higher during the months indicated in parentheses. The presence of lesser sardine is almost uniform throughout the year and that of shark (September) *Thryssa* (March), mullet (May-July), mackerel (August and March), perch (August) and carangid is higher during the months indicated in parentheses.

The catch of *Hilsa* is very high during March and meagre during other months.

Fishes, such as, seer fish, shark, Carangid are landed by all the three types of gears, namely, trawl net, mechanized gill net and indigenous gears.

Threadfin bream fishery, which is constituted by five species, is dominated by *Nemipterus japonicus* (53.0 per cent), Silver belly and sciaenid fishery is formed by more than a dozen species each. *Stolephorus bataviensis, Thryssa dussumieri* and *Sardinella gibbosa* constitute the major species of anchovy, dussumier's anchovy and lesser sardine groups respectively.

Tuticorin

As an oldest fishery port, in south east coast of India, Tuticorin is well known for marine fish production in addition to the production of valuable, good quality of natural pearls and chanks.

Percentage composition of important fisheries in this region during 1911-1915 was ribbon fishes (34.3 per cent), sardines (10.9 per cent), jew fishes (9.5 per cent), perches (8.8 per cent), rock cods (5.8 per cent), seer fishes (3.9 per cent), sea breams (3.8 per cent), sharks and rays (3.7 per cent), red mullets (2.4 per cent) and anchovies (2.3 per cent).

With the gradual change in the pattern of fishing, during 1965-66, lesser sardines alone constituted the bulk of the catch (12.1 per cent) followed by Indian sprats (9.6 per cent), sea breams (6.7 per cent), rock cods (6.7 per cent), anchovies (6.4 per cent) and seer fish (5.9 per cent). There were distinct seasons for the appearance of seer fish, tunnies, barracuda and carangid fisheries caught by troll lines (June–September) Hand lining for serranids, lutjanids, and lethrinids were brisk till March commencing from October.

Post 1970 period brought about a significant change in quality and quantity of fish landings with the introduction of mechanised fishing boats in addition to traditional fishing crafts. At that period drift nets landed 61 per cent of the traditional catch, set gill net 19.5 per cent, line fishing 18.6 per cent and seines 0.9 per cent.

The important marine fishery resources of the region are (a) pelagic fishery resources, such as clupeids (lesser sardines, Hilsa, Chirocentridae), seer fish, tuna, barracudas, sharks, anchovies, carangids, belones, bill fishes. (b) major demersal resources namely perches (lethrinids, rock cods, snappers), *Scolopsis,* sciaenids, nemipterids, silver bellies, *Thrissocles,* cat fish, pomfret and rays.

Silver bellies are the most dominant in trawl net catches and the important species are *Leiognathus bindus, L. dussumieri* and *Secutor insidiator.* Lesser sardines constituted by *Sardinella fimbriata, S. gibbosa S. albella* and *S. sirm* are landed mostly by drift gill net. The king seer. *Scomberomorus commerson* forms about 90 per cent of total seer fish catch and the rest is contributed by the streaked seer, *S. lineolatus* and spotted seer *S. guttatus.* This resource is exploited mostly by drift gill nets, hand lines, long lines, trolling and to some extent by trawlers. Though seer fish occur throughout the year, the peak period of abundance coincides with the tuna season during July to September. *Auxis thazard, Euthynnus affinis, Sarda orientalis* and *Thunnus tonggol* support the tuna fishery. Perches of the area are lethrinids, *Epinephelus* spp, *Diagramma* spp, *Drepane* spp, serranids *Scolopsis* spp, *Pomadasys* spp, *Nemipterus* spp. The last three groups along with lethrinids are landed in good quantity by trawlers and the bigger perches are caught by drift gill nets and hooks and line. *Sphyraena obtusata* is landed abundantly by trawlers from the inshore water, whereas from deeper waters drift gill nets and hooks and line land *S. jello* and *S. picuda* (barracuda).

With the introduction of high opening bottom trawl for pair trawling operations, the catch is constituted mostly of valuable large sized quality fishes like lethrinids, rock cods, snappers, seer fish, pomfrets, horse mackerels and carangids.

Port Blair

The group of Andaman and Nicobar Islands situated between Lat. 6° and 14° N. and Long. 92° and 94° E. of Indian Ocean consists of 384 main islands having a coastline of nearly 1500 km.

Nineteen group comprising of 62 species form the commercial fisheries of the area. The most important group of the catches are in order of importance are perches, caranx, seer fish, mullets, mackerel, anchovies, sardines, silverbellies, sharks and rays and tunnies. Fishing activities are mainly concentrated in Middle and South Andamans. Fishing activities are low in the North Andaman, Little Andaman and great Nicobar Islands.

More than 30 per cent of the catches is made by gill nets. Hooks and lines account for about 25 per cent, boat seines and shore seines 20 per cent and the cast net and drag net 25 per cent of the total catch.

Marine fish production of the islands from 1803 tonnes in 1980 rose to 4300 tonnes in 1983 and nearly 16,000 tonnes in 2000 of which perches, mackerels, caranx, seer, anchovies, sardines, silverbellies, tunnies, and sharks and rays constitute 16.73 per cent, 10.13 per cent, 8.16 per cent, 6.52 per cent, 5.47 per cent, 13.49 per cent, 5.68 per cent, 3.06 per cent and 3.12 per cent of the total catch respectively. Miscellaneous fishes form 5.2 per cent of the total catch; besides hilsa (2.98 per cent), pomfrets (0.83 per cent), half beak (2.26 per cent), cat fishes (1.75 per cent) and prawns (2.97 per cent). Highest catches of 124.1 kg/hour in 50-59 m depth range followed by 102 kg in 40-49 m, and 95.4 kg in 30-39 m depth ranges were observed off South Andaman during 1980 in trawl fishing. The highest catch rate was observed in respect of silver bellies (45 kg/hour) follwed by sciaenids (25 kg/hour), goat fishes (5 kg/hour), sharks and rays (4kg/hour), perches (2kg/hour) and others (39kg/hour) in trawl net. The hooking rates of tuna worked out to be 3 per 100 hooks.

Mandapam

The Ramanathapuram district with a coast line of 261 km, covering the sea-front of Gulf of Mannar and Palk Bay, ranks foremost in marine fish production of Tamil Nadu. Sustaining high levels of organic production the seas along the coast have lucrative fishing grounds with greater profusion of finfishes of economic importance. The fishery resources of the region in the fifties and sixties were mainly exploited by the indigenous crafts and gears in the near shore waters extending to a depth of 10-15m.

Mechanized fishing boats are operated from 16 centres with technological advancement in fisheries and at present about 1000 trawlers and 5000 non-mechanized boats are employed in this area to catch marine fishes. In recent years, several non-mechanized boats are fitted with either inboard or outboard engines. Bottom trawl nets and the indigenous gears, such as drift gill nets, hooks and line, perch traps, shore seines, fixed bag nets and other nets are employed in the fishery.

The fish catching activities of the region are greatly influenced by the monsoon in the Palk Bay and Gulf of Mannar. During the south-west monsoon period in June-September the fishing operations are mainly concentrated in the Palk Bay as it remains calm and offers favourable conditions.

During the north-east monsoon in October-February, however the fishing activities are shifted to the Gulf of Mannar region as the sea on the Palk Bay side gets rough. The Gulf of Mannar is open and deeper as compared to Palk Bay. The inshore region of the Gulf of Mannar (5-25m depth) is beset with rocky patches with intervening sand and mud flats, while that of Palk Bay (5-15m depth) is mostly muddy.

Over twelve important groups of fishes constitute the commercial fisheries of Mandapam and Rameswaram region. The most important of these are silver bellies, elasmobranchs, croakers, clupeids, goat fishes, perches, catfishes, lizard fishes and carangids. Silver bellies account for 48 per cent of the total trawler catches. Croakers and elasmobranchs contribute to the tune of 4-5 per cent of the total yield of fish from the centre. Goat fishes, carangids, cat fishes, flat fishes, clupeids and other miscellaneous fishes together account for about 30 per cent.

In Rameswaram also, silver bellies form the major group exploited, contributing to over 51 per cent of the trawler catch, though there is a year wise fluctuations in the concentrations of silver bellies in this region. The other important group of fishes exploited at the centre are elasmobranchs (13.3 per cent), croakers (9.5 per cent). The remaining groups of fishes constituting the catch comprise mainly of goat fish, carangids, it fish, flat fish, clupeid, contributing 16.6 per cent of the total catch.

The indigenous gears such as drift gill nets, hooks and lines, perchtraps, shore seines, fixed bag nets, operating from Mandapam and Rameswaram catch by and large fishes from pelagic and columnar regions. The important groups of fishes exploited in this sector are clupeids formed by *Sardenella albella, S.* gibbosa, mackerel and carangids. Anchovies and seer fishes support to a seasonal fishery in the region.

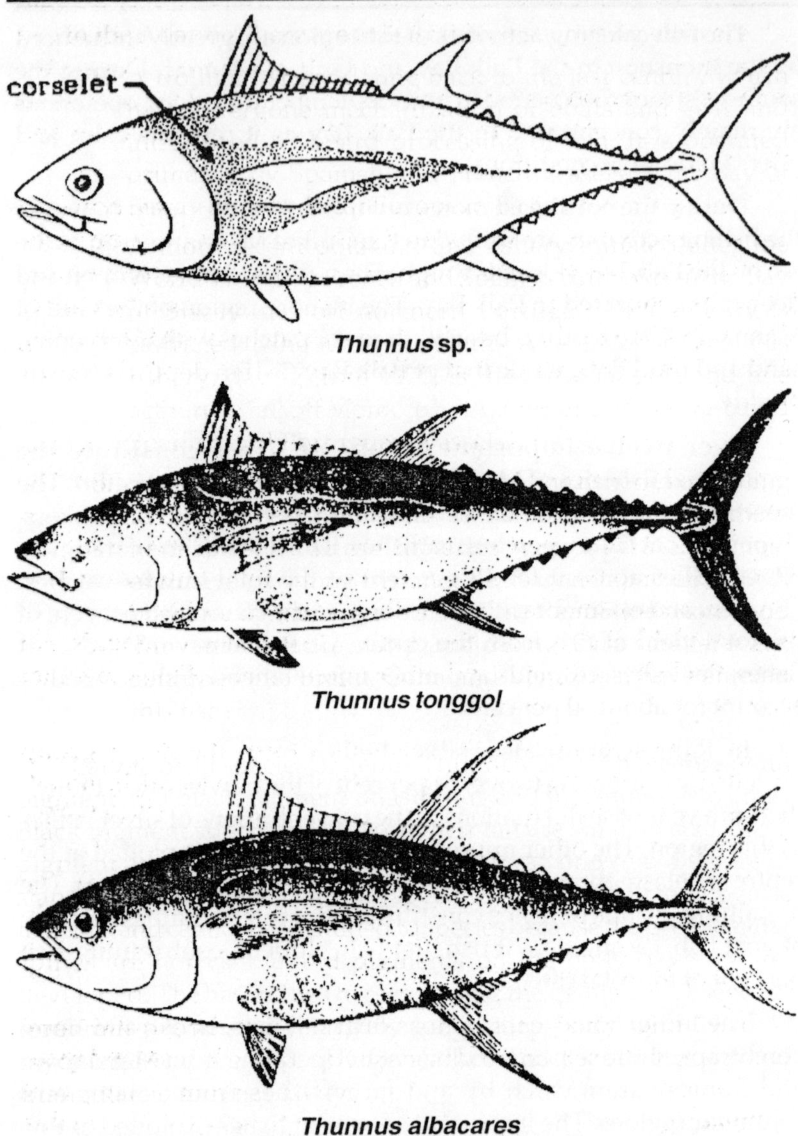

Thunnus sp.

Thunnus tonggol

Thunnus albacares

Calicut

Fish resource characteristics of Indian Oceans off Kerala coast incude pelagic fishes like tuna, seer fish, bill fish, pomfret and demersal fishes like catfish, sole etc. besides major pelagic fish species

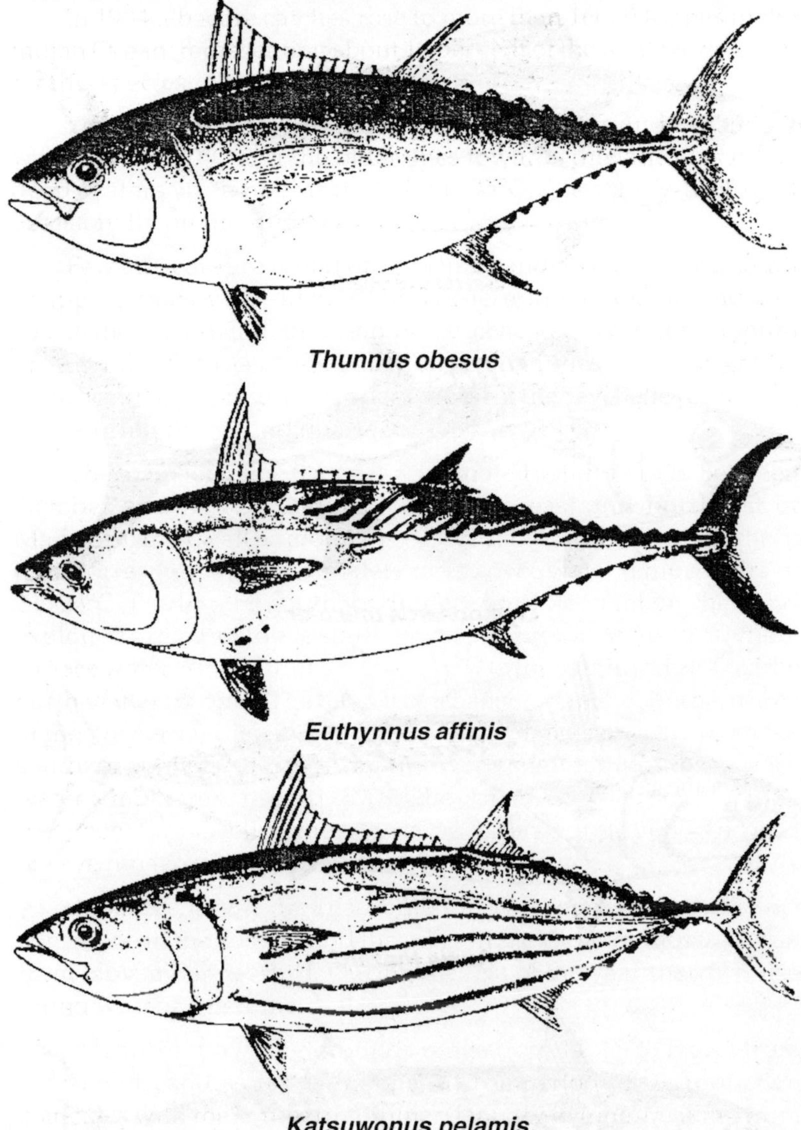

Thunnus obesus

Euthynnus affinis

Katsuwonus pelamis

like oil sardines and mackerels. The fish resources of the area are mainly exploited by country crafts with indigenous gears like boat seines, drift net, gill net and hook and lines and only 13.2 per cent is

Sarda orientalis

Gymnosarda unicolor

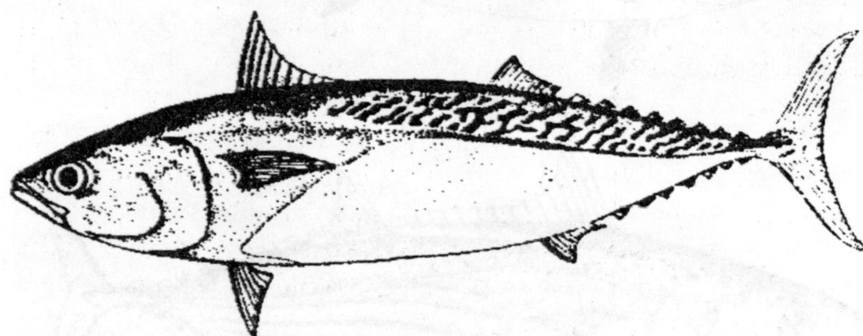

Auxis thazard

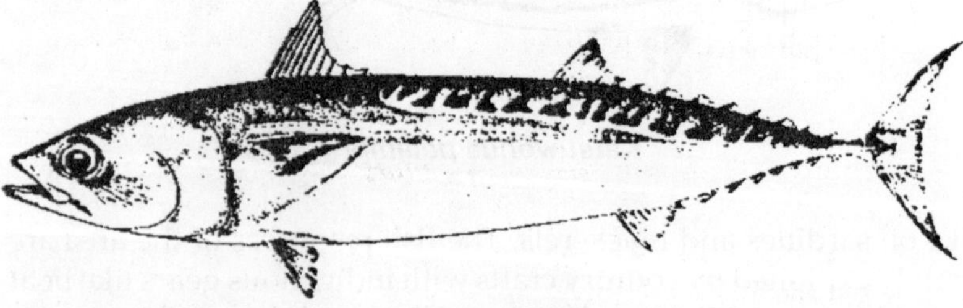

Auxis rochei

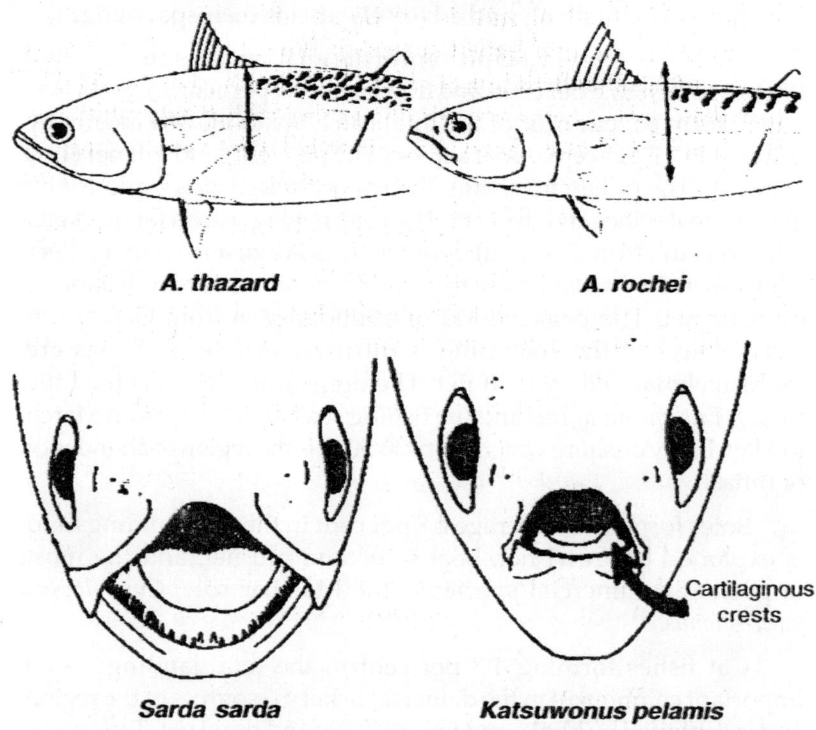

exploited by trawling and mechanised boats. About 70 per cent of total landing is obtained from boat seines alone, drift net contributing 6.6 per cent, hooks and line 5.8 per cent and gill net 4.4 per cent

The boat seines contribute 96.1 per cent of exploited clupeid fishes (54.9 per cent of total landings) and the rest 3.9 per cent is caught by gill net. The annual yield of oil sardine, *Sardinella longiceps,* the most important species among clupeids show wide fluctuations from year to year and has been showing a declining trend in recent years. Fishes of 0–year class contribute to a major portion of the catch. As maturity progresses to a size of 14 cm, with the onset of monsoon there is a seaward breeding migration. The fishery being restricted to the narrow coastal belt extending to about 10 km from the shore is exclusively exploited by artisanal fishermen using indigenous crafts and gears. The widespread operation of purse seiners along both south and north of this part of the coast restricts movement of fish shoals and adversely affects their spawning.

Anchovies forming 6.3 per cent of the total landings are exploited by boat seines (83.6 per cent) and trawl net (16.4 per cent). Scombroids which form 6.0 per cent of the total catch are exploited mostly by drift net (45.3 per cent), boat seines (34.4 per cent) and gill net (19.9 per cent). The Indian mackerel, *Rastrelliger kanagurta* is exploited by gill net, boat seines and drift net. The peak fishing season for mackerel which occurs from 5 to 40 m depth is from August to October. Seer fishes, *Scomberomorus commerson* and *S. guttatus* are mainly landed by drift net. The peak period of abundance is from October to December, but the spawning is during April–May. Tunas are exclusively landed by drift net. The dominant species is the little tunny, *Euthynnus affinis* and the fishing season extends from March to May. Fairly offshore waters from 30-50 m is the region of abundance of tunas.

Soles form on an average 4.8 per cent in the total landings and is exploited by trawl net, boat seine and gill net, and the most important commercial species is the Malabar sole, *Cynoglossus macrostomus.*

Cat fishes forming 4.3 per cent in the total landings is an important component in the demersal fishery resources of the region and is exploited by hooks and line, drift net and trawl net. *Tachysurus dussumieri, T. thalassinus, T. tenuispinis* and *T. serratus* are the most important species forming the commercial fishery. Catfishes constitutes 57.0 per cent in the landings of hooks and line and 15.2 per cent in drift net landings. The trend of this resource showed fluctuations from year to year with a general decreasing trend in the past few years. The peak period of the fishery generally coincides with the peak breading season, often with mass destructions of gesting males leading to mass mortality of eggs and embryos.

The pomfrets forming 1.1 per cent of the total landings are exploited by drift net, boat seine and trawl net, The black pomfret, *Parastromateus niger* and silver pomfret, *Pampus argenteus* are the species involved and the former forms about twice that of the latter in the fishery. Pomfrets constitute 9.0 per cent in the landings of drift net, 0.8 per cent in the trawl net and 0.5 per cent in boat seines.

Sharks, skates and rays together form only 0.42 per cent in the total landings and are mainly landed by hooks and line and drift net.

Other demersal resources like ribbon fishes, threadfin breams, sciaenids, silver bellies and lizard fishes form a fishery of some importance in the Calicut region.

Cochin

In the Indian Ocean off Cochin Harbour, the pelagic fish resources form almost 64 per cent and demersal fish 36 per cent of total fish population. Among pelagic fishes, oil sardine rank first, forming about 44 per cent of total fish population, followed by mackerel (11 per cent), tunas (3 per cent) and carangids (2.5 per cent). The demersal group is dominated by threadfin breams (12 per cent) followed by cat fishes (3 per cent), lizard fishes, flat fishes and sharks and rays.

Fishing activities in Cochin are restricted upto a maximum depth of 80 m. between Alleppey in the south and Chawghat in the north. Trawlers generally operate in depth range of 25-50 m and drift gill nets in 25-80 m; whereas, purse-seine operates beyond 30m, even though very often this gear encircles shoals of fishes in shallow depths (10-15m) also depending on the seasonal availability of fish shoals.

Purse seine make the bulk of landings and the most dominant fishes are oil sardine (74 per cent), mackerel (15 per cent), other sardines (3 per cent), carangids (3 per cent), catfishes (1.2 per cent), pomfrets (1.1 per cent). The peak fishing season for purse seine is post monsoon (42 per cent) followed by summer (32 per cent), premonsoon (14 per cent) and monsoon (12 per cent). The important fish groups from trawler catch are threadfin breams (33.7 per cent), catfishes (5.2 per cent), sciaenids (4 per cent), soles (4.6 per cent), anchovies (3 per cent), clupeids (2.3 per cent), lizard fishes (2.9 per cent). Penaeid prawns and crabs contribute a substantial catch (12.5 per cent) in trawl nets. The fishing season for the trawl gear commences in monsoon period and almost 40 per cent of landings is made during this period. The premonsoon period contribute 30 per cent, summer 20 per cent, and the post monsoon 10 per cent. In the drift gill net, tunas are the most abundant fish (31 per cent), followed by elasmobranchs (18 per cent), catfishes (15 per cent), seer fishes (16 per cent) carangids (7 per cent), pomfrets (5 per cent). Monsoon months contribute the maximum landings (41 per cent) followed by premonsoon (29 per cent), post-monsoon (18 per cent) and summer (1.2 per cent).

The average annual catch of oil sardine comprise 98 per cent, caught by purse seine, the peak months being November and December. The next abundant fish, mackerel, is 98 per cent contributed by purse seine during September to November. Tunas and carangids, the next important groups, caught by drift gill nets, with major landings during April to December. The fishing season of seer fishes is mainly from August to December followed by white bait in October and November and pomfrets in August to November.

Among demersal fishes, threadfin breams form bulk landings with peak from June to October. Catfishes, the next dominant resource is harvested mainly in June to February. The maximum harvesting of lizard fishes and flat fishes are made from June to September. The major fishing season of sharks and rays being April to December for sharks and November to March for rays. Perches are mainly caught by hooks and line and are highly seasonal in their occurrence during January and February.

Trivandrum and Kanyakumari

The catamaran fishermen of Trivandrum and Kanyakumari coasts migrate along the Malabar coast for fishing during October to January. They even go upto Karwar in their catamarans driven by sail and catch quality fishes using hooks and line. The hooks along with the bait of bright coloured fishes (oil sardines, mackerel or *Stolephorus*) are tied to a monofilament of 1mm thickness and released as the catamaran sails. The northeast wind is utilised for sailing. The fishing is done mainly along the rocky beds about 12 to 15 km away from the shore at a depth of 20-30m. These areas are proved to be rich fishing ground for quality fishes like *Scomberomorus commerson*, *Sphyraena* spp, *Rachycentron canadus* and *Sepia pharoonis*. Their length range and composition to the total catch are, 72-152cm, 45.3 per cent ; 51-92cm, 46.8 per cent ; 132-150cm, 5.0 per cent and 19-26 cm, 2.9 per cent respectively. These catamaran fishermen land their catches from the last week of October to December.

Vizhinjam

Vizhinjam, 16 km, south of Trivandrum in Kerala State is an important fish landing centre in the fishery zone extending Kollangode in the south to Valiaveli in the north spreading a distance of 50 km on the south-west coast of India. June to October presents the main fishing season with best return in July in Vizhinjam area, when nearly 71.3 per cent of the annual catch is landed. A large variety of fishes support the fishery at Vizhinjam and the landings of each gear have a characteristic species composition.

Of the ten types of traditional gears, employed to catch fish in this area, boat seine, hooks and line, hand line with smaller hooks and shore seine together contribute to the bulk (74.87 per cent) of the total fish landings and the rest (25.13 per cent) by gill nets.

Mechanized crafts, mostly with out board motor fitted boats, generally go about 20-25 km off Vizhinjam to area of 60-80 m depth, whereas the non-mechanized crafts are confined to about 10 km from the shore at a depth range of 40-50m. Both mechanized and non-mechanized units do fishing for 20 to 25 days in a month.

The juveniles of almost all major commercially important fishes form a minor fishery, which contribute to 0.03 per cent of the total landings,. The fishery season extend from October to August.

Karwar

At the beginning of September, after the cessation of south-west monsoon, large quantities of oil sardines, lesser sardines comprising *Sardinella fimbriata S. albella, S. gibbosa,* and *S. dayi,* anchovies, mackerel, tunas (*Euthynnus affinis, Auxis thazard, A. rochei*), black pomfret, carangids, cat fish (*Tachysurus tenuispinis, T. serratus*) and silver bellies gather at the northern most tip of Karnataka coast of Arabian Sea.

The period from January to May is known for moderate catches of the above groups of fishes in the region. Oil sardine is caught in plenty in most of the months of the fishing season and mackerel and lesser sardines to a lesser extent. Tunas generally appear in the seiners from September to December and again in small quantities from March to May. Bumper catches of cat fishes and pomfrets, especially the black species are caught during September and

October, causing on many-an-occasion the glut condition. Other fishes are landed throughout the fishing season on varying proportions.

Minicoy

Abundance of tuna fishery at Minicoy (Long. 73°E, Lat. 8°17' N) is seasonal. Pole and line and surface trolling gears are responsible for about 95.85 per cent and 4.14 per cent respectively of the total fish landed. Catch by hooks and line is negligible (0.01 per cent). The pole and line operations are more prevalent during post and pre-monsoon seasons; whereas maximum effort by troll line is during monsoon months. Productive period for pole and line fishery is during October–May and for surface trolling June to September with a secondary maximum in February.

In the pole and line fishery, skipjack tuna (*Katsuwonus pelamis*) contribute on an average about 90 per cent of the total marine fish catch. Yellowfin tuna (*Thunnus albacares*) respresented by sub adults (young) occur mainly during the post-monsoon period. Other components of the catch include rainbow runner (*Elagatis bipinnulatus*), little tuna (*Euthynnus affinis*), dolphin fish (*Coryphaena hippurus*) and small pelagic sharks in their order of abundance.

Species diversity is high in the troll line catches. The average annual pattern of occurrence of different species in their order of abundance are yellowfin tuna (Sub-adults 38.7 per cent), skipjack tuna (23.5 per cent), *Acanthocybium solandri* (Wahoo 10.4 per cent), pelagic sharks (7.3 per cent), *Istiophorus platypterus* (sail fish 5.4 per cent) *Caranx* sp. (3.9 per cent), *Elagatis bipinnulatus* (2.3 per cent), *Sphyraena* sp (1.8 per cent), *Euthynnus affinis* and *Auxis thazard* (0.4 per cent) *Gymnosarda unicolor* (0.4 per cent), *Coryphaena hippurus* (0.4 per cent) and others (5.5 per cent). A noticeable feature in the troll line fishery is the abundant occurrence of wahoo during post monsoon months and young yellowfin tuna during pre-monsoon months.

Factors such as sea surface temperature, surface currents, wind pattern, availability and abundance of live-baits, tuna behaviour in the fishing ground and presence of flotsam objects exert considerable influence on the fluctuation in the availability of tunas which is the mainstay of marine fishery at Minicoy.

Veraval

In the Arabian Sea off Gujrat coast, Bombay duck and pomfrets occur enormously. In the inshore waters of Veraval sciaenids and eels are prevalent. The sciaenids form 16 per cent of the trawl landings and 3.4 per cent of the gill net catches. The cat fishes constitute 5.8 per cent of the gill net catch and 0.77 per cent of the trawl landings. The thread-fin breams constitute 3.5 per cent of the fish population in the area.

About 800 trawlers, 300 OBM fitted boats and about 50 non-mechanized crafts are engaged in the area for fishing activities. The trawlers account for about 90 per cent catch and the rest by gill nets and OBM boats. The fishing operations are mainly carried out at about 80 m depth from Dwarka (250km) towards north to Nawabunder (100 km) towards south. The peak landing period is from October to January by the trawlers. During monsoon, gill netlers operate to catch sciaenids. At the fag end of fishing season many fishermen go for shark fishing.

The pelagic fishes form majority of the landings in the gill net catch (65.13 per cent); whereas 64 per cent of the demersal resources is landed by the trawlers. By and large the landings of demersal resources are comparatively higher (60.4 per cent) than the pelagic fish stock in Veraval area.

Mumbai

Bombay duck is the predominant fish forming 25.65 per cent in the 'dol' net catch. Other varieties like clupeids, sciaenids, pomfrets, eels and ribbon fishes from less than 4 per cent each in the catch.

Pomfrets both silver and black, forming 28.6 per cent dominate the gill net catch. This is followed by clupeids, tunas and seer fishes constituting 20.2 per cent, 20.0 per cent and 14.8 per cent respectively. Sciaenids, sword fishes and horse mackerel form between 3.7 per cent and 7.4 per cent.

The trawl catch has many groups of fishes occuring in different magnitude. The major groups are sciaenids, cat fishes and clupeids forming 5.01 per cent, 4.83 per cent and 4.09 per cent respectively in the catch. Sharks, ribbon fishes, threadfin breams, lizard fishes and rays contribute 3.43 per cent, 2.70 per cent, 2.11 per cent and 1.24 per cent respectively to the total catch. The remainings are minor groups forming less than 1 per cent in the total catch.

All these gears operate round the year, the trawl at 60-90 m, 'dol' at 10-30m, and gill net at 30-60 m depth zones. In addition there are hooks and lines operating from September to April at a depth of about 30 m. Their landings are very meagre and comprise of catfish, shark and occasionally eel.

Chapter 7

Larger Oceanic Pelagic Fishes

Twelve months sample survey in Arabian Sea, Bay of Bengal and North-west Indian Ocean between Lat. 7° and 12° N in south west coast and east coast of India upto Lat. 18°N. reveals that there are three species of tuna, namely, big eye tuna (*Thunnus obesus*), yellow fin tuna (*Thunnus albacares*) and skipjack tuna (*Katsuwonus pelamis*) form 25 per cent of the total catch in Indian Oceans. Yellowfin tuna was the dominant species contributing 407 numbers and big eye tuna 32 numbers out of a total catch of 1900 fishes weighting 70 tonnes. Among bill fishes, which made up 9.4 per cent of catch, striped marlin (*Tetrapturus audax*) and Indian sail fish (*Istiophorus platypterus*) were more frequent ones. Blue marlin (*Makaira nigricans*) and broadbil sword fish (*Xiphias gladius*) were caught in small numbers.

Pelagic sharks that represented 64.2 per cent of catch included different species, namely, black tip shark (*Carcharinus melanopterus*), white tip shark (*C. longimanus*) hammar head shark (*Sphyrna zygaena*), blue shark (*Prionace glauca*) thresher shark (*Alopias vulpinus*), silky shark (*Carcharinus brachyurus*) and tiger shark (*Galeocerdo* sp) in their order of abundance. Other varieties hooked were dofin fish (*Coryphynae hippurus*), barracuda (*Sphyraena barracuda*), seer fish (*Scomberomorus* sp.), Caranx, sun fish (*Mola mola*)

The catch from three regions, namely, west coast, east coast of India and equatorial seas (0°-6°N) are seperated and the percentage composition from each of the area has been worked out and furnished below.

Sl.No.	Name of Species	Aggregate	Percentage of Catch		
			West Coast	East Coast	Equatorial Seas
1.	Big eye tuna	1.69	–	–	10.42
2.	Yellow fin tuna	21.50	2.03	43.82	73.62
3.	Skipjack tuna	2.01	–	8.99	1.95
4.	Marlin	4.38	2.52	11.52	3.58
5.	Sail fish	4.23	3.09	9.27	2.93
6.	Sword fish	0.79	0.08	3.37	0.65
7.	Pelagic sharks	64.24	91.30	21.34	5.54
8.	Other varieties	1.16	0.98	1.69	1.31

Tunas were harvested in higher percentage (85 per cent) from equatorial sea followed by east coast (53 per cent). Tuna from west coast was negligibly low at 0.03 per cent. The big eye species, most esteemed among tunas was obtained only from the equatorial sea. Bill fishes were found to be more abundant in east coast contributing 24 per cent of the catch. Pelagic sharks formed the major component (91.3 per cent) in west coast and recorded the minimum share (5.5 per cent) in equatorial seas.

The equatorial waters support comparatively high quality resources with respect to three most valuable species, namely, big eye tuna, yellow fin tuna and striped marlin. The availability of the above species is very much comparable in Bay of Bengal within the Indian Exclusive Economic Zone.

Tunas

Tunas make a major contribution to the fish production of Sri Lanka, the average annual production of 35000 tonnes making up 17 per cent of all fish landed in the country. The fishes are available within a range of about 25 miles from the coast. Direct and indirect evidence points to the availability of tunas in the off shore range, beyond the presently exploited areas.

Tunas dominate the gill net fishery (62-76 per cent). Skip jack tuna (42-66 per cent) is the most dominant tuna species followed by yellowfin tuna (9-16 per cent) in the gill net fishery. In the long line, a small percentage of yellowfin (1-4 per cent) and big eye (0-0.8 per cent) are caught.

Skipjack catches in the gill net fishery show a decrease in relative abundance from south to north, while a reverse trend is seen in the case of yellowfin. Both skipjack and yellowfin show a greater abundance in the off shore range, beyond 80 km than in the coastal areas during 1st (Feb-April) and 4th (Nov-Jan) quarters. Higher catch rates are observed during the 2nd and 3rd quarters (south-west monsoon period) than in the rest of the year.

Except for target species, such as skipjack and yellowfin, most other varieties have been found to caught by entangling rather than gilling in the net. They were entangled by the mouth and teeth (small tuna), tail (cetaceans and dolphin fish), bill (bill fish) or a combination of the above. A small percentage of skipjack and yellowfin were also entangled by their snouts and fins.

While 62 per cent of the skip jack caught were gilled, only 37 per cent of the yellowfin were seen gilled. The optimum length of fish caught in the 125, 150 and 180 mm mesh sizes of nets to be 46, 56 and 65cm for skipjack tuna and 47, 53 and 61 cm for yellowfin tuna respectively.

Twelve schools of single species sighted were of skipjack, yellowfin or frigate tuna. Three schools of mixed species, comprised of skipjack and yellowfin.

In the gill net, over 90 per cent skipjack tuna caught were within 40 to 70 cm range. Small skipjack of less than 34 cm length were found mainly in the south-west during south-west monsoon suggesting possible recruitment of skipjack off the south-west, followed by a movement northwards and southwards.

Yellow fin in gill net catches measured 32 to 152 cm with a majority within 44 to 76 cm range. Troll catches comprised of immature fish (48 to 64 cm) as well as adults (104 to 160 cm). Long line catches were all adults (120-145 cm). The juvenile yellowfin, showing a seasonal shift northwards, commences to spread into deeper water and becomes available to the long line fishery.

Male to female sex ratio of skipjack tuna was of 1 : 0.9. Females out number the males during the south-west monsoon months of June to October. Size distribution of skipjack by sex showed that the proportion of females was greater in sizes less than 57 cm and that of males was more in sizes over 59 cm. The gonado-somatic-index (GSI = Weight of gonad/Total length of fish^3 x10^5) for female skipjack, correlated to the maturity stages was established as; I Immature–1.58; II Early developing–4.6; III Late developing–9.73; IV Mature–14.25; V Gravid or running–16.17, VI Partially spawned–7.98 and VII Spent–3.83.

Ova diameter showed five distinct groups of ova; immature (0–0.08 mm); developing (0.08–0.38 mm); mature (0.32–0.52 mm); gravid (0.54–0.68 mm) and running (1.02mm). The presence of a relatively high percentage of mature, spawning and spent fish was observed in most months. Their spawning commences just before the south-west monsoon (March-April) and would prevail until the end of the monsoon (September–October). The species have a long spawning period and 50 per cent of female skipjack attain maturity at 45.4 cm; while male skipjack attain maturity at 42.0 cm length.

Yellowfin tuna showed a male to female ratio of 1 : 0.91. Unlike in the case of skipjack, males out number the females during the south-west monsoon months. Females are dominant in February, April, October and November. More yellowfin males beyond a length of 112 cm is observed in the population. There is no marked difference in the sex among the juveniles. Fish below 100 cm are generally immature and the smallest fish with mature ova measured 112 cm.

Major food items of skipjack tuna are fish (35-36 per cent), squids (28-31 per cent), shrimps (32-33 per cent), crabs (1-3 per cent) and larvae of crustacean, copepods etc. (0.6-0.9 per cent) Small shrimp (Acetes spp) is a very significant food item during south-west monsoon period. Small frigate tuna (20 per cent) was the major fish food item the skipjacks feed on.

Tuna Ecology

Tuna have a high metabolism and an internal temperature higher than that of environment, a characteristic of these species. Tuna are constantly engaged in the search for suitable areas in which to ensure their survival. Consequently they spend their whole lives swimming and their resulting high energy consumption requires them to absorb daily large quantities of food which can attain as much as 15 per cent of their total weight. As a result, the areas of tuna concentration are by no means casual, and migration takes place according to "hydrological routes" in which each species finds the optimum environment for survival in every stage of its existence.

Tropical Tuna

Physiological tolerances differ from one tuna species to another. Laboratory tests carried out on skipjack tuna suggest that temperature and oxygen aspects alone could suffice in determining an ideal habitat.

The lowest critical water temperature for skipjack is at 15°C. An analysis based on sea temperature and skipjack catches, in the Eastern Pacific, indicated that catches decreased in temperature below 20°C and nil catches were obtained in waters below 17°C. Highest water temperatures for skipjack are 33°-34°C. These temperatures are not found in natural environments. The minimal dissolved oxygen value in which skipjack are known to survive is between 2.5 and 2.8 ml/litre.

The limits of temperature and oxygen necessary for the survival of other three tuna species are;

Tuna Species	*Optimal Temperature Range in °C*	*Dissolved Oxygen Values Tolerated for 10 Minutes by Fish Measuring Between 50 and 75 cm*
Skipjack	20 to 32	2.5 to 3.0 ml/l
Yellowfin	23 to 32	1.5 to 2.5 ml/l
Bigeye	11 to 23	0.5 to 1.0 ml/l
Albacore	15 to 22	1.7 to 1.4 ml/l

The level of dependence on these two parameters will vary in the course of their life span according to their morphological and physiological condition.

The first problem confronting the tuna is how to remain in the water column. Their behaviour varies according to whether they have a swimbladder (genus *Thunnus*) which reduces body density or not (skipjack, for example). Skipjack are obliged to swim increasingly faster as they grow in order to maintain constant depths, a procedure which requires increasing intakes of oxygen. Yellowfin tuna have swim bladders which grow with age. This enables them to reduce speed as well as oxygen intake. Bigeye tuna retain their swim bladders throughout their life span and consequently require less oxygen.

The second problem is water temperature in that juvenile tuna species do not have a fully developed thermoregulation system and therefore have to swim within narrower thermal boundaries. On the other hand the older individuals can swim deeper. Juvenile skipjack are limited to swimming within the warm and oxygen saturated mixed layer of the thermocline whereas the older skipjack will move in the layer which is less oxygenated, due to organic decomposition and poor renewal of water.

The above data highlight only optimal habitats and not exclusive ones. Sonic tagging experiments showed that yellowfin and skipjack can swin deeply into theoretically lethal layers. Duration and frequency of these dives are not well known. In the western Indian Ocean purse-seine tuna fishery through ecosounding could follow fish escaping under the net, within waters where temperatures are 5°-8°C, lower than the minimum temperatures limit reported.

Distribution of Tuna

The hydrological structures that can affect tuna distribution are generally discontinuities, especially those relating to temperature and salinity.

Surface water temperatures can show strong gradients (thermal fronts) especially around upwellings or during cooler winter periods or for example when colder subtropical waters reach equatorial areas. In the Western Pacific, yellowfin abundance showed a seasonal cycle in relation with winter cooling when thermal fronts are created. Up to now, in the Western Indian Ocean, the southwest monsoon is the period when substantial catches of tuna were made close to thermal fronts.

Another kind of thermal discontinuity between the warm mixed layer and the subsurface layer is the thermocline, which delimits two water masses with different densities. Its characteristic is a vertical thermal gradient depicting the water column stratification. In some areas of the Eastern Tropical Atlantic the purse-seine yield increased along with the vertical thermal gradient. In the Western

Indian Ocean significant increase in capture was found in the divergence areas where maximum vertical thermal gradients (4°-5°C/10m) were reached during the previous month. This divergence led to an accumulation of nutrients in the euphotic layer. Later, the mixing due to strong winds at the onset of the southwest monsoon generated an intense organic production in the subsequent weeks.

Salinity levels as well as their variations in the ocean do not seem to be factors affecting the distribution of tuna. But they do characterize water masses which may help to identify more potentially favourable fishing grounds. South of Japan the best skipjack catches occured during years when water temperatures were predominently higher than average and lower in salinity (34.81 ppt against 34.87 ppt). Skipjack in the tropical Pacific tended to spawn in low salinity surface waters floating above the subsurface layer with maximum salinity. Inspite of very active sampling few larvae were found outside of this hydrological pattern. Finally in the Western Pacific Ocean the salinity level of 34 ppt delimits convergences along 5°N and 5°S, when easterly winds prevail. Skipjack are caught mainly along this isohaline within the lower salinity waters.

Mention should be made of another type of discontinuity which can cause noticeable changes in the physico-chemical composition of the water column and is indirectly linked to the productivity of a given area. These are bathymetric gradients due to islands, banks and ridges. Also to be considered is the intensity of the bathymetric gradient which can cause cooling of the surface water even if the shelf is situated in waters several thousands of metres deep. In the Indian Ocean the Mascareignes ridge is crossed by the south equatorial current which can be responsible for enrichment of the slopes facing west. The deep slopes is situated east of the Seychelles plateau.

A number of interdependent hydrological factors must be considered together in order to determine the potential areas of tuna concentration. An environmentalist approach is equally necessary when assessing population structure in so far as it helps to explain and perhaps even foresee the inter-annual variations in stock abundance. These variations usually originate from anomalies in the stages preceding recruitment. Studies carried out on yellowfin in the Western Pacific and on albacore south of New Caledonia showed that stronger cohorts were the result of first life stages spent in waters cooler than average. On the other hand, for the Eastern Pacific skipjack, the most important age classes were those that previously inhabited warmer waters. It was also during the warm period that abundance was increased. A correlation was found between capture and a mixing index due to wind in the Eastern Pacific, a high index value, that means low temperature, leads to an increase of larval mortality and therefore reduces the strength of the cohort concerned. It is obvious that forecasting these abundance variation of resources can have great economic impact.

Probable Areas of Tuna Congregations

Tropical tuna species may be found any where north of 30°S–35° S in the Indian Ocean. An attempt is made to identify the specific areas of tuna concentration, where temperature and dissolved oxygen levels are compatible with physiological requirements of the species.

The favourable areas with regard to concentration and vulnerability of yellowfin tuna are those where minimal oxygen levels (1.5 ml/l) found in waters below 80 m in depth. In this environment, it is more difficult for yellowfin schools to dive deeper and consequently they would be more vulnerable to the surface gears. North of 5° N the ecological volume of yellowfin schools is reduced. South of the Equator in the Eastern Indian Ocean and south of 10°N in the western part, this volume is greater. Yellowfin tend to concentrate in larger groups along the border lines where the ecological volume decreases as they do in the areas with strong thermal gradients.

With respect to skipjack tuna, the areas delimit minimal oxygen level (2.5ml/l) between 50 and 80m depth conducive to high vulnerability if the fish schools are present. As for the yellowfin it is also possible that high concentration is obtained in the border of the unfavourable areas.

The areas, where the upper part of the thermocline is found in the first 100 m with vertical thermal gradient levels comprise between 2°C and 3° C/10m and more than 3°C/10m, tuna schools concentrate leading to fishing success.

The areas where rough seas and strong winds occur have been excluded as seine fisheries would not be able to operate easily. The minimal level has been set to a 30 per cent occurrence of wind force equal to or greater than force 4 (Beaufort scale) Then, enrichment areas due to upwelling and divergence, as well as convergence aggregating the preys have been considered.

It appears that on a yearly basis, the area west of 80°E in the Indian Ocean is a better fishing ground for surface fisheries. The most favourable areas reach their greatest extent from January to April are north of the Mozambique Channel, north and south east of Seychelles, Somalia, Maldives, along the north west coast of India, north of the Andaman Sea. In May the potentially favourable areas are reduced in the east and in July-August particularly near the Seychelles and Somalia. In September–October, potentially favourable areas reapper in the east, west of Sumatra, around the Sunda Islands and north west of Australia. In the western part, during the same period, the potentially favourable areas stretch from north of Madagascar extending as far as India.

Environment–production relationships are complex, and comprise between favourable combinations and therefore they are essentially of a provisional nature.

Schooling Habits of Tuna

The gregarious nature of tuna causes them to swim in schools which can be observed from the surface. This behaviour is considered generally as being a protection measure, and can take different forms according to area or even according to the hour of the day. Precise knowledge of the schooling pattern help to determine the fishing gear to be used.

Tuna schools mainly in surface and deep-sea. The former school can be easily spotted by simple observations by trained personnel and the later by the acoustic devices such as echosounders, sonar etc.

Surface schools swim freely or aggregate around floating objects. Free swimming schools swim close to the surface or subsurface and are often accompanied by birds. To qualify a fish school according to the behaviour of the individuals it comprises, they are termed as;

Breezer; where the presence of tuna is indicated by a rippling on the surface, caused by fish swimming sub-surface in the same direction. This situation signals very often the presence of large schools.

Finner; where the fish do not surface entirely, only dorsal fins can be seen surfacing from time to time.

Jumper; where single fish jump out of the water and dive head first. Fish behaving in this way are generally thought to have momentarily lost contact with the rest of the school.

Smoker; where the fish collectively jump out of the water, generating a choppy sea. This behaviour is characteristic of mixed schools made up of small tunas. Boiler, where large fish preying on anchovies, or euphausids. Even from a distance the fish can be seen to jump in a disorderly fashion causing the water to foam.

Schools can also aggregate around floating objects, or anchored objects and mammals (whales, dolphins) or sharks. These schools may be or may not be accompanied by birds, and may be composed of one or more species.

Deep-sea schools can be free-swimming or may aggregate with the floating objects. Three distinct patterns have been detected acoustically in the Western Indian Ocean.

Type I, only detected below floating object. It is compact in form. Catch analysis show the schools in this formation are comprised of adult skipjack mixed with small yellowfin and bigeye.

Type II, characterizes the presence of both free-swimming and aggregated schools. It comprises two parts, one above the other. The upper one is compact, the lower is longer. Sometimes, the two sections are seperated. Catches from these schools are made up of tuna mixed in size and species. The smaller individuals form the upper stratum in a compact formation and the larger ones, the second deeper one.

Type III typical free swimming formation comprised of large fish may be occupying a thick layer of water and registering an oblong diamond shape on the echosounder.

Fish Size in Different Schools

Single species schools generally comprise fish similar in size. Fish sizes in aggregated schools vary according to the speed at which the object of aggregation moves. Tuna which swim by drifting wreckage, rafts and floatsam will vary more in size than those associated with sharks or whales. Contrary to what has been observed in other oceans tuna and marine mammals association is

School of Tuna

School of *Lutjanus ambilis*

rare in the Western Indian Ocean. Yellowfin tuna, in single-species schools, are generally larger than those swimming in mixed schools (skipjack–bigeye-yellowfin mix). However, the size of the skipjack in mixed schools is close to that of single-species skipjack schools.

Volume of the tuna schools can be estimated through visual observation or by catch per positive seine nets. In the Western Indian Ocean the sizes of schools observed vary considerably from one season to another. During the north-east monsoon (December-March) and during the inter-monsoon periods (April-May and October-November), the yields obtained per positive sets are similar ranging from 21 to 25 tonnes/set. During the south-west monsoon (June-September) levels rose to 35 tonnes/set.

Best catches in seine sets are obtained on aggregated schools. The 1982-1984 average rises to 26 tonnes and nearly 29 tonnes/ positive set. Yields from free swimming schools (large single species generally) registered 13 tonnes and 23 tonnes/positive set. Whatever the type of school, yields by positive sets differ only slightly. The main variation consist in the success ratios of number per set. For instance, the average 1982-84 results were as follows; 54 per cent success registered on free-swimming schools and 90 per cent average on aggregated schools.

Tuna Resources

The potential density of tuna schools as per arial evaluation, show for the first period that the strata crossed by the equatorial current have a higher density than those crossed by the two currents flowing west (north equatorial current and south equatorial current). During the second period (October) the single stratum retained was highly concentrated in tuna.

Stratum I

Areas of high abundance are localized around the Mahe plateau, whose west, south and southeast slopes were good fishing grounds, when exploited by the Spanish pole-and-line fleets, as well as to the west and to the south of the Amirantes Islands. During the first and second survey period 1° squares with high biomass were constantly observed, west and south of the Seychelles. Howevers, while no tuna are found in the areas south of the Amirantes Islands in October, the shallower waters of the Mahe plateau show an increase in tuna potential.

Stratum–II

Covers the zones of convergence with concentration of forage organisms north of the Seychelles. The deep thermocline creates a large ecological volume for tuna and the schools may be rather scattered within the water column.

Stratum–III

The productive zone (between 3°S and 5°S) is limited to the counter current area where the thermocline at the time of survey was at its shallowest (30 m on average). The richest squares are situated from 3°S to 4°S where the thermocline gradient is strong (more than

4°C/10m). The other less productive squares (4°S to 5°S) are characterized by a weaker thermal gradient in the thermocline, varying between 1.5°C and 4°C/10m.

Stratum–IV

The area comprised from 8°S to 10°S of this stratum is crossed by a current drift rich in floating debris acting as aggregating devices for tuna schools.

Stratum–V

Gave no indication of high biomass.

Potential Tuna Surface Fishing Grounds

The potential fishing grounds of tuna is summarized country by country bordering Indian Oceans.

Australia

The most commonly found species is the southern bluefin. The adult population is exploited by long line fisheries south of Australia from June to September. Trolling, pole-and–line and purse-seining are practised in coastal inshore waters to catch the younger specimens.

Along the north-west coast, yellowfin are plentiful from September to November and skipjack from June to August.

Indonesia

Skipjack tuna is the main objective of fishing activity. They are found throughout the year west of Sumatra and south of Java from May to August.

Yellowfin are caught in smaller quantities off Sumatra throughout the year as well as south of Java. Yields are higher between November and June.

Little tuna are widely represented in the Sumatra and Java landings. Meterological conditions between June and September during the southern winter monsoon make fishing difficult due to strong winds and hurricanes. The thermocline is pronounced but deep (100–150m). Temperature gradient is highest from March to July. Currents coverage between April and October in areas 5°S to 10°S and 85°E to 100°E. During this period much drifting debris and floatsam found in the area provide favourable conditions for seining, aimed at aggregated tuna schools, before and after the monsoon.

Thailand and the Andaman Sea

Little tuna species constitute almost the entire catch of the region. Long tail tuna account for 70 per cent and are caught all year round mainly with gillnets and purse-seine by the artisanal fisheries.

Meteorological conditions are favourable from October to May during the northeast monsoon. Dissolved oxygen levels are low throughout the year which is the cause of scarcity of yellowfin and skipjack.

Sri Lanka

The skipjack and yellowfin are the species most commonly caught by the artisanal fisheries. Skipjack 40 per cent, yellowfin 25-30 per cent and the total catches average 35000 tonnes per year. Skipjack tuna is caught all year round with gillnets between June and August and pole-and-line from September to March. Yellowfin tuna are caught between October and February and from June to August.

Meteorological conditions do not favour off shore commercial fishing off the west coast between May and September and off the east coast in December–January.

India

Little tuna are the main species caught around peninsular India, whereas the catches off the Laccadives Islands consist mainly of skipjack and yellowfin tuna. The total annual catch is close to 20,000 tonnes.

Tuna fisheries around India are artisanal. Off continental India the most commonly used gear is the gill net, while off the Laccadives, fishing is mainly by pole-and-line. Around the Laccadives, fishing for skipjack is from October to December and from April to May. Yellowfin are caught mostly from December to February. But meteorological conditions during the south-west monsoon, from June to September are unsuitable for fishing tuna.

In the northern region of the Bay of Bengal low salinity levels in surface waters are caused by the heavy rains during the south-west monsoon (from June to September) and dissolved oxygen levels are low throughout the year. During the south-west monsoon, conditions are unsuitable for offshore fishing.

The most favourable conditions for commercial surface fisheries are from November to February, north of the Laccadives, and from February to April along the coast of penninsular India off Andhra Pradesh.

Maldives

Annual catches of tuna are approximately 30000 tonnes, made up of 60 per cent skipjack, and 20 per cent yellowfin. Ninety five per cent of the effort is carried out with pole-and-line and the rest with trolling gear.

Due to the wide north-south extension of the archipelago, three areas have been seperately characterized.

In, North Maldives, area north of 4°N, skipjack are plentiful from December to March and in July-August, Some yellowfins are also found during these two periods. Meteorological conditions between June and September are often unfavourable. March–April and November–December are the best periods for commercial fishing of tuna.

In Central Maldives, areas from 2°N to 4°N, skipjcak are plentiful from May to July and yellowfin in August-September.

From January to June, the thermocline lies in the first 100m, with a vertical thermal gradient in excess of 3°C/10m. The best occanographic conditions for catching tuna range from May to June.

In south Maldives, areas south of 2°N, tuna is caught all year long without great seasonal variation except for July to September when some large yellowfin tuna are caught.

The meteorological and hydrological conditions in south Maldives are favourable for fishing all the year round.

The Chagos Archiplago

This area is exploited only by the commercial purse-seine fleets. Good yellowfin catches are in principal recorded during the first half of the year. Skipjack can be caught all year round in association with floating debris.

Weather conditions are generally good except for June to September when strong winds make fishing difficult.

Arabian Sea

Little tuna species (frigate tuna, kawakawa, long tail tuna and king fish) represent the bulk of the landings and are taken throughout the year in the Gulf area along the Arabian coasts, in the Red Sea and the Gulf of Aden. The most common fishing gear are gillnets and troll lines.

Two seasonal upwellings occur in the area from May to September along the coasts of Oman and Pakistan. A converging eddy extends between these upwellings.

Meteorological conditions during the southwest monsoon can hinder offshore fishing. The oxygen saturated surface layer is thin (restricted to 30-50m). Deeper, oxygen levels are low (average of 0.5ml/l at 200m) which limits the presence of yellowfin and skipjack tuna except in the Gulf of Aden where they can be found seasonally.

The north region is not suitable for commercial surface fisheries.

East African coast

No significant artisanal fisheries operate along this coast (Somalia, Kenya, Tanzania, Mozambique). Each of these countries catches only a few hundred tonnes of tuna every year, mainly yellowfin and kawakawa.

Strong winds make coastal navigation difficult from June to September during the south-west monsoon.

The somali coastal waters are characterized by a very active upwelling developing in June which enriches the upper layers in nutrients. Large concentrations of tuna may be found offshore at the border of the enriched areas.

Madagascar, Comores and the Mozambique Channel

Commercial fisheries are beginning to develop in the fishing grounds situated along the Mozambique Channel.

In Madagascar only trolling lines are used. Catches generally yield about a hundred tonnes and consist mainly kawakawa (little tuna)

Handlining and trolling take 1800 tonnes per year off the Comores Islands. Some gillnets are used south of Grande Comore.

Navigation is easy throughout the year in the Mozambique channel. However, cyclones may sometimes occur during the summer months north of the channel.

The thermocline is generally deep (around 100m) due to a convergence eddy centred across the Comores. The temperature gradient in the thermocline is low and the water is clear. Inspite of this, purse-seining is possible due to patches of floatsam and drifting debris.

From an occanographic point of view, the area north of the Mozambique Channel could be exploited by purse seine fisheries with good probabilities of success, from January to June and in September-October.

Reunion-Mauritius

Artisanal trolling and sport fishing are the only forms of fishing carried out around these islands. Catches of tuna vary between 800 and 1000 tonnes mainly consisting of little tuna. From May to October during the trade winds season, the conditions of the sea hinder smooth navigation and make fishing difficult. The thermocline is deep and not well pronounced. Tuna resources do not seem very important and, in any case, are insufficient to enable the development of a commercial surface fishery.

Seychelles

Artisanal fisheries land approximately 100 tonnes of tuna from the plateau areas of Mahe and Amirantes.

Commercial purse-seining has been developing since 1982 reaching 1,20,000 tonnes in 1985. Yellowfin tuna account for 50 per cent of the catches and skipjack about 45 per cent. The remainder consists of bigeye and albacore. Occasionally little tuna species are also caught.

During the first quarter the seiner fleets extended from the Seychelles and the Chagos Islands. From April to July they exploited the areas between the Seychelles and the east coast of Africa (between 10°S and the Equator). From August to October the main effort is concentrated on the areas between the Equator and 10°N. In November–December, the fleets move towards the fishing grounds north of the Seychelles.

Navigation is difficult during the south-west monsoon (June to August), but easier during the same period north of the Equator (from 0° to 5° N).

Hydrological conditions are favourable for tuna concentrations throughout most of the year.

Exploitation

Tuna fishing throughout the tropical belt has remained least intensive in the Indian Ocean. Of the total world tuna catch of 3.1 million tonnes in 1984, some 410000 tonnes (14 per cent) was landed from the Indian Ocean.

Traditional artisanal fisheries at subsistence level was concentrated in the Eastern Indian Ocean in respect of tuna.

The commercial fishery for tuna and tuna like species began operations at the beginning of the 1950s and developed gradually. The first technique employed was long lining, carried out exclusively by non-coastal Asian countries, such as Japan and later Taiwan and Republic of Korea. Catches increased rapidly from 12000 tonnes in 1952 to 94000 tonnes in 1956 and finally exceeded 150000 tonnes in 1968. Catches levelled out to about 100000 tonnes at the start of the 1970s. In the following decade there was a steady decline in the longline catches. In 1971, longline catches alone represented as much as 46 per cent of total tuna catches from the Indian Ocean, but dropped to approximately 30 per cent in 1983.

This evolution was brought about by the development of the commercial surface fisheries in the early 1970s. The first commercial scale pole-and-line fishing trials were shortlived. From the start of the 1980s, the purse-seiners carried out successful surveys, and there were some 40-50 vessels operating in 1984. The new form of commercial surface fishing is still limited to the western part of the Indian Oceans.

Numerous coastal states have only recently become aware of the importance of tuna resources in the Indian Ocean. Some of them are trying to develop their own fisheries, but are still lacking sufficient information on the species availability, fishing grounds offering greatest potential, best fishing seasons and the most appropriate fishing techniques.

In successful tuna fishing, one of the main requisites is adequate knowledge of the conditions prevailing on the fishing grounds. Apart from the oceanographic data, such as, direction and rate of current and water temperature, it is necessary to study biological factors, such as the kind of tuna, their schooling habits, migration direction and size and density of schools. To obtain high catches one must know the daily changes in these factors to be able to forecast the daily movement of fish. Positioning the long line (direction of shooting) is important, since the long lines should be set not only at right angles to the current direction, but also in approximately the middle of the tuna schools. In tuna long line fishing, much depends on setting the lines smoothly, since the depth at which the hooks operate is based on the speed of the vessel while steaming out the lines and the speed with which the crew set out the gear. This calls for both individual initiative and good team work from the crew.

The type of bait used and its condition is of the highest importance in tuna longline fishing. Since the bait is nearly always frozen, the fish should be carefully chosen and properly arranged in their containers, not only to provide well-preserved bait, but also whole and undamaged fish.

The construction of tuna long lines depends on the fishing conditions in the area, the species of tuna and the schooling habit of the species.

Some species, such as, yellowfin tuna normally move in dense schools so that more branch lines should be operated for a unit of mainlines than, for instance, in Japan for black tuna, which normally swim highly dispersed. This behaviour as well as the operational depth of the hooks has a large influence on the construction and the length of branch lines, since two hooks must never be able to tangle with each other when the line is in operation. The branch lines should be spaced as close together as practical, to have a maximum of hooks working the concentrated yellowfin tuna schools.

Yellowfin in gillnet catches caught during 1987 and 1988 from Indian Ocean, south-west of Sri Lanka measured 32 to 152 cm with a majority within 44 to 76 cm range. Troll catches comprised of immature fish of 48 to 64 cm as well as adults of 104 to 116 cm length. Long line catches were all adults size of 120 to 145 cm in length.

A seasonal shift in modal size was evident in the North-West area only. The continuity in the size groups taken by trolling lines, gillnet and long line tends to indicate that the juvenile yellowfin, showing a seasonal shift northwards, commences to spread into deeper water and becomes available to the long line fishery.

Examination of 408 yellowfin caught in Indian Oceans showed a male to female ratio of 1 : 0.91. Unlike in the case of skipjack, males outnumber the females during the south-west monsoon months. Females were dominant in February, April, October and November. Size distribution of yellowfin by sex show more males beyond a length of 112 cm. There is no marked difference in the sex among the juveniles. Fish below 100 cm were generally immature and the smallest fish with mature ova measured 112cm.

Only 17 per cent of 315 yellowfin stomachs examined were found to be full. Gut contents in others were in advanced stages of digestion and only large prey species like frigate tuna could be identified.

A brown coloured (2.0-2.5 cm long) parasite located in the gill cavity of yellowfin, with its head burried in gill filaments was identified as a copepod of the genus *Pseudoeyeus*. These were observed mainly from October to January and mainly on Juvenile yellowfin tuna of 35-58cm length range.

The fishes of the genus *Euthynnus* are widely distributed in the warm and tropical waters of the World's Oceans. Of the three species, *E. lineatus* has the most restricted distribution, occurring only in the eastern tropical Pacific Ocean. *Euthynnus affinis* and *E. alletteratus* are more widely distributed in the world's oceans. *E. affinis* is widely distributed in the warm waters of the Indo-Pacific region. In the Indian Ocean the distribution of *E. affinis* is essentially continuous along the entire east African coast including the Red Sea and the Gulf of Aden, along the coast of Pakistan and both coast of India, Mayanmar, Malaysia and Indonesia and to the west coast of Australia. It occurs around Madagascar, Sri Lanka, and the offshore islands and archipelagos of Mauritius, Reunion, Comoro, the Seychelles including Aldabra, Laccadive, Maldives and Andaman and Nicobar.

The species of *Euthynnus* are fusiform, streamlined and have a lunate caudal fin similar to all tuna like fishes. The body is naked except for the scaly corselet. A series of oblique or longitudinal black markings above the lateral line distinguishes *Euthynnus* from *Katsuwonus*. which has black longitudinal stripes on the belly. The contiguous or nearly contiguous dorsal fins distinguished *Euthynnus* from *Auxis* which has widely seperated dorsal fins. The concave distal margin of the first dorsal fin seperates *Euthynnus* from *Sarda* which has a nearly straight first dorsal margin. The species of *Euthynnus* usually have eight dorsal and seven anal finlets. Palatine teeth are present in all and vomerine teeth in two of the species. The intestine is straight. There are 37-39 vertebrae. Unlike *Auxis, Euthynnus* has a common trunk for the dorsal and ventral branches of cutaneous artery. However *Euthynnus* is more primitive than *Katsuwonus* in that the cutaneous artery's ventral branch is short and dendritic and much less developed than the dorsal branch. *Euthynnus* also differs from *Auxis* wherein the dorsal cutaneous artery lies dorsal to the corresponding vein and not ventral. In *Euthynnus* the aorta is moved ventrally a distance greater than the depth of the centrum owing to the development of the trelliswork, a distance less than in *Auxis* and more than *Katsuwonus.*

The dorsal black markings that are superimposed over a blue to indigo background are most accentuated in the anterior region and are more oblique in *E. affinis* than in the other two species. Black or dusky spots are scattered over a relatively wide area on the belly below the pectoral. These spots are variable and cannot be regarded as specific character. The crest or rise of the lateral line above the pectoral fin is minimal. The crest or rise, when present, is always gentle at its anterior slope. In many specimens the lateral line is almost horizontal from the opercular margin over the anterior third of the pectoral fin, from where it slopes gently downward. The caudal keels are not conspicuously developed and are evident externally only by a dorsoventral flattening of the peduncle. The operculum is relatively straight at the posterior margin and a small dark or black elongated spot is generally present at the dorsal extremity of this margin. The posterior outline of the preoperculum is gently and smoothly rounded. A small irregular dark spot or area is present below the posteroventral margin of the eye. The maxillary extends to or slightly beyond the vertical through the middle of the eye and, depending on the size of the fish, 24 to 25 teeth are present on each side of the lower jaw. Vomerine teeth are present but palatine teeth, though always present, may not be as prominent. The dorsal fin is contiguous, is about twice the height of the second dorsal fin, and is roughly half the head length. Its dorsal outline is strongly concave. The insertion of the anal fin is on a vertical through the origin of the first dorsal finlet, or between this and the end of the second dorsal base. There usually are eight dorsal and seven anal finlets. The gill rakers are fully developed on the first arch only and there is invariably a raker at the angle of the arch including a basal process extending both dorsally and ventrally. Gill teeth (posterior rakers) are present on all arches and are most developed on the first arch.

Larval and juvenile *E. affinis* are widely distributed in the Indo-Pacific area. It appears that larval and juvenile *E. affinis* are generally taken close to land masses and less frequently around oceanic islands. In the fishery for *E. affinis* off the southwest coast of Sri Lanka, fish smaller than 240 mm appear regularly in the commercial landings.

The three species of *Euthynnus* seldom occur more than a few hundred miles from land. They are usually found around land masses and island chains generally between lat. 35°N and 35°S. Around the Maldive Islands in the Indian Ocean *E. affinis* are found close to the islands and they often venture into shallow water in pursuit of bait. *E. affinis* are also found in waters around oceanic islands and archipelagos although here too they are found close to shore. Off the coast of Japan *E. affinis* may occur in waters with salinities ranging from 31.22 to 33.80 ppt. and water temperature ranging from 18° to 28°C. In Indian Ocean the tropical and subtropical waters are generally characterized by a scarcity of plankton off shore except in areas where vertical mixing of water masses occur, *e.g.*, off Somalia, where upwelling takes place strong development of phytoplankton and the ensuing zooplankton is confined to the immediate coastal areas.

All of the species of *Euthynnus* are heterosexual, males and females do not differ in external appearance, and there is no record of hermaphroditism in these fishes. *E. affinis* attains sexual maturity at a relatively small size, at least in certain areas of its distributional range. In Philippine waters 49 cm long female was found with mature gonads. The smallest mature female was around 38.5 cm long.

Euthynnus affinis in the Indian Ocean reach sexual maturity later than those occurring in the Philippines. *E. affinis* taken near the Seychelles attained sexual maturity between 50 and 65 cm total length, probably in the third year of life. In Mauritius waters *E. affinis* of 55 cm long are mature. Off the southwest coast of India, near Vizhingam (lat. 08°22′N, long. 76°59′E) the smallest female with ripe ovaries was found to be 48 cm. In the South China Sea off Kwangtung, China, *E. affinis* attains sexual maturity at around 50 cm. Fertilization is external in all the species of *Euthynnus.* In the Indian Ocean, *E. affinis* around Vizhingam spawned 210000 to 680000 ova per spawning and 790000 to 2500000 ova during the spawning season. The number of ova produced by *E. affinis* increased with the size of fish.

Fork Length (cm)	*Weight of Fish (kg)*	*No. of Ova per Spawning (millions)*	*No of ova per Spawning Season (millions)*
48.0	1.37	0.21	0.79
52.5	2.06	0.31	0.88
55.5	2.35	0.30	1.31
58.2	3.20	0.50	2.14
65.0	4.57	0.68	2.50

Euthynnus affinis in the Philippines spawn all year round, more than one batch of eggs during a spawning season. The ripe eggs of *E. affinis* would probably be between 0.8775 and 1.1050 mm in diameter. In the western Indian Ocean near the Seychelles the spawning season for *E. affinis* is during the period of the northwest monsoon, from October-November to April-May, including a peak from January to March. Off the coast of east Africa the spawning season extends from January to July, the middle of the north-west monsoon to the beginning of the south east monsoon. Off the southwest coast of India, near Vizhingam *E. affinis* were found to spawn from April to September. The capture of juveniles up to 200 mm in the months of January, May, June, September, October and November near Vizhingam and Calicut, India suggests rather lengthy spawning season for *E. affinis* in these waters. In the eastern equatorial Indian Ocean, larval *E. affinis* have been captured from August to October in Indonesian waters which suggest spawning during these months. Larval and juvenile *E. affinis* have been recorded from many other areas in the Indo-Pacific. These records suggest that *E. affinis* probably spawns throughout its range. The eggs of

E. affinis took less than 24 hours to hatch. At hatching the larva was approximately 2.1 mm with 39-41 "myotomes". The smallest specimen identified as *E. affinis* measured 4.6 mm in total length. The head length is moderate (33.9 per cent TL), there are 40 myomeres, and the abdominal sac is nearly triangular. The mouth is fairly large (71.9 per cent of head length) and contains 12 teeth on the upper jaw and 9 teeth on the lower jaw. Three long spines are present on the posterior edge of the preopercle, the longest spine is the one at the angle. Generally a single chromatophore is found at the middle of the lower jaw and at the symphysis of the pectoral girdle. The median fins show very little development, there are 4 or 5 rays in the caudal fin and only strong striations representing rays in the incipient anal and dorsal fins. The pelvic fins are undeveloped and the pectoral fins show only striations. Five short spines are present in the first dorsal fin and about 9 rays in the caudal. The other median fins and the pectorals still show only strong striations and the pelvic fins are just emerging as small buds. The single chromatophore at the middle of the mandible and that at the pectoral symphysis are still evident. A 7.6 mm long fish larva exhibits some striking changes, particularly in pigmentation. A series of 4 to 5 chromatophores is present on the anterior half of the mandible and the first dorsal fin is almost completely pigmented.

E. affinis in Japanese waters grow to 60 cm and 3.5 kg and on rare occasions, fishes over 100 cm long and 10 kg in weight are found. The largest *E. affinis* caught during survey in the Mauritius-Seychelles area was 87 cm weighing 8.6 kg. *E. affinis* of 0.4 to 1.4 kg become established in cage enclosures more readily than yellowfin tuna. The species had been in captivity for two years. They are hardy enough to be successful in their natural environment, but they can only stand a minimum of handling. The competitors of *E. affinis* include skipjack tuna, small yellowfin tuna, *Auxis thazard* and a carangid, *Megalaspis cordyla.* These fishes school together with *E. affinis* off east Africa in the Indian Ocean. In waters around Sri Lanka. *E. affinis* are found in schools together with *A. thazard* and presumbly these two species are direct competitors for food.

E. affinis are voracious feeders and their feeding method is darting swiftly into a school of small fish and scattering them. Their feeding habits may vary throughout the year. *E. affinis* travelled an average of 5.9 body lengths while feeding and can attain a maximum speed of 10.0 body lengths. The feeding mechanism is drag free and that *E. affinis* feeds by swimming over its prey rather than by sucking it into its mouth. The species in general is facultative filter-feeder in that they are able to feed in open water upon swimming organisms that are minute in comparison to the size of the predator. In the Indian Ocean squids (56.5 per cent by volume) are most important food of *E. affinis* followed by fishes (38.3 per cent by volume) and crustaceans (5.2 per cent by volume) are least important. *E. affinis* grow at the rate of 1.0 to 1.3 mm/day and attain 150 and 250 mm in the first year of their life. *E. affinis* attain an average length of 345 mm, 456 mm, 511 mm, 544 mm and 554 mm at 1, 2, 3, 4 and 5 years of their age. The corresponding weight gain during the periods were 0.9, 2.1, 2.8, 3.3 and 3.5 kg respectively. Aspects of the schooling behavior of *Euthynnus* are relative to feeding. In the Indian Ocean off the east coast of Africa. *E. affinis* are found in schools with small *T. albacares, K. pelamis, Auxis* sp and a carangid, *Megalaspis cordyla*. All individuals in the schools are of same size. In small *E. affinis* the schooling is strong and disciplined. Around Sri Lanka in Indian Ocean mixed schools of *Auxis spp and E. affinis* are common. Schools of *E. affinis* are composed of 100 to 1000 or more individuals in east Africa. In Sri Lanka as many as 5000 individuals (mixed *E. affinis* and *Auxis* spp) may be taken in one haul of a beach seine.

Euthynnus affinis in the experimental tanks, significantly increased their side-to-side spacing and atltered their diagonal to abeam position ratios so that abeam orientation assumed increased importance when the hydrodynamic field between orienting fish was blocked by a transparent partition. Hydrodynamic contact is essential for the typical spacing and positional orientation in schooling, and that the lateral line plays a prime regulatory role as the water turbulence detector.

E. affinis are attracted of continuous white light over a range of moderate intensity. The species are not attracted to a light of weaker intensity and are repelled by a light of stronger intensity. White lures are slightly more attractive than red, black or silver *E. affinis* has a well-developed sense of smell or taste in that they are strongly attracted to clear colour-less extracts of tuna flesh. The attractant is contained in the protein rather than the fat fraction of the clear extract. Visual acuity, the ability to see clearly fine details of objects especially when the objects become smaller and closer together, is greater in *K.*

pelamis at higher luminances than *E. affinis*. At lower luminance the visual acuity of two species, *E. affinis* and *K. pelamis* are similar. *E. affinis* can perceive sounds from 100 to 1000 Hz. The thermal sensitivity of *E. affinis* is no more acute than that of inshore fishes and appear inadequate for direct sensing of weak horizontal temperature gradients at sea. *E. affinis* travel an average of 5.9 body lengths while feeding and a maximum of 10.0 body lengths. The non-feeding swimming speed with food present averaged 4.5 body lengths and ranged from 2.9 to 12.5 body lengths. The swimming speed was highest after a meal and decreased when the fish are deprived of food. *E. affinis* swim at a speed of 0.30-1.27 m/s during the day and 0.33–0.75 m/s under artificial lights in the experimental tank of 4m in diameter and 0.6 m deep. *E. affinis* exhibited black spots ventral to the pectoral fins, faint vertical bars on the flanks and a yellowish middorsal stripe while feeding. For *E. affinis* from 38 to 49 cm the sexes appeared to be represented in equal numbers but for fish from 50 to 73 cm. 67 per cent of the fish were males. The *E. affinis* taken by the commercial fishery ranges from 20 to 65 cm. Occasionally fish in the 15 to 20 cm and 65 to 68 cm groups are also caught.

A little over 600 tonnes of *E. affinis* were landed in 1969 on the southwest coast of Sri Lanka. *E. affinis* occur in commercial quantities in Australian waters of Indian Ocean. There are large unexploited schools of *E. affinis* in east African waters. *Euthynnus* and *Sarda* represent about 0.5 per cent of all fish by weight in the Indian Ocean.

Off the coast of South Africa and south-west Australia *E. affinis* occur and fished during the southern summer when surface water temperature are at a maximum. In the Seychelles, the fishery is highly seasonal and takes place during the northwest monsoon (October-November to April-May), when the surface water temperature is highest (29°–30°C). In northern Somalia the fishery is conducted during the northeast monsoon at the time of minimum sea-surface temperature. In Pakistan, west coast of India and Sri Lanka, *E. affinis* is taken throughout the year with slight local seasonal variations. Good landings of *E. affinis* occur during the eight months from October to May and the period of low catches is from June to September in the Vizhingam fishery (south-west coast of India). The period of low catches coincides with the southwest monsoon when the seas are rough and very little fishing effort is expended. Off east Africa and the Malagasy Republic, *E. affinis* is taken throughout the year. *E. affinis* is present around Philippines throughout the year but the highest catches are made from October through December and April through May. Around Japan *E. affinis* is taken from May to November. Young *E. affinis* of 15 to 25 cm have been caught from August to October.

Thunnus obesus (Lowe)

Big eye tuna, *Thunnus obesus* (Lowe) is distributed, world wide in tropical and subtropical waters of the Atlantic, Indian and Pacific Oceans, but absent from the Mediterranean.

A large species, deepest body near the middle of first dorsal fin base. 23 to 31 gill rakers are present on first arch. Pectoral fins are moderately long, 22 to 30 per cent of fork length in large individuals over 110 cm fork length, but very long (as long as in *T. alalunga*) in smaller individuals though in fish shorter than 40 cm they may be very short. In fish longer than 30 cm, ventral surface of liver is striated. Swimbladder is present. A total 39 vertebrae is present. The lower side and belly of the fish is whitish in colour. A lateral iridescent blue band runs along sides in live specimens. The first dorsal fin is deep yellow. The second dorsal and anal fins are light yellow. Finlets are bright yellow edged with black.

Big eye tuna is epipelagic and mesopelagic in oceanic waters, occurring from the surface to about 250 m depth. Temperature and thermocline depth seem to be the main environmental factors governing the vertical and horizontal distribution of big eye tuna. Water temperatures in which the species has been found range from 13° to 29° C, but the optimum range lies between 17° and 22°C. This coincides with the temperature range of the permanent thermocline. In fact, in the tropical western and central Pacific, major concentrations of *T. obesus* are associated with the thermocline rather than with the surface phytoplankton maximum. For this reason, variation in occurrence of the species is closely related to seasonal and climatic changes in surface temperature and thermocline.

Juveniles and small adults of big eye tuna school at the surface in mono-species groups or together with yellow fin tuna and/or skipjack. Schools may be associated with floating objects.

In the eastern Pacific some spawning is recorded between 10°N and 10°S throughout the year, with a peak from April through

September in the northern hemisphere and between January and March in the southern hemisphere, Kame (1967) found a correlation between the occurrence of sexually inactive big eye tuna and a decrease of surface temperature below 23° or 24°C. Mature fish spawn at least twice a year, the number of eggs per spawning has been estimated at 2.9 million to 6.3 million.

The food spectrum of big eye tuna covers a variety of fish species, cephalopods and erustaceans, thus not diverging significantly from that of other similar sized tunas. Feeding occurs in day time as well as at night. The main predators of big eye tuna are large bil! fish and toothed whales.

The species attain a maximum fork length over 200 cm, but common to 180 cm, corresponding to an age of at least 3 years. The all-tackle angling record for the Pacific is a 197.3 kg fish from off Cabo Blanco, Peru in 1957. This fish was 236 cm long. For the Atlantic, the all-tackle angling record is a 170.3 kg fish with a fork length of 206 cm taken off ocean city, Maryland, USA in 1977. Maturity seems to be attained at 100 to 130 cm fork length in the eastern Pacific and in the Indian Ocean, and at about 130 cm in the central Pacific.

Catch statistics are reported by 17 countries for 14 fishing areas. Yearly catches of more than 10000 metric tons are taken in Fishing Areas 34, 51, 61, 71 and 77 with more than two thirds of the total taken in the Pacific upto 1980. Among the countries reporting big eye tuna catches Japan ranks first, followed by the Republic of Korea with much lower landings. The world catch increased from about 164000 metric tons in 1974 to 201000 metric tons in 1980 reaching a peak of 214000 metric tons in 1977. For 1981 a decrease to about 167000 metric tons was estimated. In the Indian Ocean, the big eye tuna fishery was dominated by Japanese fleets upto the end of the sixties, but subsequently operations of vessels from the Republic of Korea became more important, and have accounted for more than 60 per cent of the catch in the late seventies.

The most important fishing gear, at least in the Pacific, are long lines, which comprise some 400 "baskets", consisting of 5 branch lines, each with a baited hook, extending over up to 130 km. Species commonly used as bait include frozen Pacific saury (*Cololabis saira*), chub mackerel (*Scomber japonicus*), jack mackerel (*Trachurus*) and squid. Day and night-time operations are common throughout the year, but there are seasonal variations in apparent abundance reflected in changes of fishing effort. In the seventies, deep long lines employing between 10 and 15 branch lines per basket were introduced. This new type of gear is theoretically capable of fishing down to 300 m depth, as compared to the usual 170m reached by traditional long line gear. Catch rates increased for about 3 years and then declined to previous levels again, suggesting that only a portion of the big eye resources are exploited.

Big eye tuna is exploited in increasing quantities as associated catch of the spring and summer pole-and-line fishery in the north western Pacific, and of the purse seine fishery in the eastern Pacific, both directed primarily at skip jack and yellowfin tuna. In Japan its meat is highly priced and processed into sashimi in substitution for blue fin tuna.

Black pigment spots are absent on dorsal edge of tail of big eye tuna larvae. One or two pigment spots are present on the ventral edge of the tail. The ventral spots are very small and difficult to see, usually located on the ventral midline about 7 to 8 myomeres anterior to caudal fin base. The caudal fin may have 0-2 black pigment spots, usually one (and rarely a caudal fin pigment spot dorsal to the notochord, leading to confusion with *T. maccoyii*). Red pigment–spot pattern on the tail region of small fresh specimens is diagnostic.

This species is very similar to *T. maccoyii* and *T. albacares*. It can be separated from *T. albacares* by (i) absence of body pigment in *T. albacares* and (ii) by the pigment spots on the under-side of the lower jaw tip, which are usually a pair of distinct spots in *T. obesus* and usually a single, smaller and fainter spot in *T. albacares*

The dorsal tail pigments on the body of *T. maccoyii* larvae are very tiny even in larger specimens and especially difficult to see if the specimen has become opaque in the preservative. If they are missed, the larvae will key out as *T. obesus*. Approximately 15 per cent of *T. maccoyii* larvae will have no dorsal body pigment spots and will key out as *T. obesus*. However, approximately 11 per cent have pigment mid-laterally on the body and 16 per cent have body pigment internally, usually in the caudal peduncle region, and the presence of either of these body pigments will distinguish *T. maccoyii* from *T. obesus*. *T. obesus* has only 1 or 2 small faint ventral pigment spots, while *T. maccoyii* have upto 4, and they are usually quite distinct. The caudal fin pigment in *T. maccoyii* can be diagnostic with upto 4 black spots (0-1 in *T. thynnus*, and 0-2, rarely 3 in *T.*

obesus) and except for rare occurrences in *T. obesus*. *T maccoyii* is the only species that may have black caudal fin pigment on the dorsal side of the notochord, with the possible exception of *T. tonggol*. However *T. tonggol* is a coastal species and therefore unlikely to occur in samples with *T. maccoyii* larvae. The only tuna besides *T. maccoyii* that may have internal (deeply embedded) melanophores is *T. thynnus*, never *T. obesus*, *T. albacares* or *T. alalunga*.

Big eye tuna, like the yellowfin are present throughout the intertropical zone. They are also found in bordering areas where the yellowfin are scarce, such as, the Arabian Sea with its low oxygen levels and also in subtropical areas where water temperatures are lower. Big eye is found in greater depths, its vertical distribution seems to be closely linked to the thermocline. Young big eye schools are found frequently below drifting wreckage in 50-100 m of water depth.

The horizontal and vertical distribution of the larvae is little known. The only concentrations now have been observed from November to April south of Indonesia. Lesser quantities have been found south of the Bay of Bengal, northeast off the Maldives and from May to October, north of Australia.

Young big eye are found between 10°N and 10°S only.

The smallest sexually mature big eye female caught by long line fisheries in the Indian Ocean measured 92 cm. Big eye tuna in the Western Indian Ocean reach sexual maturity at the age of 3 years, which would correspond to an average length of 67 cm. Big eye tuna spawn from January to March, over a wide area extending to the Eastern and Western Indian Ocean. A sexually mature female in March was obtained near Timor, Eastern Indian Ocean. In February sexually matured male and female big eye were observed to come from Eastern Indian Ocean (95° to 97° E and 9° to 10° S) and was observed that this stock was sexually matured only between January and March.

The food assimilation rates of big eye were found to be lowest off the eastern coast of Africa and on both sides of the Equator between 2°N to 3°S during the period May to October. High assimilation rates were found in the area from 42° to 75° E and from 3° to 10°N, where big eye prey mainly on fish and squid. Big eye also prey on flying fish in the area 5° to 10°S. Near Maldives, large numbers of pelagic crab species are assimilated. Although big eye and yellowfin prey on the same species, there is no competition between them as they hunt at different depths.

In the Cape area, the big eye have been found to grow between 30 to 35 cm per year. The growth rates in young big eye situated northwest of Madagascar are approximately 18 cm per year as big eye measuring 51 cm in February reached 60 cm in September. Big eye originating in the Northeast Indian Ocean grow faster than those from other regions.

Catches of big eye in the Indian Ocean amounted to 40000 tonnes in 1984, representing approximately 20 per cent of the total world catch for this species. Big eye are the prime target of Asian longliners which catch the large deep-swimming individuals. The small ones are often caught by pole-and-line or by purse seining. They are very similar in appearance to young yellowfin. Adult big eye are much broader and do not have the long fins (ventral and second dorsal), characteristic of adult yellowfin tuna.

Based on the catches by longline fisheries, the proportion of males is greater than that of females. This proportion changes not only according to area, but also according to size of individuals. Among larger individuals there are generally more males.

The size structure of the catches is conditioned by the selectivity of the gear used and is therefore not represented of the population as a whole. Sizes of 120-150 cm dominate in the long line catches even though important variations may be observed according to the areas exploited. The big eye caught by pole-line are only young individuals. Seiners also catch a portion of young big eye together with a fair quantity of 4 to 5 year olds.

Thunnus alalunga (Bonnaterre)

Tunas are mostly oceanic species with epipelagic to mid-water (more than 500 m depth) distributions depending on species and size. They do not inhabit the polar seas. They are unique among bony fish for their high metabolic rate (resulting in an extraordinary growth pattern) and in their vascular heat exchanger system (retia mirabile) permitting them to maintain body temperatures several degree higher than the ambient water. As muscles are more powerful when warm, this gurantees steady swimming required to maintain sufficient gas exchange via gills, which in turn is indispensable to sustain their high metabolism. As juveniles they must swim upward at 50 km per day and are capable of remarkable bursts of speed. The ability to regulate body temperature increases with size and is of particular importance to albacore, yellowfin tuna, big eye tuna and is best developed in bluefin tuna.

Albacore, *Thunnus alalunga* (Bonnaterre) is cosmopolitan in tropical and temperate water of all oceans including the Mediterranean Sea, extending north to 45° to 50°N south to 30° to 40°S, but not at the surface between 10° N and 10°S.

The genus *Thunnus* is characterized with fusiform, elongated and slightly compressed body. Teeth are small and conical, in a single series. Gill rakers 19 to 43 situated on first arch. Two dorsal fins, seperated only by a narrow interspace, the first with 11 to 14 spines. Anterior spines are much higher than posterior spines giving the fin a strongly concave outline. The second dorsal fin with 12 to 16 rays is shorter, as high as or higher than first dorsal fin followed by 7 to 10 dorsal finlets. Anal fin with 11 to 16 rays, about as high as second dorsal fin is followed by 7 to 10 finlets. Pectoral fin of variable length depending on species and age with 30 to 36 rays are present, which is more than in any other genus of Scombridae. Interpelvic process is small and bifid. Body is covered with very small scales, corselet of larger scales developed but not very distinct. Caudal peduncle is very slender, bearing on each side a strong lateral keel between two smaller keels. Swimbladder is present in most species. Vertebrae 39. Liver in ventral view either without striations and the right lobe is largest or with prominent striations and the central lobe is largest. The back of the fish is metalic dark blue in colour, lower sides and belly whitish. No dark stripes or spots on the sides are visible. Finlets bright yellow, edged with black in several species.

Albacore is a large species, deepest at a more posterior point than in other tunas. Gillrakers 25 to 31 on first arch. Second dorsal fin is clearly lower than first dorsal. Pectoral fins are remarkably long, usually 30 per cent of fork length or longer in 50 cm or longer fish, reaching well beyond origin of second dorsal fin, usually upto second dorsal finlet. Fish smaller than 50 cm will have proportionately smaller pectorals than other tunas. *i.e., T. obesus.* Ventral surface of liver is striated (vascular network). Swim bladder is present, but poorly developed and not evident in fish smaller than about 50 cm fork length. A total of 39 vertebrae is present. A faint lateral iridescent blue band runs along sides in live fish. The first dorsal fin is deep yellow, second dorsal and anal fins are light yellow. Anal finlets are dark in colour and posterior margin of caudal fin is white.

As an epipelagic and mesopelagic oceanic species it is abundant in surface waters of 15.6° to 19.4° C. The deeper swimming large albacore are found in waters of 13.5° to 25.2° C. The temperature as low as 9.5°C may be tolerated for short periods. In the Atlantic, the larger size classes (80 to 125 cm) are associated with cooler water bodies, while smaller individuals tend to occur in warmer strata. According to available data, the opposite phenomenon occurs in the northeastern Pacific. Albacore tend to concentrate along thermal discontinuities (oceanic fronts such as the transition zone in the north Pacific and the Kuroshio Front, east to Japan) where large catches are made. The transition zones are preferred to cooler upwelling waters which are richer in forage organisms but poorer in oxygen content. Minimum oxygen requirements are probably similar to those of yellowfin tuna, that is 2ml/l. Albacore migrate within water masses rather than across temperature and oxygen boundaries.

Throughout its range, the albacore migrates over great distances and appears to form seperate groups at different stages of its life cycle. Several diverging, sometimes contradictory models have been suggested to portray these migrations. At least two stocks (northern and southern) are believed to exist in both the Atlantic and the Pacific Oceans, each with distinct spawning areas and with little or no interchange across the warm equatorial waters.

The depth distribution in the Pacific ranges from the surface down to atleast 380 m and is governed by the vertical thermal structures and oxygen contents of the water masses. In the Atlantic, for the same environmental determinants, albocore are believed to occur as deep as 600 m. Like other tunas, albacore form schools with fewer fish, hence more compact units when composed of larger fish. They may also form mixed schools with skipjack tuna, yellowfin tuna and bluefin tuna. Schools may be associated with floating objects, including sargassum weeds.

Although fecundity does increase with size generally, there is no close relationship between fork length and ovary-weight and hence, number of eggs. A 20 kg female may produce between 2 and

3 million eggs per season, which are released in two batches. The sex ratio in catches is about 1 : 1 for immature specimens, but males predominate among mature fishes, which is possibly due to both differential mortality of sexes and differential growth rate after maturity.

Tunas are agile, opportunistic predators feeding on a great variety of suitably sized forage fishes, crustaceans and squids. Because of their size, adult tuna have few predators, mainly bill fishes, sharks and toothed whales.

Albacore attain a maximum fork length of 127 cm. The all-tackle angling record is a 40 kg fish with fork length of 123 cm. caught in the Canary Islands. In the Pacific surface fishery, pole-and-line and troll fishery, smaller sizes (between 55 to 80 cm fork length) predominate, while long line fisheries take bigger fishes of 95 to 115 cm length. In the Indian Ocean, common size ranges are from 40 to 100 cm fork length, while males upto 109 cm and females upto 106 cm are not exceptional in the Atlantic. In the Pacific, maturity may be attained at about 90 cm fork length in females and at about 97 cm in males. In the Atlantic it is reached at about 94 cm in both sexes.

The world catch of *Thunnus* species remained relatively stable around 750000 metric tons in the period between 1975 and 1981, but exceeded one million tons in 1978. Japan caught more than 300 000 metric tons in 1981; while the USA and the Republic of Korea caught about 100 000 metric tons each in 1981, accounted more than 60 per cent of the world catch. Apart from being taken by big game sports fishermen on hook and line and trolling lines, tunas are caught predominantly by industrial gear including purse seines, live bait hook and-line, conventional long lines and deep long lines. Presently the tuna fishing industry faces at least two major problems apart from market fluctuations;

1. Rising fuel costs that have already forced adoption of less energy-intensive fishing methods *i.e.* by reducing searching time through introduction of a variety of fish-locating and aggregating aides;
2. Insufficient supply of suitable bait fishes for potential surface fisheries which has induced aquaculturists to investigate the feasibility of economic mass rearing of convenient species. There have been several trials to test

the utility of cultured bait since the behaviour, growth, hardiness and shape of bait have a great bearing on its efficient use in the tuna fisheries. However, the bait fish rearing operations do not appear to be viable in any industrial context, particularly in island areas where land and labour are at a premium.

Another aspect deserving attention is the by-catch of the tuna long line fisheries, which is particularly composed of various sharks, but includes dolphin and other finfishes. This is mostly discussed in terms of the damage inflicted to the tuna catch and not so much as an additional source of food. In the southeastern Arabian Sea and the Laccadive Sea sharks constituted 63.8 per cent by number and 57.8 per cent by weight of the total catch during March to May in 1981. Part of the by-catch is discarded, but another part is locally marketed. There is also a substantial rejection rate of tunas caught in most long line fisheries, because they may be too small to sell at reasonable prices in the major high-quality, size-specific market places. Reported landings are believed to be underestimating catches by as much as 25 per cent in numbers in some fisheries.

In general, however, all species are highly appreciated and marketed fresh, deep frozen or canned.

There are important fisheries for *T. alalunga* in the Atlantic and Pacific Oceans. Catches have been reported by 15 countries in the period from 1974 to 1981. Along with increasing effort in the major fisheries, the world catch has been gradually declining from a peak of about 245000 metric tons in 1974 to a low of about 181000 metric tons in 1981. More than half of the catch in the last years was taken in the Pacific in fishing areas 61, 77 and 81. The landings in Area 61 were almost exclusively made by Japanese vessels. More than 10,000 metric tons were reported in 1981 from two other fishing areas, namely Areas 27 (predominantly by Spain, while the French catch collapsed to less than one tenth of its previous level) and 47.

Albacore fisheries involve four basic types of fishing operations, longlining, live-bait fishing, trolling and purse seining. Surface methods (trolling, live-bait, purse-seining) tend to take smaller fish than longlining. In recent years, boats and gear have been improved by introduction of longer vessels (trollers upto 22m length), more modern boat building materials (fibreglass, aluminium), larger ice storage or brine freezing capacities, better navigational aids and fish locating devices, and larger bait-holding capacities that increase the autonomy of the vessels.

The most important albacore fisheries are as follows ;

In the Pacific there are five major fisheries which are operational at various times of the year;

1. The Japanese live-bait fishery originates south of Japan and then develops off shore into the area of the Kuroshio Front. It extends from March through July, with a peak in June.
2. The Japanese long line fishery operates across the North Pacific throughout the year, although the best catches are obtained from December to February.
3. The US surface fishery from off Baja California to Canada attains its peak in August and September. Fishing activities extend from June to December in the northern part of this area and from May to January in the southern part. 90 per cent of catches are taken in waters of 15.6° to 19.4°C. Catches of this fishery are believed to include fishes as young as one year of age, with only few mature adults.
4. Long-line operations in the South Pacific between 10° and 40°S from Samoan and Japanese bases extend throughout the year with the peak season from August to February.
5. The New Zealand surface fishery, mostly in waters from 18.5° to 21.3° C, extends from January to April, with best catches usually in February.

Albacore is also caught as a by-catch in the Hawaiian longline fishery for yellowfin and big eye tuna.

In the Indian Ocean, the fishery is barely developed, but areas of potentially successful exploitation, as derived from an assessment of favourable hydrographical conditions are given by Sharp (1979). Upto the mid-sixties, catches in this area were taken exclusively by Japanese vessels, while in the late seventies, vessels from Taiwan, Province of China, were the most abundant, followed by boats from the Republic of Korea and Japan.

In the Atlantic, there are at least three fisheries for albacore;

1. A trolling fishery dating back to the last century which has undergone mechanization of boats and gear and introduced on-board processing of fish. It is operated primarily by Spanish and French vessels in the Bay of Biscay and the West European Basin.
2. A more recent pole-and-line fishery initiated after the World War II by French and Spanish bait boats in the Bay of Biscay and off northern Portugal. This activity is restricted to the summer months. A recent offshoot of this fishery, dating from 1970, is the seasonal pole-and-line activity in fall off Morocco by Azores-and Maideira-based Spanish and Portuguese vessels.
3. Seasonal summer and winter long line fisheries operating in different offshore areas in the northern and southern hemispheres. These fisheries were operated almost exclusively by long-distance fleets from Japan upto 1970, but since that time vessels from the Republic of Korea and Taiwan Island are also participating. On a smaller scale, countries like Brazil, Cuba and Venezuela have entered this fishery.

Larvae of all species of genus Thunnus do not have fore brain pigment and black pigment on isthmus and just anterior to anus. Black pigment spots is present on their tail are small in size. Black pigment appears on first dorsal fin in larvae of 6 mm or more in standard length. Newly hatched specimens are difficult to recognise to genus and cannot be identified to species because the preopercular spines are not visible and pigments on the body have not yet developed. The basic clues to recognising the genus are; absence of forebrain pigment, a relatively long, deep body with 39 to 40 myomeres, and presence of large head, compact eye and mouth and triangular abdominal cavity.

In larvae of albacore tuna, no black pigment spots on tail or tip of lower jaw, inside or outside, is seen. Centre of the eye in head profile is situated more or less in line with the body axis. Red pigment-spot pattern on the tail region of small fresh specimens is diagnostic. Albacore larvae are not usually found in equatorial regions (0°-10° N and S).

In 1984 albacore catches rose to more than 16000 tonnes in the Indian Ocean, representing about 10 per cent of the total world catch for this species.

Young albacore are found in subtropical areas and in shallow waters of 15°-20°C and the older ones towards the latitudes in the deep waters in temperatures of 14°-25°C. They may be found occasionally on the surface.

Few albacore are found of 10°N in the Indian Ocean, and south of approximately 35°-40°S. in the Western Indian Ocean and 25°-30°S in the Eastern Indian Ocean. Some albacore may be found south of Australia as far as 40°S. Fairly large quantities were caught by French and Spanish surface fisheries west of the Seychelles economic zone during the second quarter of 1984 and 1985.

The main long line fisheries are situated south of 10°S. Between October and March, activities moves to east and northeast of Madagascar as well as along the east coast of Africa. The southern limit corresponds approximately to a surface water temperature of 23°-24°C. During the southern winter, April-September, the areas exploited are more to the south toward approximately 35°S where surface water temperatures register 16°C to the south and 20°C to the north of that latitude. The habitat of albacore from the southern area of the Western Indian Ocean is the same as that of albacore from the southern part of the East Atlantic. In the summer months surface water temperature rises to 21°C in the Cape area. This would enable the albocore, as well as the tropical species, such as, yellowfin and big eye tuna to migrate from one ocean to another.

Albacore larvae are distributed in two seperate and distinct zones, one in the Eastern and the other in the Western Indian Ocean from November to April. During the rest of the year they are also found west of Australia.

Albacore spawning grounds range from 14° to 28°S and from 55° to 80°E, and as far as Indonesia. The period of reproduction coincides with the southern summer, January to June, peaking from February to April.

There is little information on growth rates of the species or on size structures in the Indian Ocean. Albacore caught in the western area by seiners are usually large, averaging 116-120 cm.

Thunnus maccoyii (Southern Bluefin Tuna)

The adults of this species are larger than all the other species found in the Indian Ocean. Adult bluefin can attain as much as 200 kg in weight. 225 cm in length and live for upto 20 years.

The Japanese and Australian fisheries land most of the world southern bluefin catch of 40000 tonnes. Along the south coast of Australia the exploitation of southern bluefin is ensured by pole-and-line, purse seine and to a lesser extent, by trolling fisheries. Long line catches represent a little more than half the total catch, which are exclusively carried out by the Japanese fleets.

Southern bluefin is found in the southern hemisphere. It is widely distributed throughout the three oceans from 30° to 50°S and in relatively cold waters, 5°–10°C. However, this species can be found as far as 10°S in the Indian Ocean, south of Indonesia where they spawn. Long line fisheries exploit adult southern bluefin south of Tasmania from June to September and from May to August off South Africa.

The area between 10° and 20° S east of 100° E is the only area where southern bluefin is known to breed. Maximum density of larvae is found south of Java. Its distribution in the Indian Ocean varies throughout its existence. During the first two years schools move progressively from their place of birth towards southwest Australia. During the following two years, at 3 and 4 years age, they migrate in easterly direction remaining for some time in the southern part of the continent. During these 4 years southern bluefin is exploited by the Australia surface fisheries. Later the schools spread out in many different directions, some reaching west of South Africa and the South Atlantic. At 7 or 8 years of age southern bluefin return to the place where they were born to spawn.

Southern bluefin become sexually mature at 7 years when they measure approximately 140 cm in length. The spawing season extends from October to March with more than one spawning per year. The species is highly fecund, since a fully mature female can lays as many as 14 to 15 million eggs at a time.

Thunnus albacares (Bonnaterre)

Yellowfin tuna, *Thunnus albacares* (Bonnaterre) is distributed world wide in tropical and subtropical seas, but absent from the Mediterranean Sea.

The species is epipelagic, oceanic, inhabit above and below the thermocline. The thermal boundaries of occurrence are roughly 18° and 31°C. Vertical distribution of the fish appears to be influenced by the thermal structure of the water column, as is shown by the close correlation between the vulnerability of fish to purse seine capture, the depth of the mixed layer and the strength of the temperature gradient within the thermocline. Yellowfin tuna are essentially confined to the upper 100 m of the water column in areas with marked oxyclines, since oxygen concentrations less than 2ml/l encountered below the thermocline and strong thermocline gradients tend to exclude their presence in waters below the discontinuity layer. Larval distribution in equatorial waters is transoceanic the year round, but there are seasonal changes in larval density in subtropical waters. It is believed that the larvae occur exclusively in the warm water sphere, that is above the thermocline.

The species is recognized by its large size and deepest body near the middle of first dorsal fin base. There are 26 to 34 gillrakers located on first arch. Some large specimens have very long second dorsal and anal fins, which can become well over 20 per cent of the fork length. Pectoral fins are moderately long, usually reaching beyond second dorsal fin origin, but not beyond end of its base, usually 22 to 31 per cent of fork length. No striations on ventral surface of liver. Swimbladder is present. Vertebrae 18 precaudal plus 21 caudal. The back of the fish is metallic dark blue in colour changing through yellow to silver on belly. Belly is frequently crossed by about 20 broken, nearly vertical lines. Dorsal and anal fins and dorsal and anal finlets are bright yellow. The finlets are with a narrow black border.

Schooling of the fish occurs more commonly in near-surface waters, primarily by size, either in monospecific or multispecies groups. In some areas, *i.e.* eastern Pacific, larger fish, greater than 85 cm fork length, frequently school with porpoises. Association with floating debris and other objects is also observed.

Although the distribution of yellowfin tuna in the Pacific is nearly continuous, lack of evidence for long ranging east-west or north-south migrations of adults suggests that there may not be much exchange between the yellowfin tuna from the eastern and the central Pacific, nor between those from the western and the central Pacific. This hints at the existence of subpopulations.

Spawning occurs throughout the year in the core areas of distribution but peaks are always observed in the northern and southern summer months respectively. Joseph (1968) gave a relationship between size and fecundity of yellowfin tuna in the eastern Pacific.

The species attain a maximum fork length over 200 cm. The all-tackle angling record was 176.4 kg fish of 208 cm fork length taken off the west coast of Mexico in 1977. Common size is 150 cm fork length.

Off the Philippines and Central America, the smallest mature fish were found within the size group from 50 to 60 cm fork length at an age of roughly 12 to 15 months, but between 70 to 100 cm fork length the percentage of mature individuals is much higher. All fish over 120 cm attain sexual maturity.

There are important yellowfin tuna fisheries throughout tropical and sub tropical seas. Catch statistics for this species include reports from 14 fishing areas by 35 countries. The most important catches, well over 100 000 metric tons were recorded from Fishing Areas 71, 77 and 34. Japan and the USA were the two countries with the largest catch, about 100000 metric tons each per year. Landings were relatively stable over the period from 1975 to 1981, varying only between about 496000 and 545000 metric tons. The world catch for 1981 totalled 526340 metric tons. The above level of production could be maintained due to an increase in fishing effort, but reduction in catches per unit effort suggests decreasing abundance of some stocks.

Near surface schooling yellowfin tuna are captured primarily with purse seines and by pole-and-line fishing, while trolling and gillnetting are of much lesser importance. In 1979 eastern Pacific surface fleet numbered 259 purse seiners. 45 bait boats and 17 other vessels flying 16 flags. The carrying capacity of this fleet amounted to 169149 metric tons. Purse seining is increasing in the western Pacific, initially taking mainly skipjack and bluefin tuna. In 1982, the yellowfin tuna catch by US purse seiners in this area probably exceeded that of skipjack tuna and the total purse seine catch of yellowfin by all vessels may have been higher than that of bluefin tuna.

Pole-and-line fishing is still one of the major surface fishing techniques for yellowfin tuna in the Pacific, even though this method is declining in overall importance throughout the world.

The most important fishing method for deep swimming yellowfin tuna is long lining, primarily by vessels from Japan, the Republic of Korea and Taiwan Island. Although these fisheries operate virtually throughout the geographical range of the species, the largest catches are made in the equatorial waters of the Pacific.

For the purpose of assessing maximum sustainable yield of yellowfin stocks, the Inter-American Tropical Tuna Commission has established a yellowfin Regulatory Area and some experimental fishing areas in the eastern Pacific. The fishery has been under regulation since 1966. However, since 1979, the I-ATTC has been unable to administer effectively any conservation measures due to lack of agreement on a quota system in the CYRA among its member countries. Overfishing was suspected a couple of years ago, but the 1982 catch was so low that this is no longer the case.

In the Indian Ocean, yellowfin tuna were caught exclusively by Japanese vessels upto the early sixties, but thereafter, boats from the Republic of Korea and Taiwan, Province of China, started operating in this area and accounted for more than 50 per cent of the total catch by the late seventies. Nevertheless, Indian Ocean yellowfin tuna fisheries are not yet fully developed. Based on hydrographical data, Sharp (1979) has suggested the existence of certain areas of potential exploitation.

Larvae of yellowfin tuna do not have black pigment spots on tail. Underside of lower jaw usually has a single, small, black pigment spot (however, this spot may not develop until larvae reach 4-4.5 mm standard length). Centre of the eye in head profile is situated above the line of the body axis. Red pigment-spot pattern on the tail region of small, fresh specimens is diagnostic.

Pigments on the forebrain area begin to appear in most tuna post larvae by the time they reach 8-10 mm standard length, which could cause confusion to identify species with forebrain pigment. However, the forebrain pigment used to distinguish *Katsuwonus pelamis* and *Euthynnus affinis* from *Thunnus* species can still be discriminated because it is internal (on the surface of forebrain) while the other pigment on the forebrain area develops on the outer surface of the head above the forebrain. Also *Katsuwonus* and *Euthynnus* have much more pigment on the forebrain area than do *Thunnus* species larvae of comparable size.

Juvenile stages (15-60 mm standard length) are almost impossible to identify to species because the larval characteristics become obscured by the development of general body pigmentation and meristic counts are almost identical for all species. Juvenile stages of *Auxis, Scomber* and *Rastrelliger* have a gap between the first and second dorsal fins, unlike tunas which have the first dorsal fin continuous with the second dorsal. Matsumoto *et al* (1972) evaluated internal characters for seperating four species of *Thunnus* juveniles.

Larger stages can be identified from the adult characteristics of gill raker counts, shape of liver and vertebral column characteristics, such as the position where the first closed neural arch occurs. *K. pelamis* and *E. affinis* have a peculiar trellis or complex basket–work structure of the vertebral column, as well as distinct vertebral counts. *K. pelamis* has 41 vertebrae, *E. affinis* and all the *Thunnus* species have 39 vertebrae.

Yellowfin tuna is widely distributed throughout the Indian Ocean. Indian Ocean catches attained 100000 tonnes in 1985, as against world catch of the species of 610000 tonnes in 1984. However the species is not found south off Australia where the influence of Antartic waters is strongly felt. It is less abundant in the north of the Arabian Sea (Gulf of Oman) and in the south of the Gulf of Aden. Young individuals are rather concentrated in the upper waters near the Equator (10°N to 10°S) where they are caught by commercial purse-seining fleets and by artisanal fisheries using pole-and-line and trolling techniques. The adults disperse more, covering approximately the first 150m. They are known to live to greater depths, and are generally caught by longlining and purse-seining.

Greatest larval concentration has been reported from November to April, south off Java, Maldives, Chagos Islands, the Seychelles as well as off Madagascar. The largest quantities have been registered always on the east side of these islands. West off Sumatra and around the Chagos larvae have also been found from May to October, but in lesser quantities. Maximum density in the Mozambique Channel has been registered between November and April.

Fifty per cent of yellowfin tuna observed reach first sexual maturity when measuring 120 to 140 cm, but some rare individuals display the same maturity when measuring only 80 cm. The distribution of sexually mature yellowfin tuna indicates that spawning intensifies between January and March in the Central Indian Ocean, from 5° to 15°S and from 70° to 85°E, as well as off Sumatra and south off the Seychelles. Later, between April and June, only the yellowfin tuna caught in the vicinity of Sri Lanka are spawning. Between October and December, yellowfin spawn again in areas north off Australia and north off Madagascar.

Yellowfin tuna from the Mozambique Channel, southern tip of India, Maldives and the Chagos Islands feed on almost exclusively flying fish, small young tunas and squid. Northwest of this area, off the Somalian coasts, squid and scombroids predominate in the stomach contents. The proportion of shellfish found in yellowfin tuna stomachs increases as they appear closer to the shore.

The species attain 54 cm at first year, 92 cm in 2 years, 120 cm in 3 years, 140 cm at 4 years and 154 cm at 5 years. Yellowfin juveniles measuring 45-70 cm fork length have a growth rate of 3 cm per month. The growth rate of Indian Ocean yellowfin is 3 cm/month at 57-76 cm fork length and reaches 4.3 cm/month at 88-101 cm fork length.

The sex ratio of Indian Ocean yellowfin tuna is equal (1:1). In yellowfin measuring more than 140 cm, they registered a 52 per cent male predominance. A decrease in sex ratio have been observed moving from the Arabian Sea towards the Equator, male predominance decrease from 80 per cent to 60 per cent.

There are two stocks of yellowfin tuna in the Indian Ocean, located both sides of 100°E. During their life-span, yellowfin are exploited by various surface fisheries both commercial and artisanal. Longliners catch mainly large yellowfin of average size of 120 cm. Live bait pole-and-line fisheries exploit smaller fish averaging 60 cm. In the western Indian Ocean purse seiners catch both small and large individuals. The large yellowfin caught in this instance generally come from free-swimming schools, whereas the smaller ones are often caught near drifting wreckage.

Katsuwonus pelamis (Linnaeus)

Skipjack tuna, *Katsuwonus pelamis* (Linnaeus) is cosmopolitan in tropical and warm-temperate waters, except the Black Sea.

The fish have a fusiform, elongated and rounded body. Their teeth are small, and conical in a single series. The fish have numerous gillrakers, 53 to 63 on first gill arch. Two dorsal fins seperated by a small inter space, not larger than eye. The first dorsal with 14 to 16 spines and the second dorsal followed by 7 to 9 finlets. Pectoral fins are short with 26 or 27 rays. Interpelvic process is small and bifid. Anal fin is followed by 7 or 8 finlets. Body of the fish is scaleless except for the corselet and lateral line. A strong keel on each side of caudal fin base is present between two smaller keels. The species do not have swimbladder, but have 41 vertebrae. The back of the fish is dark purplish blue, lower sides and belly silvery with 4 to 6 very conspicuous longitudinal dark bands which in live specimens may appear as discontinuous lines of dark blotches.

An epipelagic, oceanic species with adults distributed within the 15°C isotherm, overall temperature range of occurrence is 14.7° to 30°C., while larvae are mostly restricted to waters with surface temperatures of at least 25°C. Aggregations of this species tend to be associated with convergences, boundaries between cold and warm water masses, upwelling, and other hydrographical discontinuities. Depth distribution ranges from the surface to about 260 m during the day, but is limited to near surface waters at night.

Skipjack tuna spawn in batches throughout the year in equatorial waters and from spring to early fall in subtropical waters, with the spawning season becoming shorter as distance from the equator increases. Fecundity increases with size but is highly variable. The number of eggs per season in females of 41 to 87 cm length range between 80000 and 2 million.

The fish predominantly feeds on fishes, crustaceans and molluscs. Even though fishes of families Carangidae and Balistidae are part of the diet of skipjack tuna in all oceans, the wide variety of species caught suggest it to be an opportunistic feeder preying on any forage available. The feeding activity peaks in the early morning and in the late afternoon. Cannibalism is common among skipjack tunas. The principal predators of skipjack are other tunas and bill fishes.

It is presumed that the skipjack tuna in the eastern central Pacific originate in equatorial waters, and that the pre-recruits, upto 35 cm length, split into a northern group migrating to the Baja California fishing grounds, and a southern group entering the central and south American fishing areas. Having remained there for several months, both groups return to the equatorial spawning areas. A similar migration pattern has been observed in the north western Pacific. Studies of the local movements of skipjack tuna showed that small fish under 45 cm length made nightly journeys of 25 to 106 km away from a bank but returned in the morning, while big individuals moved around more independently.

Skipjack tuna exhibit a strong tendency to school in surface waters. Schools are associated with birds, drifting objects, sharks, whales or other tuna species and may show a characteristic behaviour of jumping, feeding and foaming.

In absence of reliable age determination methods, estimates of longevity vary at least between 8 and 12 years.

Skipjack make up about 40 per cent of the world's total tuna catch and have come to replace yellowfin as the dominant tuna species since a few years. In the period from 1978 to 1981 catches of *K. pelamis* were reported by 43 countries from 15 Fishing Areas. The yearly world catch in this period fluctuated between 697760 (1981) and 796034 (1978) metric tons. Nearly half of the annual catch (47-50 per cent) in this period was landed by Japan. Other countries landing over 10000 metric tons were the USA, Indonesia, Papua New Guinea, Solomon Islands, the Philippines, France, Senegal, Spain, Maldives, Sri Lanka and Ecuador. In the Indian Ocean, skipjack fisheries are not yet well developed, but on the basis of the disrtribution of hydrological factors it offers areas of potential exploitation.

Skipjack tuna is caught at the surface, mostly with purse seines and pole and line gear, but also incidentally by longlines. Other artisanal gear include gillnets, traps, harpoons and beach seines. The importance of man made aggregating devices has increased greatly in recent years. Supporting exploration techniques, such as, arial spotting find increasing application in skipjack fisheries and utilization of remote sensing is being tried. In the pole-and-line/bait boat fishery, availability of suitable bait-fish presently represents one of the major constraints and hence efforts to culture bait-fishes are receiving more attention. It appears, however, that bait rearing is hardly feasible on large enough scale to support a major fishery. Skipjack tuna are marketed fresh, frozen and canned. In Japan they are also dried.

Black forebrain pigment is present in larvae of skipjack tuna of more than 4 mm standard length. Black pigment is absent from isthmus and just anterior to anus. Distinct black pigment spots are noticed on ventral edge of tail. Black pigment at the tip of lower jaw is visible. On the dorsal edge of caudal peduncle black pigments are rarely visible. Snout sharply pointed and projects beyond the lower jaw. Black pigments does not appear on first doral fin until the larvae are more than 8 mm in standard length. Red pigment spot pattern of the tail region of small, fresh specimens is diagnostic.

Skipjack tuna are found throughout the Indian Ocean from as far as 40°-45°S in the west as well as off south Australia. In the Indian Ocean, inspite of recent development of the commercial surface fisheries, catches of the species were only 150000 tonnes in 1985, though the world catches were nearly one million tonnes in 1984. The species is wide spread throughout the Indian Ocean and is significantly under exploited.

Katsuwonus pelamis larvae can be found in most parts of the Indian Ocean from 36°S on the western side to 30°S in the east. They are not found further north than 11°N to 15°N. Large larvae concentrations have been observed in northwest off Madagascar, south of Sri Lanka close to Laccadive Islands as well as off South Somalia. In the east of Indian Ocean larval abundance is highest in February and minimal in June-July. In north of Mozambique Channel maximum concentration was observed between November and April.

Skipjack juveniles are seldom seen in the open sea and consequently are rarely caught there. Their area of distribution is limited between 15°N and 35°S.

Skipjack tuna is heterosexual except a few cases of hermaphrodism found among the species off Laccadive Islands. Off Minicoy Island, the species attain its first maturity at 40-50 cm length. But species caught in northwest coast of Madagascar this size is 41-43 cm.

Central Indian Ocean skipjack tuna spawn from January to April and June to early September with maximum activity in January and June. Peak gonadosomatic index levels are registered from

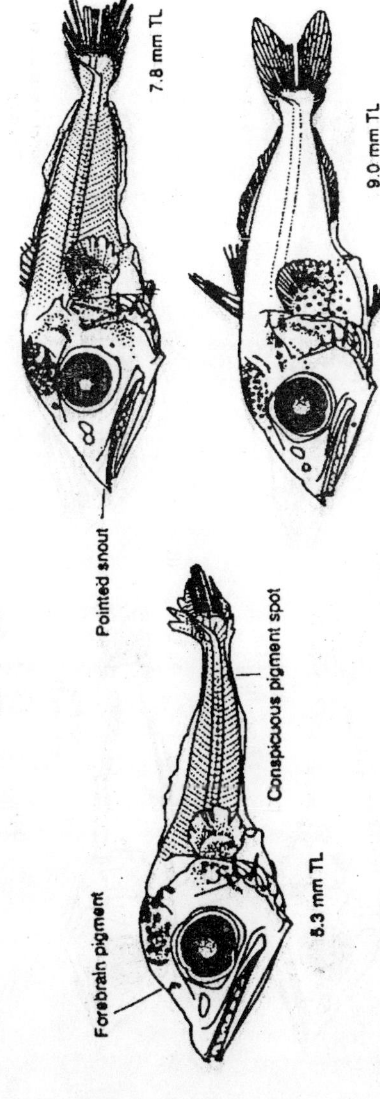

Larvae of Skipjack Tuna, *Katsuwonus pelamis* (from Yabe, 1955)

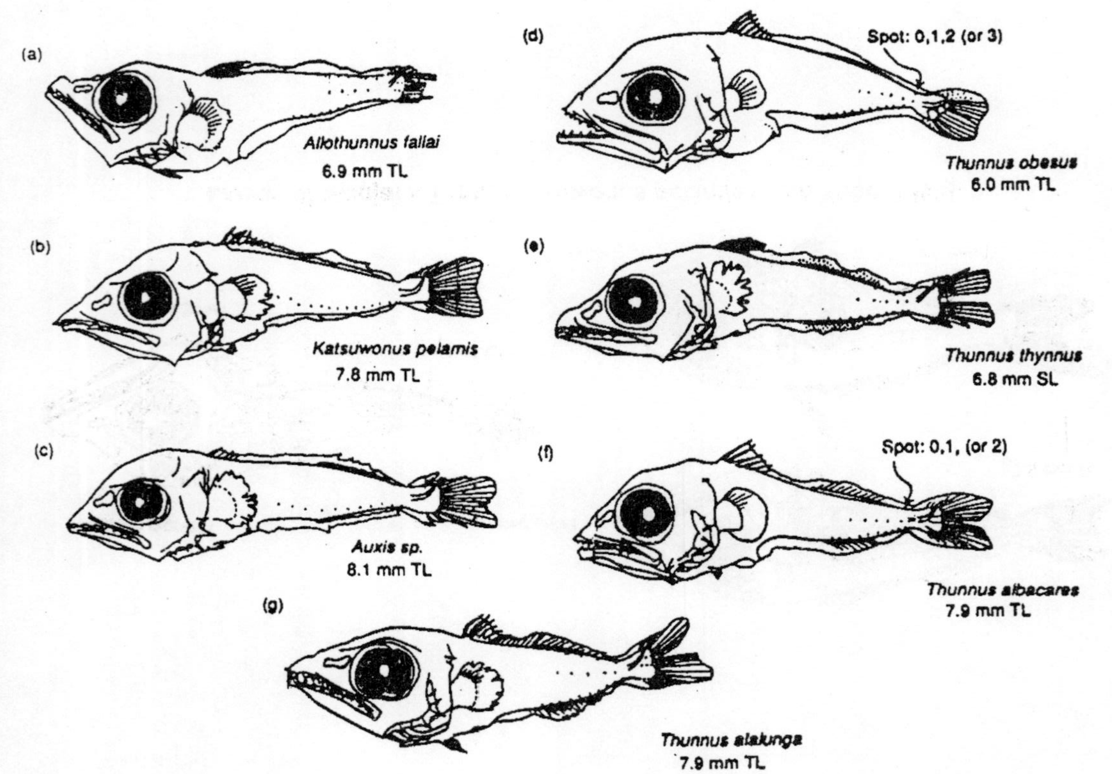
(a)
Allothunnus fallai
6.9 mm TL
(b)
Katsuwonus pelamis
7.8 mm TL
(c)
Auxis sp.
8.1 mm TL
(d)
Spot: 0,1,2 (or 3)
Thunnus obesus
6.0 mm TL
(e)
Thunnus thynnus
6.8 mm SL
(f)
Spot: 0,1, (or 2)
Thunnus albacares
7.9 mm TL
(g)
Thunnus alalunga
7.9 mm TL

October to April between Indonesia and Australia in the Eastern Indian Ocean. Peak spawning thus corresponds to the southern summer. There is no defined peak of spawning periods, the species appear to spawn all the year round. Even though eggs are released all the year round off the north-west coast of Madagascar, there is maximum activity from December to March during the southern summer. Observation of larvae abundance in this particular area suggests a continuous spawning season throughout the year with peak from November to May.

Skipjack caught in the Indian Ocean by longlining in deep waters are generally larger than those caught by trolling, pole-and-line fishing or seining in shallow waters. However, large fish have been caught occassionally in surface waters off the Laccadive Islands and Sri Lanka.

Skipjack tuna, weighing less than 5 kg are found throughout all intertropical waters. This species are found throughout the Indian Ocean as far as 40°-45° S in the west as well as off South Australia, a vast area with an extremely heterogenous environment.

Skipjack larvae can be found in most parts of the Indian Oceans, from 36°S on the western side to 30°S in the east. They are not found further north than 11°N to 15°N. Large larvae concentrations have been observed in the north west of Madagascar, south off Sri Lanka close to the Laccadive Islands as well as off South Somalia. In the east of the ocean larval abundance is highest in February and minimal in June-July. In north of the Mozambique Channel maximum concentration was observed between November and April.

Skipjack juveniles are seldom seen in the open sea and consequently are rarely caught there. On examining 200 skipjack juveniles of 3-35 cm length, their area of distribution was determined between 15°N and 35°S.

Apart from a few cases of hermaphrodism found among skipjack tuna off the Laccadive Islands, the species is heterosexual. At 40 to 50 cm total length off Minicoy Island, they have been found to attain first maturity. For specimens caught off the northwest coast of Madagascar, this size is 41-43 cm.

Gonadosomatic levels of skipjack tuna have been studied in three different areas of the Indian Ocean.

Central Indian Ocean skipjack tuna spawn from January to April and June to early September with maximum activity in January and June. Peak gonadosomatic index levels are registered from October to April between Indonesia and Australia in the Eastern Indian Ocean. Peak spawning thus corresponds to the southern summer. There is no defined peak of spawning periods, except possibly for December. Sexual rest seems to last from May to September, but some observed skipjack tuna to spawn all the year round. Even though eggs are released all year round of the north-west coast of Madagascar, there is maximum activity from December to March during the southern summer. Observations of larvae abundance in this particular area suggests a continuous spawning season throughout the year and peaking from November to May.

Variations in fecundity from 151900 to 1977900 eggs for fish ranging between 41.8 cm and 70.3 cm (total length) have been observed off the Laccadive Islands. While skipjack tunas of 44.1–56.5 cm fork lengths obtained from Madagascar produced between 87600 and 8,24,000 eggs per batch.

The principal prey of skipjack tuna are shellfish, fish and molluscs. They grow 40-45 cm in 3 years and 40-60 cm in 4 years. Monthly growth rates of species, north west off Madagascar were 34 cm at 1 year, 45 cm at 1.5 years, 52 cm in 2 years, 56 cm in 2.5 years and 58.5 cm at 3 years.

Skipjack tuna are essentially caught by surface gear (purse seine and pole-and-line). World catches were nearly 1 million tonnes in 1984 or nearly 50 per cent of total catches of large tunas. In the Indian Ocean, inspite of recent development of the commercial surface fisheries, catches were only 1,50,000 tonnes in 1985. This species is wide spread throughout the entire ocean and is undoubtedly significantly under exploited.

Based on the catch of pole-and-line fisheries alone the north west coast of Madagascar from February 1974 to March 1975 female predominence with the average sex ratio for the entire period as 0.83:1 (males : females) was observed.

Fish caught in Indian Ocean by long lining in deep waters are generally larger than those caught by trolling, pole-and-line fishing or seining in shallow waters. However, large specimens have been caught occasionally in surface waters off the Laccadive Islands and Sri Lanka.

In Minicoy Island, Laccadive Sea, skipjack tuna is caught entirely with pole and line and live bait fishes. Appreciable fluctuation has been noticed in the catches of tuna in the Island from year to year.

The seasonal character of the skipjack fishery at Minicoy is shown by the monthly catch average. Usually the poorest catches were observed in the months of June to October. The catches tend to increase gradually from November to March with peak in March. From April again the catches decline till October. Though the skipjack is the major fishery of the Island and fishing for it continues all round the year, the five-month period of November to March is considered as the peak season. June to October is the lean period. The reason for the low catches during this period is mainly due to the suspension of or decrease in the fishing activity due to the onset of southwest monsoon in the Island.

The annual catches of skipjack tuna also vary widely. The average catch with non-mechanized boats at Minicoy for skipjack was about 338.6 tonnes. The highest catch recorded was in 1969 being 522 tonnes. Even though '0' year group was observed in the catch, *K. pelamis* enters the main fishery only when it is in the 1 year group. The II year group formed the mainstay of the fishery followed by the 1 year group and III year group in the order of abundance. IV year group was found in almost all the months in lesser quantities. Like the 0-year group, the V year group is very insignificant in their occurrence in the commercial catches. This indicates that the fishery of *K. pelanis* is dependant mainly on individuals belonging to the I, II and III year age groups.

It is found that the species attains 402.2, 494.0, 562.6 and 620.5 mm respectively when it is 1-year, 2-year, 3-year and 4-year old. By length-frequency study five year classes were traced, having the size of 410, 500, 570, 630 and 680 mm respectively for 1-, 2-, 3-, 4– and 5–year classes. The size group of 700-800 mm total length was very rarely available in the fishery and that there was high percentage of males with the increase of size over females. The maximum length of 872 mm and 784 mm for males and females respectively were recorded for this species.

Length frequency distribution of skipjack tuna in the gillnet fishery covering western and southern off shore seas of Sri Lanka showed that a size range of 30 to 82 cm has been sampled with over 90 per cent fish within 40 to 70 cm range, during 1987 and 1988. Small skip jack of less than 34 cm were found mainly in the south-west. There are also signs of modal progressions from south-west to northwest and south. The quarterly length frequency distribution shows that small skipjack of south-west were observed mostly during second and third quarters, suggesting possible recruitment of skipjack off the south-west, followed by a movement northwards and southwards. Recruitment seems to be occuring predominantly in the southwest during the second and third quarters as observed from the catch data of 1988.

On examining 1440 skipjack tuna during 1987 and 1988, a male-to-female sex ratio of 1:0.97 has been observed. Females outnumbered the males during the south-west monsoon months of June to October. Size distribution of skipjack by sex showed that the proportion of females was greater in sizes less than 57 cm and that of males was more in sizes over 59 cm.

The Gonado-Somatic index for female skipjack, correlated to the maturity stages established is given below.

	Maturity Stage	Description
I.	Immature	Ovaries are small threadlike, much elongated, round in cross section with no apparent vascularization, fairly turgid. Max. ova diameter 0-0.08 mm and mean gonado-somatic index 1.58.
II.	Early developing	Ovaries increase in size and are shaped like hollow tubes, pinkish yellow, Vascularization is slightly visible. Max. ova diameter 0.08-0.24 mm and mean gonadosomatic index 4.6
III.	Late developing	Ovaries are increasingly rounded, tightly packed, cream or yellow in colour. The branching of blood vessels increases and they are clearly visible. Ova is visible through the ovarian wall. Max. ova diameter 0.16-0.38 mm and mean gonado-somatic index 9.73

Contd...

	Maturity Stage	*Description*
IV.	Mature	Ovaries are yellow or yellowish pink in colour. Well vascularized and have a thin wall through which the much larger eggs can be seen. Eggs are translucent and light orange in colour. Max. Ova diameter 0.32-0.52 mm and mean gonado-somatic index 14.3
V.	Gravid	Ovaries are orange or yellowish pink in colour. Well vascularized and have a thin wall, soft and delicate. The branching of blood vessels increases and they are fuller with blood. The thin skin shows many free large eggs inside. If the wall is damaged, eggs disperse. Max. ova diameter 0.54-0.68 mm and mean gonadosomatic index 16.2
VI.	Partially spawned	Ovaries are reddish brown or orange brown in colour. Blood vessels still prominent. Ovaries are soft, some what flabby and contain some large degenerating eggs. Still large ova can be seen through the ovarian wall. Max. ova diameter 0.47-0.58 mm and mean gonadosomatic index 8.0.
VII.	Spent	Ovaries are soft and flabby. Brown or reddish brown in colour. May still have a few degenerating ripe eggs. Ova cannot be seen through the ovarian wall. Max. ova diameter 1.02 mm and mean gonado-somatic index 3.8.

Frequency distribution of ova diameter showed five distinct groups of ova. They are immature (0-0.08 mm), developing (0.08-0.38mm), mature (0.32-0.52 mm) gravid (0.54-0.68 mm) and running (1.02mm). Maturity studies indicated the presence of a relatively high percentage of mature, spawning and spent fish in most months. Separation of female skipjack into "immature" and "mature" fish suggest that spawning commences just before the south-west monsoon in March/April and would prevail until the end of monsoon in September/October.

Occurrence of partially spent fish during this period may be indicative of spawning pulses within the long spawning period. Length at maturity studies indicated 50 per cent maturity at 45.4 cm for female skipjack and 42.0 cm for male skipjack.

A total of 614 stomachs of skipjack tuna were examined during 1987 and 1988. During the two years 35 per cent of them were categorised as "full" stomachs. The percentage composition of the major food items observed were as below.

Major Food Item	*Percentage Composition*	
	1987	*1988*
Fish	35.9	34.7
Squids	30.7	28.0
Shrimps	31.5	33.3
Crabs	1.3	3.1
Other items (larvae of crustaceans and copepods)	0.6	0.9

A fairly consistent pattern of composition was observed for the two years. A higher percentage of full stomachs were observed during the peak fishing season which is the south-west monsoon period. Small shrimp (Acetes spp) was a very significant food item during the peak fishing season. Small frigate tuna (20 per cent) was the major fish food item identified in the stomachs.

In skipjack, an external parasite was found on the membrane between the spines of the first dorsal fin, as a 'yellow dot' and has been identified as a sessile parasitic trematode of the genus *Didymocystis*. More than 70 per cent of this parasite was found between 1st and 2nd and 2nd and 3rd spines. Fish sampled from the west and south-west had more of this parasite, compared to fish sampled from other areas. Relatively higher numbers were also found on fish sampled during the south-west monsoon period.

A round, whitish and worm-like parasite, a larval form of cestode belonging to the genus *Tentacularia*, was found on gills, in the mouth, body cavity, gut and in muscles. The largest proportion was located within the body cavity. Large numbers of this parasite were found during the south-west monsoon period. Further, presence of large numbers of this parasite coincided with the presence of larger amounts of shrimps (*Acetes* spp) in the stomachs of skip jack.

Euthynnus affinis (Cantor)

Kawakawa (*Euthynnus affinis*), present throughout the warm waters of the Indo-West Pacific, including oceanic islands and archipelagos, is a medium-sized fish with a robust, elongate and fusiform body. It has conical small teeth in a single series, about 25 to 35 on each side of lower jaw including palatine teeth. 29 to 33 gill rakers are present on first arch,. On posterior surface of first gill arch 28 or 29 gill teeth are present. Vomerine teeth is absent in the species.

Two dorsal fins, the first with 10 to 15 spines. Both fins are seperated by only a narrow interspace, not wider than eye. Anterior spines of first dorsal is much higher than those mid-way, giving the fin a strongly concave outline. The second dorsal fin is much lower than the first and followed by 8 to 10 finlets. Pectoral fins are short with 25 to 29 rays, never reaching the interspace between the dorsal fins. The interpelvic process is small and bifid. Anal fin with 13 to 14 fin rays is followed by 6 to 8 finlets. Body of the fish is naked except for corselet and lateral line. A very slender caudal peduncle with a prominent bony lateral keel is present between two small keels at the base of caudal peduncle on 33rd and 34th vertebrae. Swimbladder is absent in the species and the vertebrae are 39 in numbers. Dark blue colour on the back of the fish with a complicated striped pattern which does not extend forward beyond middle of the first dorsal. Dorsal markings are composed of broken oblique stripes. The lower sides and belly is silvery white without dark longitudinal stripes. Several characteristic dark spots between pelvic and pectoral fins are present, though may not always be very conspicuous.

An epipelagic, neritic species inhabiting waters temperatures ranging from 18° to 29°C. The species is rarely encountered in waters where surface temperatures fall below 20°C. Spawning occurs in batches when the water is warmest. Like other scombrids, *E. affinis* tend to form multispecies schools by size, that is, with small *Thunnus albacares, Katsuwonus pelamis, Auxis* sp and *Megalapsis cordyla* (a carangid), comprising from 100 to over 5000 individuals.

Although sexually mature fish may be encountered throughout the year, there are seasonal spawning peaks varying according to the region, that is, March to May in Philippine waters, during the period of north west monsoon (October-November to April-May) around the Seychelles, from the middle to the north west monsoon period to the beginning of the south-east monsoon (January to July) off East Africa and probably from August to October off Indonesia. The only available information on fecundity to Indian Ocean species is that a 1.4 kg female of 48 cm length spawns approximately 0.21 million eggs per batch, corresponding to about 0.79 million eggs per season, whereas a 65 cm long female weighting 4.6 kg spawn some

0.68 million eggs per batch, that is 2.5 million per season. The sex ratio in immature fish is about 1 : 1, while males predominate in the adult stages.

E. affinis is a highly opportunistic predator feeding mainly on fish, crustaceans, squids, tunicates and other forage. They compete for food with other species they school with and also with dolphins and cetaceans. In turn, they are preyed upon by large yellowfin tuna, marlins and sharks.

The species attain a maximum length of 100 cm weighting 13.6 kg, but common size is 60 cm. In Indian Ocean the fish attain maturity on reaching a length between 50 and 65 cm in the third year of age; while in Philippines waters the species are matured for spawning at 40 cm length.

The reported world catch of the species by eight countries fluctuate between 67500 metric tons (1981) to 84000 metric tons in 1977. In India, *E. affinis* is an important species in local drift gill net and hook and line fisheries. Typically the multispecies fisheries also take *E. affinis.* Besides gillnetting, trolling is the major fishing technique in use to catch the species. Occasionally beach seines and longlines are deployed to catch the species. Some gear types are rather size-selective, that is, trolling lines catch smaller fish than gill nets.

The meat of *E. affinis* is of good quality when fresh, but it deteriorates very fast, if not preserved adequately.

There is no information on the embryonic or larval stages of *E. affinis* from Indian waters. However, a fairly large number (82 specimens) of juveniles (24.5-74.0mm) and young ones (75.0-321.5 mm) of *E. affinis* were obtained from shore-seines from Vizhinjam, a fishing centre about 16 km south of Trivandrum and from boat seines from Calicut.

Eggs of *E. affinis* have a diameter of 0.85-0.95 mm with an oil globule of about 0.21-0.24 mm. The newly hatched larva measures about 2.05 mm

In specimen of 24.5 mm standard length, the eyes are large and conspicuous and the mouth has small teeth. All the larval characters are lost except the vestiges of preopercular spines. Lateral line is not distinct. The general colour is brown dorsally gradually becoming white below.

In specimen of 56.6 mm standard length, the eyes are comparatively smaller and the body is more elongated than in the previous stage. The lateral line is distinct in the anterior region. The caudal peduncle bears a keel on each side. The general colour of the body is brown, but there are dark bands, which are broad at the dorsal aspect but taper and disappear as they come downs to the sides.

The general colour of 96 mm (standard length) young ones is somewhat deeper than the previous stage and the bands of which 13 on each side could be distinguished are more distinct. The lateral line is complete with corselet scales developed at the base of the pectoral fins.

In specimens of 162 mm standard length, except for the rather elongated appearance of the body, most of the external features of the adult are present at this stage. The corselet of scales have developed but they extend to only upto the vertical to the ninth dorsal spine on each side. Oblique irregular bands are present on the dorsal aspect above the lateral line, but these have not resolved into the distinct patterns as in the adult. Two diffused dark areas on either side between the pectoral and pelvic fins evidently indicate the position of the so-called thumb impressions or blotches in the adult.

The young fish of 222 mm SL has all the unmistakable features of the adult except for the characteristic colour pattern between the dorsal fin and the lateral line. The blotches below the pectorals are clear though not so intense as in the adult.

Euthynnus affinis populates the coastal waters and can be found in the tropical and subtropical waters of the Indian Ocean as well as along all the coastal countries from South Africa to Indonesia and around the islands of Madagascar, Reunion, Mauritius, Seychelles and Sri Lanka. It is also present along the west Australian coast but only as far as 25°-30°S. Larvae and juveniles are generally caught near the coast, but also some times offshore. In some countries artisanal fisheries land substantial quantities of juveniles. Off the southwest Sri Lankan coast the catch is also marketed when the individual size is below 24 cm. The species are also found near beaches. Off the Maldives for example, quite large schools sometimes prey on small fish in the shallow waters of the lagoon.

In the Indian Ocean, sexual maturity occurs rather late. Specimens taken in the Seychelles, found that sexual maturity is reached at 3 years when *E. affinis* measure 50-65 cm in length. Off Mauritius, the species measuring 55 cm. are sexually mature. On the East African coasts sexual maturity is reached when the fish is 55-60 cm long. However, off the southwest coast of India females obtain sexual maturity when measuring 48 cm and between 210 000 and 680000 eggs are spawned by fish measuring 48-65 cm. Spawning periods off the Seychelles occur between November and March and between April and September along the southwest coast of India. Along the equatorial area spawning takes place between August and October.

E. affinis has no preferential prey and this species will eat whatever food is available. The diet of this species consists essentially of crabs and fish and occasionally cephalopods. In Indian Ocean, the species grow to 25 cm in 1 year, 25-45 cm in 2 years, 45-65 cm in 3 years and 65–80 cm in 5 years. The average size of the fish, caught around Sri Lanka is 20-65 cm. Sometimes smaller individuals of 15-20 cm are also caught. On the southeast coast of Africa, around the Seychelles and Madagascar, trolling fisheries catch large individuals of 40-85 cm. *E. affinis*, frigate tuna and young yellowfin often found together in mixed schools.

The eggs of *Euthynnus affinis*, collected from Java Sea have a diameter of 0.85-0.95 mm with an oil globule of about 0.21-0.24 mm. The newly hatched larva measures about 2.05 mm.

In all 84 juveniles and young stages of *E. affinis* were obtained from shore seines from Vizhinjam, a fishing centre about 16 km south of Trivandrum and in boat seines from Calicut. In the larvae of *Euthynnus affinis*, black forebrain pigment is present. Black pigment spot is also present on isthmus and just anterior to the anus. One to three black pigment spots are also seen near the skin surface on ventral edge of tail (unlike *Sarda* where pigment spots are embedded in the skin). Distinct black pigment appears on first dorsal fin in larvae of 6mm or more standard length. Black pigment spots appear on underside of lower jaw.

In specimen of 29.5 mm total length, the eyes are found to be large and conspicuous and the mouth has small teeth. All the larval characters are lost except the vestiges of the preopercular spines.

Lateral line is not distinct. The general colour is brown dorsally gradually becoming white below.

Their eyes become comparatively smaller and the body is more elongated when they attain 66 mm total length. The lateral line is distinct in the anterior region. The caudal peduncle bears a keel on each side. The general colour of the body is brown, but there are dark bands, which are broad at the dorsal aspect but taper and disappear as they come down to the sides.

On growing to a total length of 106 mm., the general colour becomes deeper and the bands, of which 13 on each side could be distinguished more distinctly. The lateral line is complete and corselet scales have developed at the base of the pectoral fins.

Except for the rather elongated appearance of the body most of the external features of the adult are present on reaching 179 mm total length. The corselet of scales have developed but they only extend to the vertical to the ninth dorsal spine on each side. Oblique irregular bands are present on the dorsal aspect above the lateral line but these have not resolved into the distinct patters as in the adult. Two diffuse dark areas on either side between the pectoral and pelvic fins evidently indicate the position of the so-called blotches in the adult.

The young fish of 248 mm total length has all the unmistakable features of the adult except for the characteristic colour pattern between the dorsal fin and the lateral lines. The blotches below the pectorals are clear though not so intense as in the adult.

The fish grows to a length of about 75 cm and is the commonest thunnid on the coasts of the mainland of India and forms a fishery of appreciable importance during certain seasons.

Auxis thazard (Lacepede)

The species is cosmopolital in warm water habitat including Indian Oceans with a very few documented occurrences in the Atlantic Ocean. Commonly known as frigate tuna, the species also occur in the Mediterranean and the Black sea.

With robust, elongated and rounded body the frigate tuna have two dorsal fins, the first one with 10 to 12 spines, seperated from the second by a large interspace, at least equal to the length of first dorsal fin base. The second fin followed by 8 finlets have a large single interpelvic process, longer than the pelvic fins. The anal fin is followed by 7 finlets. Teeth of the fish are small, conical and in a single series. Pectoral fins of the fish are short, but reaching past vertical line from anterior margin of scaleless area above corselet. Body of the fish is naked except for the corselet, which is well developed in its posterior part. A strong central keel on each side of the caudal fin base is present between two smaller keels. The species have a total 39 vertebrae, but without swimbladder. Black bluish in colour turning to deep purple or almost black on the head, belly white, the fish have a pattern of 15 to more narrow, oblique to nearly horizontal, dark wavy lines in the scaleless area above the lateral line. Their pectoral and pelvic fins are purple without any stripes or spots on the belly.

Auxis thazard is an epipelagic, neritic as well as oceanic species inhabit in warm waters with strong schooling behaviours. Though larvae have a high temperature tolerance, between 21.6 and 30.5°C, the widest among tuna species studied, their optimum temperature is between 27.0 and 27.9°C and the species is usually confined to oceanic salinities.

Larval records indicate that the species spawns throughout its distribution range. In correlation with temperature and other environmental changes, the spawning season varies with areas, but in some places it may even extend throughout the year. In the southern Indian Ocean, the spawning season extends from August to April; north of the equator it is reported from January to April. Fecundity was estimated at about 1.37 million eggs per year in a 44.2 cm long female. Fecundity of fish in Indian waters ranged between 200000 to 1.06 million eggs per spawning in correlation with the size of females.

Food is primarily selected by the size of the gill rakers and consists of fishes, crustaceans, cephalopods and others. In turn, the species are preyed upon primarily by large tunas, billfishes, barracudas, various sharks and others. Cannibalism is wide spread among the species. Because of their abundance, they are considered an important element of the food web, particularly as forage for other species of commercial importance.

Maximum length from driftnet records in the Indian Ocean is 51 cm, but off Shri Lanka it is 58 cm. The common size is catches ranges between 25 and 40 cm, but depends on the type of gear used and may also vary seasonally and by region. The size of first maturity is reported at about 29 cm in Japanese waters, but about 35 cm around Hawaii.

Catches of *Auxis* are usually not identified to species. Indian Ocean catches are predominantly *A. thazard*. The species are caught most commonly with pole and line. Other commercial and artisanal gear include trolling lines, hand lines, small-scale long lines and a wide variety of nets, including traps, gill or drift nets, ring nets, beach seines, otter trawls and purse seines. In some of these gears the species are taken incidentally to other species sought. In the purse seine fisheries for yellowfin and skipjack tunas, the species being smaller and hence getting "gilled" in the webbing.

The dorsal head profile of larvae of frigate tuna is rounder and blunter than the other tunas. Their black forebrain pigment is absent. Black pigment spot present on isthmus and just in front of anus. Three dotted stripes of black pigment are present on tail towards caudal peduncle (along dorsal edge, along mid-lateral and along ventral edge). The first dorsal fin is not pigmented. Juveniles have a distinct gap between first and second dorsal fins. Red pigment-spot pattern on the tail region of small fresh specimens is diagnostic.

Auxis thazard can be found in the tropical and subtropical waters of the Indian Ocean. It is caught in variable quantities along coastal countries from Australia to South Africa. Exploited only by artisanal fisheries this species is small, rarely measuring more than 60 cm. Average sizes in the catch range from 25 to 40 cm. Larvae are more usually found closer to the coast. Large quantities of larvae have been caught off the coasts of Somalia. Their presence was also observed in the north Mozambique Channel from November to May and from May to October between 5°S and 58°E during the southern winter.

Females begin to develop sexually when measuring 33 cm and reach sexual maturity at 38 cm in length. Maximum concentration of sexually mature females occurs from December to March.

Along the coast of India frigate tuna feed on fishes (42.4 per cent), crustaceans (24.4 per cent) and squid mainly (22.7 per cent). Among fish diet clupeids form 13.4 per cent, leognathids 5.3 per cent and other fishes 23.7 per cent. Crab larvae is the main food item (21 per cent) among crustacean diet.

An annual growth rates of 11 cm in *Auxis thazard* measuring 25-36 cm and an annual growth rate of 6 cm in fish measuring more than 40 cm. have been observed. As such fish grow 26 cm in 1 year, 37 cm in 2 years and 43 cm at 3 years.

Thunnus tonggol (Bleeker)

Long tail tuna, *Thunnus tonggol* (Bleeker) inhabit in the Indo-West Pacific Ocean from Japan south through the Philippines to Papua New Guinea, New Britain, the northern three quarters of Australia (Twofold Bay, New South Wales to Freemantle, Western Australia) west through the East Indies to both coasts of India, southern Arabian Peninsula, the Red Sea and the Somalia coast.

The species in characterized by its small size, deepest near middle of first dorsal fin base. Gill rakers are few, 19 to 27 located on first arch. Second dorsal fin is higher than the first dorsal. Pectoral fins are short to moderately long, 22 to 31 per cent of fork length in smaller specimens (under 60 cm fork length) and 16 to 22 per cent in larger individuals. Ventral surface of the liver is not striated. Swimbladder is absent or rudimentary. Vertebrae 39 of which 18 are precaudal and 21 caudal. Lower sides and belly of the fish is white in colour with colourless elongate oval spots arranged in horizontally oriented rows. Dorsal, pectoral and pelvic fins blackish. Tip of second dorsal and anal fins are washed with yellow. Anal fin is silvery. Dorsal and anal finlets are yellow with greyish margins. Caudal fin is blackish with streaks of yellowish green.

An epipelagic, predominantly neritic species the fish avoid very turbid waters and areas with reduced salinity, such as estuaries. Longtail tuna may form schools of varying size. Being an opportunistic feeder, its diet includes many species of crustaceans, cephalopods and fishes, at varying percentages.

The fish attain a maximum fork length of about 130 cm. In the Indians Ocean, common fork lengths range between 40 and 70 cm. The all-tackle angling record is a 35.9 kg fish of 136 cm fork length taken at Montagne Island, New South Wales, Australia in 1982.

This species in known to be fished off Japan (but is very rare), the Philippines, Australia, Papua New Guinea, Indonesia and India. But catch statistics were only reported for Australia and Papua New Guinea, ranging between only 9 and 59 metric tons per year in the period from 1975 to 1980. In 1981, catches of 350 metric tons were for

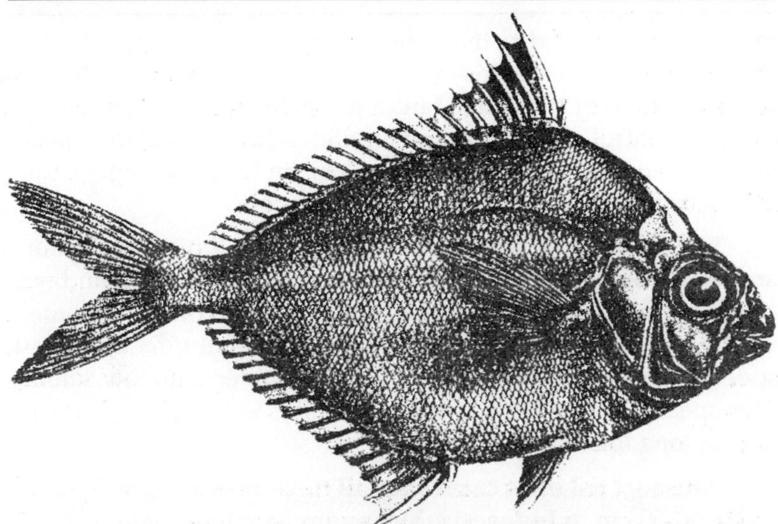

Equula splendens

The Indian Anchovy
Engraulis purava

Hilsa Ilisha
Tenulosa ilisha

the first time reported by the United Arab Emirates bringing the total to 368 metric tons in this year. This is doubtlessly a still gross underestimate of the actual landings of this species. Fishing gear comprise of trolls, drift nets and longlines. Juveniles of this species are very similar to bluefin tuna, yellow fin tuna and bigeye tuna. Misidentification of catch records are possible.

The distribution pattern of longtail tuna is limited to the north and east of Indian Ocean, Indonesia and along the north and west coasts of Australia. Some longtail tuna are found off north Somalia on the east coast of Africa. Even though longtail tuna is a coastal species, it is not found in turbid water or water with low salinity. This species is caught by artisanal fisheries with trolling lines, gill nets or long lines along the coasts.

Artisanal fisheries catch longtail tuna measuring 40-70 cm in India, 30-50 cm in Indonesia. Maximum length registered are 130 cm in India and 105 cm in Indonesia.

The size at first sexual maturity is attained by individuals measuring 45-50 cm in the Northeast Indian Ocean and only 44 cm off the west coast of Thailand. Ripe ovaries are found in a 7.3 kg female measuring 81 cm caught in September off the coast of India. There seem to be two spawning seasons off the west coast of Thailand. The main one occurs from January to April during the northeast monsoon and the other during August-September. Females measuring 44-49 cm produce an average of 1.4 million eggs per spawning.

Longtail is one of the most voracious tuna species. The species grow 30 cm at 1 year, 47 cm in 2 years with a monthly growth rate of 1.5 cm during their second year.

Sailfish *Istiophorus gladius* (Broussonet)

The Indian Ocean sailfish, if considered as distinct from the sailfish of other oceans, is limited in its range to that part of the sea bounded by Africa, Arabia, India and the Indo-Australian Archipelago. The southerly limit of its range has not been established, but it certainly reaches Durban on the south African coast and Mauritius.

There has been no sailfish catch along the Malabar coast till the drift nets were operated at 20-35 m. depth. With the subsequent operations of the nylon-drift net units (towed by mechanized vessels towards the fishing grounds and back) in the offshore waters extending up to the 60 m. depth region, moderate catches of the sailfish were obtained. Sailfish schools occur usually beyond 40m. depth region. The drift net fishermen at Calicut reported that small schools of sailfish have been occassionally sighted exposing their sail (dorsal fin) above the surface film of water during certain nights in October and November.

The sail fish catch in October 1974 at Calicut was the highest both in total catch and monthly catch per gear in the drift net. The catches registered a marked fall after January, 1975. In October, 1975, however, the sailfish catch was very meagre, but it improved during November and December. The catches decreased markedly thereafter.

Sailfish catches made in drift nets in October, 1974 bear a standard length of 145 to 255 cm. The highest standard length mode was 200 cm in that month, In November 195 and 205 cm length modes appeared. In the month of December, in addition to these two modes, an equally strong modal length of 200 cm was observed. In January, 1975 two new length modes appeared, indicating relatively fast modal length progression.

The sailfish catch trends indicate that the fish tend to visit coastal waters where cold conditions prevail in winter and migrate to deeper waters with the warming up in summer.

Istiophorus gladius is highly predatory and carnivorous, feeding mainly on teleost fishes and cuttle fishes. Among the fishes, *Anchoviella heterolobus* dominated in their diet. The sailfish captures its prey by chasing it mostly from behind. Their speedy chasing after the baited trolling hooks during day time suggests that their feeding frenzy is more prominent during day time, being facilitated by better visibility at surface waters.

Drift net has been found very efficacious for the capture of sailfish and other large scombroid fishes. Usually, twenty pieces of white nylon webbing are laced together provided with floats and sinkers. Each rectangular pieces of netting is about 19 m long and 8.5 m wide and the total length of one fishing unit is about 385-400 m. The mesh size, on an average, knot to knot, when stretched in wet condition, ranges from 85 to 120 mm.

Chapter 8
Clupeidae

Fishes under the clupeidae family have very wide gill-openings and pseudobranchiae when present are well developed. Their abdomen are mostly compressed, generally with sharp edged scutes and serrated. Their eyes are lateral with or without adipose lids. Mouth may have a deep cleft with small premaxillaries and the elongated maxilla. Either their upper or lower jaw is projecting or their mouth may be transverse. The fishes have a single dorsal fin with a few or moderate number of weak and articulated fin rays. Their anal is sometimes many rayed. Scales are present on the body and none on the head.

Genus-*Engraulis*

Fishes belonging to this group have oblong or elongated, compressed and serrated body along the abdominal edge. Their cleft of the mouth is lateral, snout conical, the longer upper jaw and maxillaries of varying length, mostly long, having a membraneous attachment to the cheeks. Their teeth are small, sometimes of unequal size usually present on the jaws. The dorsal fin may be wholly or partially in advance of or entirely posterior to the origin of the anal fin. The upper pectoral fin rays may or may not be prolonged. Anal fin is with many or moderate number of rays. Scales are large or of moderate size.

Engraulis hamiltonii (*Stolephorus hamiltoni*)

Engraulis hamiltonii (*Stolephorus hamiltoni*) inhabiting in Indian Ocean has an elongated body with strong scutes, 16 anterior to the ventral fin and 10 posterior to it. Its snout is slightly in advance of the end of the lower jaw, the enlarged maxilla opposite to the mandibular joint and nearly reaching the base of the pectoral fin. Its opercle is more than twice as high as wide. The fish is bronze coloured on the dorsal side, which in fresh state is divided by a silvery band from the purple and golden tinted abdomen. Fins are yellow in colour and dorsal fin sometimes edged with black. Fine teeth are present in jaws. Dorsal fin of the fish commences about midway between the end of the snout and the base of the caudal fin. Pectoral fin reaches to above the middle of the small ventral fins. Anal fin originates behind the vertical from the last dorsal ray. Thirteen gill-rakers are present along the horizontal limb of the outer branchial arch. They are nearly as long as the eye. Pseudobranchiae are rudimentary.

Engraulis malabaricus

Engraulis malabaricus occuring in Indian Ocean are silvery shot with gold and purple colour. Their fins are yellow in colour. Dorsal and end of caudal fin are edged with dark colour. Pectoral fins are sometimes, but not usually black. Their snout is slightly in advance of the lower jaw. Maxilla dialated above the mandibular joint and extends posteriorly to the gill-opening. Fine teeth are present on the jaws and on the palate. Dorsal fin of the species originates rather nearer to the snout than to the base of the caudal fin. Anal fin commences behind the vertical from the last dorsal ray. Pectoral fin reaches the base of the ventral fin. Scales are regularly arranged, 17 scutes before and 9 or 10 behind the base of the ventral fin. Gill rakers are half the length of the eye, from 21 to 25 on the horizontal branch of the outer branchial arch. Pseudobranchae are rudimentary.

Thryssa mystax (*Engraulis mystax*)

Thryssa mystax (*Engraulis mystax*) inhabiting in Indian Ocean are characterized by black venules over the scapular region. Their dorsal and caudal fins are yellow in colour and the latter having a black upper and posterior end, at least in the young stage. The fish has a projecting snout with enlarged maxilla situated opposite the mandibular joint. It extends backwards to opposite the base of the pectoral fin. Opercle is twice as high as wide. The dorsal fin originates midway between the snout or anterior end of the orbit and the base of the caudal fin, which latter has the lower lobe usually the longer. Anal fin commences just posterior to the last dorsal ray. Pectoral fin extends to the last third of the small ventral fins. Nine strong scutes are present behind the ventral fin and 16 to 17 weaker ones before it. Gill-rakers are as long as the eye, 13 on the horizontal limb of the outer branchial arch. Pseudobranchiae are rudimentary.

Engraulis kammalensis

Engraulis kammalensis inhabiting in Bay of Bengal attain 15 cm in length. They are silvery tinged with gold colour; the upper portion of the head, dorsal and caudal fins are yellowish with numerous fine black dots. Dark venules on the shoulder extend indistinctly over the back. The species has got the protecting snout beyond the lower jaw, which latter anteriorly reaches below the front edge of the eye. Maxilla is large, opposite the angle of the mouth and extend posteriorly to opposite the gill-opening. Fine teeth are present on the jaws. Dorsal fin originates rather nearer the base of the caudal fin than the end of the snout or else midway between these points. Pectoral fin reaches to opposite the end of the small ventral, which is inserted rather in advance of the origin of the dorsal fin. Anal fin commences a short distance behind the last dorsal ray. Fourteen scutes are present before and eight behind the base of the ventral fin. Twenty two gill-rakers are present on the horizontal branch of the outer branchial arch. Pseudobranchae are rudimentary.

Stolephorus boelama (*Engraulis boelama*)

Stolephorus boelama (*Engraulis boelama*) inhabitant of Indian Ocean is bluish colour above, becoming silvery on the sides and beneath. Its head is glossed with gold. The species has compressed and pointed snout projecting considerably beyond the end of the jaws. Maxilla is oblique, truncated posteriorly and reaching slightly beyond the mandibular joint. The fish has small teeth, equal in both the jaws. Its dorsal fin originate nearer the end of the snout than the base of the caudal fin. Anal fin commences a short distance behind the last dorsal ray. Pectoral fin nearly reaches the ventral fin, which is inserted below the anterior dorsal rays. The scales of the species are regularly arranged. Scutes are very indistinct, not extending above half the way between the base of the pectoral and ventral fins.

Gill-rakers are fine, closely set, about 21 or 22 on the outer branch of the lower branchial arch. They are nearly as long as the eye.

Engraulis dussumierie

Engraulis dussumierie, inhabiting in Indian Ocean is characterized by its coppery colour, becoming silvery below. A large black shoulder spot is passing over the back. Its caudal fin is straw coloured with a dark extremity. The snout of the species projects considerably over the mouth, the angle of the mouth is far behind the posterior margin of the orbit. The maxilla dilated opposite the mandibular joint, elongated posteriorly and almost or quite reaching the ventral fin. Fine teeth are present on the jaws. The dorsal fin of the species originates rather nearer the snout than the base of caudal fin. It is situated entirely in advance of anal fin, which commences a short way behind it. Pectoral fin is as long as head without the snout. Ventral fin is inserted slightly in advance of the vertical from the origin of the dorsal fin. Caudal fin is forked, lower lobe is longer. Seven to eight strong spines are present behind the ventral fin and 13 to 14 anterior to it. Pseudobranchae is rudimentary.

Engraulis telara (Settipinna phasa)

Engraulis telara (Settipinna phasa) inhabiting in Indian Ocean attains 40 cm in length is characterized by greenish colour along back, becoming silvery dashed with gold along the abdomen. Its dorsal and caudal fins are yellow with the upper lobe of the caudal fin and upper margin of the dorsal fin stained with black colour. Pectoral fin in the young is yellowish but in the adult of a deep blue-black except the elongated ray which is usually uncoloured in its posterior three-fourths. Ventral and anal fins are also uncoloured. The abdominal profile of the fish is more prominent than the dorsal profile. Snout slightly over hanging the mouth. The maxilla extends to opposite the gill-opening. It is enlarged above the mandibular joint, from where it is cut straight to its pointed extremity. Opercle is twice as high as wide. Fine teeth are present in both the jaws. Origin of dorsal fin is slightly posterior to that of anal fin, much nearer to snout than the base of the caudal fin. Pectoral fin with its superior ray elongated to opposite the centre of anal fin (in some fish this ray is slightly produced), while the fin itself extends as far as to opposite the posterior end of the ventral fin. Lower caudal fin lobe is longer, the upper being truncated. The length of the base of the anal fin is considerably more than half the distance from the end of the snout to the caudal fin. Scales are very deciduous, some over the base of the anal and dorsal fins. Seven spinate scales are present behind the ventral fin and 15 or 16 anterior to it. Pseudobranchae are rudimentary. Gill-rakers are rather longer than the eye.

Engraulis taty (*Stolephorus taty*)

Engraulis taty (*Stolephorus taty*) inhabiting in Indian Ocean attains atleast 15 cm in length. It has got a projecting snout over the lower jaw. The maxilla extends backwards beyond the mandibular joint, where it is dilated and posteriorly truncated. Fine teeth are present in the jaws. The dorsal fin arises nearer the snout or midway between it and the root of the caudal. The anal fin is half as long as the total length excluding the caudal fin. It begins below the middle of the dorsal. The first pectoral fin ray having (but not invariably) a long filamentous ending. Caudal fin is lobed, the lower lobe being longer. Twelve scutes posterior to the base of the ventral fin and about 23 anterior to it are present. Gill-rakers are lanceolate and rather widely set. The upper surface of head and back are green or greenish-yellow in colour, sparingly dotted with black. Sides, abdomen, cheeks and opercles are silvery. Dorsal, caudal and anal fins are yellow. Some black dots are present on dorsal fin. Ventral and pectoral fins are yellowish. The latter sometimes is dark. The edges and posterior margins of the caudal fin is blackish. Iris silvery and supraorbital half bluish black.

Engraulis purava

Engraulis purava, inhabitant of Indian Ocean attain 30 cm in length. The species is characterised by slightly elevated and rather projecting snout (in adults) beyond the extremity of the lower-jaw. Smooth portion of the opercle is rather above twice as high as wide and oblique in its direction. The maxilla rather dilated opposite the mandibular joint and gradually decreasing in width. It reaches nearly to the base of the pectoral fin. Fine teeth are present in both jaws. The dorsal fin commences midway between the snout and the base of the caudal fin, or rather the latter in some fishes. Pectoral fin reaches the ventral fin which does not extend half way to the base of the anal fin. Anal fin arises beneath the posterior dorsal rays. Caudal fin is forked. Scales are arranged in regular horizontal lines, with a fine row at the base of the dorsal and anal fins. There are 15 or 16 spinate scales between the throat and the base of the ventral fin, and 10 or 11 between the ventral and the base of the anal fin. The fish is silvery in colour, steel-blue along the back and a golden tinge above the head. Dorsal and caudal fins are yellowish and the others are uncoloured. Gill-rakers are lanceolate, rather widely set, and the largest are about as long as the eye.

Stolephorus commersonianus (*Engraulis commersonianus*)

Stolephorus commersonianus (*Engraulis commersonianus*) occuring in Indian Ocean is silvery in colour, greenish above with opercles shining silvery shot with gold. A large black spot, sometimes indistinct, is present just behind the occiput. A broad silvery band passes from the centre of the posterior margin of the opercle to the centre of the caudal fin, becoming widest posteriorly. Abdomen of the fish is light buff-colour. Fins are yellowish with minute black dots. Eyes are silvery, orbital margin is dark. Caudal fin is shot with bluish colour. Its scales are very deciduous. Pseudobranchiae is well-developed. The dorsal fin of the fish commences nearer the base of the caudal fin than to the front end of the snout, and posterior to the insertion of the ventral fin. The anal fin arises below the middle of the dorsal fin. The species have fine teeth in the jaws. Snout of the fish is pointed, projecting considerably beyond and overhanging the mouth. The maxilla moderately dilated opposite the mandibular joint and continued to the gill-opening.

Stolephorus russelli (*Engraulis indicus*)

Stolephorus russelli (*Engraulis indicus*), an inhabitant of the Indian Ocean ascends the tidal rivers. They are characterized by moderately compressed body with the dorsal and abdominal profiles equally and slightly convex. The greatest height of the head equals its length excluding the snout. Snout is pointed and projecting considerably beyond the jaws. Maxilla is enlarged and truncated opposite the mandibular joint. Small teeth are present in the jaws. Its dorsal fin commences nearer the base of the caudal fin than to the end of the snout. Its first half in advance of the origin of the anal. The scales of the fish is very deciduous. Four scutes are present before the base of the ventral fin. Pseudobranchiae is present. Gill-rakers are closely set, not so long as the eye. The colour of the fish is silvery, dashed with green along the back and sometimes some dark spots behind the occiput. A brilliant silvery band passes from opposite the upper edge of the eye to the middle of the caudal fin.

Stolephorus tri (*Engraulis tri*)

Stolephorus tri (*Engraulis tri*) inhabitant of Indian Ocean attain 10 cm in length ascends the Hoogly river as high as Kolkata. The fish has a compressed body, the abdominal profile more convex than the dorsal. The greatest height of the head equals its length excluding the snout. Snout is pointed and projecting considerably beyond the jaws. Maxilla is enlarged opposite the mandibular joint, pointed posteriosly and reaching to the gill-opening. Teeth are present on jaws. Dorsal fin commences between the front edge of the eye and the base of the caudal fin. Ventral fin is inserted on a line anterior to the origin of the dorsal fin. Anal fin commences under the middle of the dorsal fin. Scales are regularly arranged and not very deciduous. Four long slender scutes are present before the ventral fin. About 25 gill-rakers are present in the horizontal branch of the outer branchial arch. The colour of the fish is silvery shot with purple. A silvery band passes from opposite the eye to the base of the caudal fin. A dark spot is present behind occiput.

Genus–*Coilia*

Branchiostegals are nine to eleven. Body elongated compressed and tapering to a pointed tail. Abdomen is trenchant and serrated. Snout pointed and projecting. Mouth cleft extend to behind the eye. The maxilla is produced posteriorly. A single rather short dorsal fin placed in the anterior portion of the back. Anal fin is elongated and confluent with the caudal fin. Some of the upper pectoral rays produced into modertely thick filaments. Scales are of moderate or small size.

Coilia ramcarati

Coilia ramcarati, an inhabitant of Indian Ocean ascends the rivers and estuaries of Bengal. The species is characterized by golden colour, with a darkish stain behind the gill-opening. The last half of the anal and the whole of caudal fins are blackish in colour. No golden spots are present on the body. Five spinate scales are present anterior and 10 or 11 posterior to the base of the ventral fin. Gill-rakers are rather fine and closely set, about 28 in the horizontal branch of the outer branchial arch. Pseudobranchiae is present.

Teeth are fine and rather widely set in jaws and also on palate. Six lower pectoral fin rays are very short and almost concealed in a scaly seath. The free ones reach to about the middle length of the fish. Ventral fin as long as the post orbital length of the head is inserted slightly before the front edge of the dorsal fin. The length of the base of the anal fin is about one third of the distance between the gill-opening and the base of the caudal fin, the upper ray of the latter fin being the longest. Snout of the fish is somewhat projecting beyond the mouth. Maxilla dilated above the mandibular joint, behind which it is truncated. The distance between the snout and dorsal fin is one fourth of the total length.

Coilia dussumieri

Coilia dussumieri inhabiting in Indian Ocean grows to 18 cm in length. The species is characterized by projecting snout beyond the mouth. The distance between the snout and the first dorsal ray is one fourth of the total length. Maxilla dilated opposite the mandibular joint extending posteriorly to the gill-opening. Fine teeth are present in jaws and palate. Six upper pectoral fin rays produced to about the middle of the length of the fish, the longest of the lower rays equalling the head excluding the snout. Ventral fin is not half as long as the head. It is inserted below the middle of the dorsal fin. The length of the base of the anal fin is nearly equal to three-fourth of the distance between the gill-opening and the base of the caudal fin, the upper ray of which last is the longest. They decrease in length to the most inferior one. Five or six spinate scales anterior to the ventral fin and eight posterior to it is present. They commence just behind the insertion of the pectoral fin. Gill rakers are fine closely set, from 21 to 25 in the lower branch of the outer branchial arch, the longest being as long as the eye. Fishes are golden in colour with two or three rows of round burnished golden spots along the lower half of the side.

Coilia borneensis

Coilia borneensis lives in Indian Ocean and is characterized by golden colour, fins yellowish without any black marks. Maxilla of the fish is dilated above the mandibular joint and continued backwards as far as the gill-opening. The distance between the snout and the dorsal fin is one-fourth of the total length. Six pectoral rays are short, the upper 12 ones extend almost of the middle of the length of the body of the fish. Ventral fin is about half as long as the head. The length of the base of the anal fin equals two fifth of the distance between the snout and the base of the caudal fin. Four or five spines are located along the abdominal edge between the bases of the pectoral and the ventral fins, five behind the latter.

Genus-Chaetoessus

Fishes of this group have oval, short, deep and moderately compressed body with a sharp, serrated abdominal edge. Their snout are overhanging and a narrow transverse mouth. The superior combs of the first branchial arch unite with those of the opposite side, forming two angles, one pointing forwards, the other backwards. The fourth branchial arch having an accessory respiratory organ. Teeth are absent from jaws. The fishes have a single dorsal fin having the last ray prolonged in some species. Ventral fin is situated anterior to or below the dorsal fin. Anal fin commence posterior to the dorsal with a moderate number of or many rays. Caudal fin is forked. Air-vessel is large, rounded anteriorly and pointed posteriorly. Pseudobranchiae is well developed.

Chaetoessus nasus

Chaetoessus nasus, inhabiting in Indian Ocean has a compressed body with abdominal profile more convex than the dorsal. Gape of mouth is nearly twice as wide as the cleft depth and overhung by the snout. The dorsal fin commences much nearer the snout than the root of the caudal fin. Its last ray is elongated and in some specimens reaching to the base of the caudal fin. Pectoral fin is as long as the head excluding the snout. Ventral fin is inserted under the anterior dorsal fin rays. Caudal fin is deeply forked. Scales with serrated edge are placed in regular rows. 28 scutes are present along the abdominal and thoracic edge, about 15 of which are anterior to the ventral fin. 18 rows of scales are situated before the base of the dorsal fin. The fish is greyish-green in colour dorsally, with the centre of each scale in the first seven rows are darkest, forming horizontal lines, the lower of which do not extend to the caudal fin. Abdomen is whitish shot with gold. A bluish spot some time present on the shoulder behind the upper half of the opercle. Preopercle has a brilliant golden tint. Dorsal fin greenish-yellow with the posterior margin stained blackish, Pectoral, ventral, anal and caudal fins are yellowish with a dark extremity at the end.

Genus-*Clupea*

Fishes of this group have oblong or sub-elongated body with the serration of the abdomen extending anteriorly into the thoracic region. Their upper jaw is not projecting beyond the lower. Mouth is anterior or antero-superior. Teeth when present are rudimentary and deciduous. Their dorsal fin is situated opposite the ventral fin. Anal with moderate or large number of rays. Caudal fin is forked. Scales are large, or moderate or more rarely of small size. Pseudobranchiae is well developed.

Sardinella brachysoma (Clupea branchysoma)

Sardinella brachysoma (Clupea branchysoma) inhabiting in Indian Ocean have more convex abdominal profile than the dorsal. Their lower jaw is prominent, maxilla reaching to below the first third or middle of the eye. Opercle is more than twice as high as wide. A few teeth is situated on either side of symphysis of lower jaw, fine ones on tounge, palatine and pterygoid bones. The fish is greenish in colour superiorly. The bases of the scales are little dark and caudal fin dark tipped. Dorsal fin of the fish commences midway between the end of the snout and the posterior extremity of the base of the anal fin. Pectoral fin reaches to above the ventral fin, the latter fin being inserted beneath the centre of the dorsal fin. Last two anal rays are prolonged. Caudal fin is deeply forked and covered with fine scales. Scales are regularly arranged. Scutes are well developed, 18 before and 12 behind the base of the ventral fin. Free portion of the tail is higher than long.

Clupea atricauda

Dorsal and abdominal profiles are nearly equal in *Clupea atricauda* and is slightly convex. The greatest height of the head equals its length excluding the snout. Lower jaw projecting beyond the upper, the posterior extremity of the maxilla reaches to beneath the first third of the eye. Numerous fine radiating lines on the cheeks. Opercle is twice as high as broad. A few small teeth are present in front of lower jaw, a narrow band along the tongue and some deciduous ones on the palatines and pterygoids. Dorsal fin of the species starts midway between the end of the snout and above the posterior extremity of the base of the anal fin. Pectoral fin is as long as the head excluding the snout and reaching two-third of the way to the ventral fin, which last is inserted beneath the middle of the dorsal fin. Last two anal fin rays are thickened and slightly elongated. Lower caudal lobe is sometimes slightly longer. Gill-rakers numerous, closely set, and is about one-third as long as eye. Scales are regularly arranged with crenulated outer margin. Scutes are small anteriorly becoming stronger under the middle of the base of the pectoral fin and 13 behind the base of the ventral fin. The fishes are dark steel blue in colour along the back and upper third of the body, divided by a yellowish line from the silvery sides and abdomen, which are glossed with blue and purple. Caudal with its extremity deep brown or black.

Clupea klunzei

Dorsal and abdominal profiles of *Clupea klunzei* are about equal and slightly convex. Head is about one fifth longer than deep. Lower jaw is prominent, maxilla reaching to beneath the middle of the eye, cheeks with numerous radiating striae. Opercle is twice as high as wide. Eyes are broad with adipose lids. Teeth are few on lower jaw, tongue, palatine and pterygoid bone. Dorsal fin starts midway between the end of the snout and above the posterior end of the base of the anal fin. Pectoral fin is as long as the head excluding the snout and reaching two-third of the distance to the base of the ventral fin, which last is inserted below the middle of the dorsal fin. Last two anal rays are rather thickened and slightly elongated. Gill-rakers are fine about one-third as long as the eye, about 38 in the outer branch of lower branchial arch. Scales are in regular rows. Thirty badly developed scutes are present, about 13 behind the base of the ventral fin. The colour of the fish is bluish along the back becoming silvery white shot with purple on the sides and below. The species inhabit in Indian Ocean.

Sardinella longiceps

Sardinella longiceps an inhabitant of Indian Ocean is distributed in Sind down the western coast of India, more rarely on the eastern coast, Sri Lanka, the Andamans to the Malay Archipelago. They attain about 20 cm in length and appear in large quantities in Malabar coast. But along the Coromandel coast they never appear in sufficient quantities. They are abundant in some years and occasionally disappear for several consecutive seasons, returning again in enormous quantities. Eyes of the fish are with broad adipose lids, lower jaw is slightly longer. The maxilla reaches to beneath the first one third or centre of the eye. The greatest height of the head equals its length behind the centre of the eye. Opercle is twice as high as wide. Fine teeth are present on the tongue, very deciduous ones on palatines and more rarely on the pterygoids, very minute or absent from the lower jaw. Dorsal fin commences rather nearer the snout than the base of the caudal fin. Its upper edge is concave and last two rays are short. Pectoral fin is as long as the postorbital portion of the head, ventral fin is inserted in a line rather behind the middle of the dorsal fin. Anal fin is short, the length of its base is not quite half that of the head. Its last two rays are thickened and rather elongated. Caudal fin is forked. Scales of the fish are indistinctly crenulated on the outer edge and regularly arranged, 13 rows before the base of the dorsal fin. About 18 badly developed scutes before and 13 to 14 behind the base of the ventral fin are present. Gill-rakers are numerous about half longer than the eye. The fish is bluish in colour along the back with golden reflections. Abdomen is silvery shot with purple and sometimes a golden line divides the colour of the back from that of the sides. Head is of the same colour as the body, with a large greenish gold spot on the upper margin of the opercle and preopercle. Dorsal fin is greenish. Caudal fin is stained with green and the other fins are transparent.

Clupea finbriata (*Kowala lauta*)

Clupea finbriata (*Kowala lauta*), an inhabitant of Indian Ocean is characterized by more convex abdominal profile than the dorsal profile. Lower jaw is slightly prominent, the maxilla reaching below the first third of the eye. Opercle is about twice as high as wide, sub-opercle nearly triangular and rather longer than height. Teeth are deciduous present in the jaws, a band along the middle of the tongue and a patch on the palatines. Scales are regularly arranged with their free edges jagged. Scutes are moderately developed, 14 or 15 posterior to the ventral fin and 16 or 17 before it. Free portion of the tail is higher than the length. Dorsal fin is rather higher anteriorly than its base length. If commences midway between the end of the snout and the posterior extremity of the base of the anal fin. Its upper border is concave. Pectoral fin reaches two-third of the way to the ventral fin, the last finray being inserted under the middle of the dorsal fin. Last two anal rays thickened and little prolonged. Caudal fin is deeply forked. The fish is bluish-green in colour with silvery sides. Dorsal fin is with numerous fine black dots, and a black mark at the base of its anterior rays. Caudal fin with bluish reflections and tipped with dark.

Clupea sindensis

Clupea sindensis, an inhabitant of Indian Ocean attain 20 cm in length. The greatest height of the head exceeds its length excluding the snout. Lower jaw is prominent. The maxilla reaches below the first one-fourth of the eye. Eyes are with broad adipose lids. Preorbital is with a raised and branched ridge. Opercles are smooth and are about twice as high as wide. Sub-opercle is short and truncated posteriorly. Teeth are present in a narrow band along the tongue and fine ones are present in lower jaw. Dorsal fin commences nearer the snout than the base of the caudal fin, its entire base being midway between these two points. Pectoral fin below the middle of the dorsal fin. Scales with little rough edges are having some vertical lines. Scutes are not well-developed, 13 to 14 present behind the base of the ventral fin. Gill-rakers are closely set, nearly as long as the eye. Gill cavity is deep brown or black in colour. The back of the fish is intensely deep blue, sides golden, shot with purple and blue. Fins are yellowish, a black spot at the base of dorsal fin anteriorly and its upper edge is dark. The end of caudal lobes are sometimes dark.

Clupea lile

Clupea lile, an inhabitant of Indian Ocean grows to 10 cm in length. The dorsal profile of the species is nearly horizontal and abdominal very convex. Snout is obtuse, lower jaw is slightly longer. The maxilla reaches below the first third of the orbit. Eyes are with broad adipose lids. Teeth are present on pterygoids, and a band along the centre of the tounge but none on jaws. Scales are adherent and in regular rows. Their edges are smooth. Free portion of the tail is as deep as long. Dorsal fin commences nearer the snout than the base of the caudal fin. Its upper edge is concave. Pectoral fin is not reaching to above the ventral, the latter being inserted under the anterior dorsal rays. Caudal fin is moderately forked. Gill-rakers are closely set and half as long as the eye. Pseduobranchiae is well developed. The fish is golden spot with purple in colour with a brilliant silvery band along the side. Caudal fin is dark tipped, shot with blue. A brilliant bronze coloured spot is present on occiput. The species appears in vast numbers along the western coast of India.

Clupea variegata

Clupea variegata, which grows to 18 cm in length inhabit Indian

Ocean and ascends Irrawaddi and its branches. Abdominal profile of the fish is more convex than dorsal profile. Jaws are of equal length and the maxilla extending to beneath the centre of the orbit. Opercles are smooth. Teeth are very minute present on the tongue. Scales are regularly arranged behind a line from the opercles to the base of the anal fin, anterior to which they are very irregular. Pseudobranchiae is well developed. Gill-rakers are closely set, and not quite as long as the eye. Serrated scales commence under the middle of the pectoral fin, ten are posterior to the ventral fin and about ten anterior to it. Dorsal fin arises rather nearer to the snout than to the base of the caudal fin. Ventral fins are situated below the first third of the dorsal fin. Anal fin is situated in the posterior third of the distance between the posterior margin of the orbit and the base of the caudal fin. The colour of the fish is silvery glossed with gold and bronze. A dark humeral spot is present. A row of about 18 bars passes across the back and descends a short way over the sides. Dorsal fin is with a black band in the lower portion of its posterior half. End of the tail is tipped with black.

Clupea chapra

Clupea chapra attain 20 cm in length, resides in Indian Ocean and ascends fresh water rivers throughout India. Their abdominal profile is more convex than the dorsal. The maxilla is straight and rather narrow, reaching to below the anterior third or centre of the orbit. Eyes are with broad adipose lids. It's scales are smooth, placed in horizontal rows except over the abdomen. Eighteen to ninteen scutes anterior and 9 to 10 scutes posterior to the ventral fin are present. Teeth are absent. Origin of the dorsal fin is opposite or slightly before that of the ventral fin, and a little nearer the snout than the base of the caudal fin has been observed. It is highest anteriorly. Pectoral fin does not quite reach the ventral fin and highest in front. Caudal fin is deeply forked, lower lobe is slightly longer. Gill-rakers are closely set, very numerous and rather shorter than the eye. Caudal peduncle is as high as long. Fish is silvery shot with golden in colour with dark back. The edge of the caudal fin is stained darkest. A dark spot may be present on the shoulder.

Tenualosa ilisha (*Clupea ilisha*)

Tenualosa ilisha (*Clupea ilisha*) inhabitants of Indian Ocean ascend the large rivers of India and Mayanmar to breed. The species is characterized by silvery colour, shot with gold and purple. No spots are found in adult fish, but a row of them along the upper third of the body is present in the immature fish. The most distinct of the spot is behind the upper third of the opercle. The young are usually of a bronze colour along the back with silvery sides and a burnished silvery band going from above the eye to the upper half of the caudal fin. The caudal fin is often deeply edged with black in its entire circumference. Scales are present in regular rows and many over the caudal fin. Gill-rakers are numerous and are as long as the eye. 16 to 17 scutes before and 14 to 15 are present behind the insertion of ventral fin. Pseudobranchiae is well developed. Eyes are with broad adipose lids, situated some distance before the middle of the length of the head. The posterior extremety of the maxilla reaches to below the middle or even hind edge of the orbit. Lower jaw is not projecting beyond the upper. The greatest width of the opercle equals about two third of its height. Sub-opercle is rounded posteriorly. Teeth are absent. Dorsal fin of the fish commences a little nearer the snout than the base of the caudal fin. Its upper edge is concave. Pectoral fin reaches to above the origin of the ventral fin. Ventral fin is inserted beneath the anterior half of the dorsal fin. Caudal fin is deeply forked and caudal peduncle is as deep as long.

Clupea kanagurta

Clupea kanagurta, inhabitant of Indian Ocean, has abdominal profile more convex than that of the back. The thickness of the body at the shoulder equals two fifth of its height. Greatest width of the smooth portion of the opercle is nearly half its depth. The subopercle tapering is behind to its posterior–superior angle. Lower jaw scarcely projects beyond the upper. The maxilla extends posteriorly below the hinder third of the orbit. Eyes are with a broad adipose membrane on either side. Teeth are absent. Scales are regularly arranged. 27 scutes in front and 11 behind the base of the ventral fin are present. Dorsal fin commences midway between the front end of the snout and the end of the base of the anal fin. Its greatest height exceeds the length of its base. Its upper edge is concave. Pectoral fins are more than half as long as the head and does not reach to above the ventral fin. Ventral fin is situated behind the middle of the dorsal and extends two-fifth of the distance to the base of the anal fin. Free portion of the tail is higher at its base than long. The fish is bluish-green in colour superiorly, gold dotted and purple on the sides and beneath.

Generally a row of about six or eight oval spots pass in a line from behind the upper edge of the opercle along the side. Upper edge of the dorsal fin is with a dark mark.

Hilsa toli (Clupea toli)

Hilsa toli (Clupea toli) an inhabitant of Indian Ocean occurs from Bombay through the seas of India to the Malay Archipelago and China. It attains a length of 90 cm and does not ascend rivers to breed like *Clupea ilisha* but is sometimes found in their mouth. The abdominal profile of the fish is more convex than that of the back. The thickness of the body at the shoulder equals two-fifth of its height. Greatest width of the smooth portion of the opercle is nearly half its depth. Sub-opercle tapering behind to its posterior-superior angle. Lower jaw scarcely projects beyond the upper jaw. The maxilla extends posteriorly to below the hinder third of the orbit. Eyes are with a broad adipose membrane on either side. Teeth are absent. Scales are regularly arranged. 27 scutes in front and 11 behind the base of the ventral fin are present. Dorsal fin commences midway between the front end of the snout and the end of the base of the anal fin. Its greatest height exceeds the length of its base and its upper edge is concave. Pectoral fins are more than half as long as the head and does not quite reach to above the ventral fin. The ventral fin is situated behind the middle of the dorsal fin and extends two-fifth of the distance to the base of the anal fin. Free portion of the tail is higher at its base than long. The fish is bluish green in colour superiorly, gold dotted with purple on the sides and beneath. Generally a row of about 6 or 8 oval spot pass in a line from behind the upper edge of the opercle along the side. Upper edge of dorsal fin is with a dark mark.

Genus–*Corica*

Fishes of the group have oblong and moderately compressed body. Lower jaw is longer. Teeth when present are rudimentary and deciduous. Dorsal fin is situated nearly opposite the ventral fin. Anal fin is moderate, its last rays may be thickened and detached. Caudal fin is forked. Scales are of medium size, abdomen serrated, scutes usually commence behind the pectoral fins.

Corica soborna

Corica soborna inhabiting in Indian Ocean attain 5 cm in length. Their body is strongly compressed and elongated. Maxilla reaches to below the middle of the orbit. Dorsal fin arises nearer the base of the caudal than the snout, and behind the origin of the ventral fin. The anal fin commences under the posterior extremity of the dorsal fin and has its last two rays detached. Caudal fin is forked in its posterior third and the lower lobe is longer. The fish is silvery in colour with a light band. Ten or eleven serrated scales are present along its abdominal edge, commencing below the pectoral fin. Anterior to the ventral fin there are 10 or 11 and 7 or 8 posterior to it, serrated scales are present.

Genus-*Pellona*

Fishes of this group have got elongated and strongly compressed body with serrated thoracic and abdominal edges. Their mouth are of moderate size, upper jaw generally emarginate and shorter than the lower one. They have got fine sharp teeth in the jaws, palatines and pterygoid bones and also on the tongue. Dorsal fin is small and median. Ventrals are small and inserted anterior to the dorsal fin. Anal fin is elongated. Scales are large or of moderate size and rarely small. They have six branchiostegals.

Pellonar filigera

Pellonar filigera inhabiting in Indian Ocean, along Bombay coast attain a length of 30 cm. Their abdominal profile are much more convex than that of the back. A slight concavity over the front end of the head is observed. The maxilla reaches to below the middle of the eye. Teeth are present in jaws, tongue and palate. The dorsal fin of the species commences about midway between the snout and base of the caudal fin. Its posterior rays are not above the anal fin. Ventral fin is very short, situated midway between subopercle and base of the anal fin. Pectoral fin reaches to above the ventral fin, its upper ray is very broad. Length of base of the anal fin equals $2^2/_3$ in the total length excluding the caudal fin. Caudal lobes, are elongated. 19 to 20 rows of scale are present anterior to the base of the dorsal fin. Scutes are well developed, 22 to 23 of them are present before the base of the ventral fin, while 10 or 12 are located posterior to the base of the ventral fin. Fishes have a coppery tinge colour along the back, while their sides are silvery with mother-of-pearl reflection. Dorsal and pectoral fins are black tipped.

Pellona elongata

Pellona elongata, inhabiting in Indian Ocean have more convex abdominal profile than that of the back. Their maxilla reaches to below the middle of the eyes. Teeth are present in jaws, tongue and palate. Dorsal fin commences midway between the snout and the base of the caudal fin. Its last rays do not extend to over the commencement of the anal fin. Ventral fins are very minute, situated midway between the hind margin of the head and the commencement of the anal fin. Length of the base of anal fin equal from 3 to $3^1/_4$ in the total length excluding caudal fin. Abdominal scutes are well-developed, 20 to 24 located anterior and 8 or 10 posterior to the base of the ventral fin. The fish is silvery in colour, shot with mother-of-pearl. Fins are yellowish, edge of dorsal fin is black tipped.

Pellona motius

Pellona motius inhabiting in coastal waters of Indian Ocean attain 10 cm in length. Their body is very compressed and abdominal profile is more convex than that of the back. The maxilla reaches to beneath the middle of the orbit. The posterior halves of the ridges on the summit of the head are nearly parallel one to the other. Teeth are present on the jaws, tongue and palate. The dorsal fin originate slightly nearer to the snout than to the base of the caudal fin and situated entirely in front of the anal fin. The length of the base of the anal fin being equal to $^1/_3$ of that of the total length excluding the caudal fin. Ventral fin is nearly as long as the eye. Scales in 16 rows are present anterior to the base of the dorsal fin. There are seven to eight scutes along the abdominal profile posterior to the base of the ventral fin and 15 or 16 anterior to it. The fish is silvery in colour with mother-of-pearl reflections and having a burnished lateral-band. There are some fine dots on the fins and along the edges of some of the scales. Caudal fin is with a dark outer edge. A dark band is present along the middle of the dorsal fin.

Pellona indica

Pellona indica, an inhabitant of Indian Ocean have more convex abdominal profile than the dorsal. The maxilla reaches to below the anterior $^1/_3$ of the eye. Opercle is more than twice as high as wide and not striated. The posterior halves of the ridges on the occiput are parallel one to the other. Teeth are present in the jaws, tongue, palatine and pterygoid bones. Dorsal fin of the fish commences nearer the snout than the base of the caudal fin. Its base being entirely, or almost so, in advance of that of the anal fin. The length of the base of the anal fin being $2^3/_4$ to 3 in that of the body length excluding the caudal fin. Ventral fin is small, shorter than the eye and inserted in advance of the origin of the dorsal fin. Scales are with rough edges and having one or two vertical lines. 25 to 26 scutes are present, 18 being situated between the throat and base of the ventral fin. Fishes are greenish along the back, becoming silvery shot with purple and gold on the sides and below. Dorsal fin is often tipped with black.

Pelona ditechela

Pelona ditechela, a resident of Indian Ocean have more convex abdominal profile than the dorsal. The maxilla reaches to below the middle of the orbit, anteriorly a bone armed with teeth instead of a ligament, connect the premaxillaries to the maxilla. Teeth are present in jaws, tongue and palate. Scutes are well developed, 23 anterior and 10 posterior to the base of the ventral fins are present. Origin of the dorsal fin is nearer the snout than the base of the caudal fin, while the fin is entirely in advance of the anal fin. Ventral fin is as long as the eye. Length of the base of the anal equal to $3^1/_2$ in the total length excluding the caudal fin. The fish is silvery in colour with a burnished lateral-band. The upper edge of the dorsal fin is dark.

Pellona hoevenii

Pellona hoevenii abounds in Coromandel coast of Indian Ocean. Its abdominal profile is more convex than that of the dorsal. The maxilla reaches below the middle of the eye. Opercle is more than twice as high as wide. Teeth are present on the jaws, tongue, palatine and pterygoid bones and also along a small accessory bone extending between the premaxillaries and the superior maxillary. Dorsal fin of the fish commences nearer the snout than the base of the caudal fin and is entirely or almost so in advance of the anal fin. The length of the base of the anal fin being $3^1/_4$ times the distance between the snout and the base of the caudal fin. Ventral fin is small and shorter than the eye. Scales are with almost smooth edges. 22 scutes are present of which 13 being, anterior to the base of the ventral fin. The fish is greenish colour along the back becoming silvery with mother-of-pearl reflection on the sides and beneath.

Pellona brachysoma

Pellona brachysoma, an inhabitant of Indian Ocean have more convex abdominal profile than that of the back. Their maxilla extends to below the middle of the orbit. The posterior halves of the ridges on the occiput are nearly parallel one to the other. They have teeth on jaws, tongue and palate. There are 18 scutes anterior to the ventral fin and 8 posterior to it. The dorsal fin commences midway between the snout and the posterior end of the base of the anal fin, over which last few rays of anal fin is extended. Ventral fin is small. Length of base of the anal fin is $2^1/_2$ to $2^2/_3$ in the total length excluding caudal fin. The fish is golden glossed with purple in colour.

Pellona megaloptera

Abdominal profile of *Pellona megaloptera* is slightly more convex than that of the back. The maxilla extends to nearly beneath the middle of the orbit. The posterior halves of the occipital ridges are parallel. Teeth are present on jaws, tongue and palate. There are 22 scutes anterior and 8 posterior to the ventral fin. The dorsal fin of the fish commences about midway between the snout and the base of the caudal fin. Its last few rays are above the commencement of the anal fin. Its first ray is over the centre of the interspace between the ventral and origin of the anal fins. Pectoral fin reaches to above the ventral fin. Length of base of anal fin equal to nearly $^1/_3$ of the length of the body excluding the caudal fin. 22 scutes anterior and 8 scutes are present posterior to the ventral fin. The species inhabit in Indian seas.

Pollona sladeni

Pollona sladeni, a resident of Indian seas attains 18 cm in length. The species is characterized by silvery colour with golden opercles and black edged caudal fin has 23 spinate scales on the abdominal profile anterior to the ventral fins and 10 posterior to them. Their maxilla reaches to below the middle of the eye. Their first two dorsal fin rays are in advance of anal fin, the remainder of the fin over it. It arises midway between the base of the caudal fin and the posterior extremity of the opercle. Pectoral fin reaches to nearly the end of the ventral fin which is small. Caudal fin is forked.

Genus-*Opisthopterus*

Fishes of this group have six branchiostegal, oblong and compressed body. Their lower jaw is projecting and the maxilla is elongated posteriorly. They have small sharp teeth in the jaws, palatines, plerygoids and tongue. Their dorsal fin is situated behind the commencement of the anal fin which has many rays. Ventral fins are absent. Their scales of moderate or small size, very deciduous and serrature along the abdominal profile is well developed.

Opisthopterus tartoor

Opisthopterus tartoor, a resident of Indian seas, attain 23 cm in length. Their maxilla reaches to below the anterior third or middle of the orbit. Its length is nearly half of that of the head. Lower jaw is very prominent. Opercle is $^1/_3$ to $^3/_4$ higher than wide. Teeth are present on jaws, tongue, palatine and pterygoid bones. Origin of dorsal fin is nearer the base of caudal fin than to the maxilla. The pectoral fin is somewhat longer than the head, reaches to over the first anal ray and anal fin commences midway between the end of the snout and the posterior extremity of its own base. Caudal fin is forked. Scales are very deciduous. 28 to 32 spines are present along the lower profile. There are 28 gill rakers, nearly as long as the eye. Pseudobranchiae are well developed. The fish is silvery in colour.

Genus-*Raconda*

Fishes of this genus have six branchiostegals, compressed and oblong body. Their lower jaw is projecting and maxilla may be elongated posteriorly or truncated. Small teeth are present on the jaws, palatines, pterygoids and tongue. Dorsal and ventral fins are absent and the anal fin is elongated. Scales are small and deciduous. Serrature along the abdominal profile is developed but weak.

Raconda russelliana

Body is strongly compressed in *Raconda russelliana* and end of lower jaw forms a portion of the dorsal profile which is concave above the orbit, whilst the anterior portion of the abdominal profile is very convex. Its snout is obtuse and maxilla extends to opposite the middle of the eye. The species have fine teeth along the whole extent of the maxilla and also on the lower jaw, palate and tongue. Upper pectoral fin ray is enlarged. It does not extend so far as to above the root of the anal fin. Caudal fin is deeply forked and the lower lobe is longer. Scales are in regular horizontal rows, covering the body and along the base of the anal fin. They are very deciduous. From 31 to 38 weak scutes are present along the abdominal edge. Gill-rakers are rearly as long as eye and about 28 in number on the lower branch of the outer branchial arch. Pseudobranchiae are well developed. Being residents of Bay of Bengal, the fish has a narrow dark bluish band along the back, succeeded by a light bronze line, which is divided by a lighter and wider band from a broad silvery one passing from the head to the caudal fin. A deep brown or black spot exists on the shoulder, which is sometimes continued on to the opercle. The young are a beautiful purple with a silvery band.

Genus-*Dussumieria*

Fishes of this group have numerous branchiostegals and well developed pseudobranchiae. Their gill membranes are entirely seperate. Body somewhat elongated, compressed with rounded abdomen without serration. They have pointed snout, upper jaw not projecting and cleft of mouth is moderately wide. Eyes are with broad adipose lids. Small fixed teeth are present in the jaws and villiform ones on the palatines, pterygoids and on the tongue. Dorsal fin originate opposite to ventral fin and anal fin is of moderate length. Scales are of medium or small size and deciduous.

Dussumieria hasseltii

Dussumieria hasseltii have spherical eggs with 1.41-1.67 mm in diameter. Yolk is colourless, transparent, minutely vacuolated and almost fills the egg capsule. Perivitelline space is narrow. Single colourless oil globule of 0.113-0.182 mm diameter is present near the vegetative pole. No pigment is found on embryo, yolk or oil globule. Ten hour old larva grows to 5.00 mm in length. Eyes have appeared at this stage. Yolk sac is found tapering posteriorly with the oil globule at the far end. Anus is located below 48th myotome. Four post-anal myotomes are present. Heart started functioning. Auditory vesicles are visible. Crossed muscle fibres are present on myotomes. Melanophores are present on the head, and along the body above alimentary canal.

Dussumieria elopsoides

Dussumieria elopsoides, inhabitant of Indian seas occur from Canara to Coromandel coast of India. It attains 20 cm in length. The species have pointed snout, the greatest height of its head equals its length excluding the snout. Its maxilla reaches to beneath the front edge of the orbit, jaws of the same length anteriorly or the lower slightly projecting Eyes are covered with broad adipose lids. Teeth are present in a single row in the jaws and on the tongue, palatines and pterygoids. Its dorsal fin commences midway between the hind edge of the eye and the base of the caudal fin. Ventral fin is inserted beneath the last half of the dorsal fin. Its scales are very deciduous. Gill rakers are widely set, the longest being about $^1/_2$ as long as the eye, and 19 on the lower branch of the outer branchial arch. The branchiostegals of the left side overlap those of the right to a slight extent. The fishes are beautifully green, shot with blue. Upper margin of the opercle and along the back is of a light blue colour, below which is a bronzed line and below that again a silvery one shots with pink. Caudal fin shot with blue, green and gold. Upper surface of the head and eyes are emerald green. Pectoral, ventral and anal fins are white except the first ray of the pectoral fin which has a little black on it. Dorsal fin is yellowish-green. End of snout is greeish and eyes are white.

Dussumieria acuta

Abdominal profile of *Dussumieria acuta* is more convex than that of the back. Height of head equals its length excluding the snout. Snout is pointed, lower jaw slightly longer when the mouth is closed. The maxilla reaches below the front edge of the eye. Opercle is twice as high as wide. Teeth are present in a single row in the jaws, tongue and palate. The dorsal fin commences midway between the front edge of the eye and the base of the caudal fin. Ventral fin is inserted beneath the last half of the dorsal fin. Scales are very deciduous. Gill-rakers are widely set, half as long as eye, and about 22 in the horizontal branch of the outer branchial arch. The species occur in the seas of India and attain 18 cm in length. The fish is of green in colour shot with blue. Upper surface of the head and eye are emerald green. Pectoral, ventral and anal fins are white in colour.

Genus-*Spratelloides*

Fishes of this group have elongated moderately compressed or sub-cylindrical body having rounded abdomen. Their snout is compressed. The mouth is anterior having a lateral left and the upper jaw is not overlapping the lower jaw. They have six branchiostegals and well developed pseudobranchiae. Gill-membranes are entirely seperated. Eyes are without adipose lids. Teeth are small and deciduous, but may be present on the jaws, vomer, pterygoids and tongue. Dorsal fin is placed opposite the ventral fin and anal fin is of moderate length or short. Scales are medium sized and deciduous.

Spratelloides malabaricus

Spratelloides malabaricus occur in western coast of India and attain 7 cms in length. Abdominal profile of the fish is more convex than the dorsal profile. Lower jaw is longer than the upper jaw. The maxilla reaches to below the front edge or first one third of the orbit. Teeth are minute and deciduous in both jaws. Dorsal fin of the fish commences slightly anterior to the origin of the ventral fin and rather nearer the end of snout than the base of the caudal fin. Caudal fin is deeply forked. The fish is light yellowish-green in colour superiorly, a silvery stripe along the side, and abdomen is silvery. Upper caudal lobe is with a bluish posterior edge. Some fine black points are along the back. Upper edge of the eye is dark green.

Genus-*Albula*

Fishes of this genus have eleven to sixteen branchiostegals. Gill membranes are entirely seperated. Body of the fish is oblong or elongated. Abdominal edge is rounded and not keeled. Snout is pointed projecting beyond the mouth. Eyes are with a broad annular adipose covering. Villiform teeth are present on the jaws, vomer and palatine bones, granular on the tongue, pterygoid and sphenoid bones. Dorsal fin is situated opposite to the ventral fins. The anal fin is shorter than the dorsal. Scales are small not deciduous and lateral line is present. Pseudobranchiae is well developed.

Albula conorhynchus

Albula conorhynchus, an inhabitant of Indian seas are silvery in colour. Their ventral fins are inserted beneath the last half of the dorsal fin. Its maxilla reaches to below the front edge of the eye.

Genus-*Elops*

Fishes of this group have numerous branchiostegals. Pseudobranchiae is also present. Gill-membranes are entirely seperated. Body is elongated and compressed. Mouth is wide, anterior and lower jaw is longer. Villiform teeth are present in the jaws, vomer, palatine and pterygoid bones, and also on the tongue and base of the skull. Ventral fins are opposite to the dorsal fin. Scales are small and lateral line is distinct.

Elops saurus

Lady fish (*Elops saurus*) has long slender body with small scales present over entire body. It has got a terminal mouth and deeply forked tail. The fish is silvery in colour.

The body of *Elops saurus* is elongated, sub-fusiform and wide and smooth abdomen. Both profiles are nearly horizontal and almost of equal curves. The lower jaw is slightly shorter than the upper jaw and rather thickened at its extremity. Cleft of the mouth is oblique. The maxilla extends to rather behind the posterior margin of the orbit. Teeth are numerous and villiform in both jaws, while they are continued along the anterior edge of the maxilla, on vomer in a triangular patch and also on the palatines where there are about ten rows. Dorsal fin commences nearer to the base of the caudal fin than to the snout. It is highest in front with a concave upper edge. Anal fin is more concave than the dorsal fin. Caudal fin is deeply lobed in its posterior three-fourths. A few fine rows of scales form a sort of seath to the dorsal and anal fins. Some scales are over the root of caudal fin. Eleven rows of scales are situated between lateral line and the base of the ventral fin. The fish is silvery in colour and fins yellowish with greenish tinge. The fish inhabits in seas of India and attains considerable length.

Genus-*Megalops*

Fishes of this genus have numerous branchiostegals. Gill-membranes are entirely seperated. Body is oblong and compressed. Mouth is anterior with prominent lower jaw. A narrow bony plate adheres to the symphysis of the mandibles and cover the space intermediate between the two bones. Villiform teeth are present in the jaws, vomer, palatine, pterygoid bones and on the base of the skull and the tongue.

Megalops cyprinoides

Megalops cyprinoides, a resident of Indian seas and estuaries have oblong and compressed body. Their lower jaw is prominent and the maxilla reaches to opposite the hind edge of the eye. Eyes have narrow adipose lids. Villiform teeth are present in both jaws, on the vomer, palatine, pterygoids and sphenoid bone. The dorsal fin commences opposite the ventral fin and about midway between the snout and base of the caudal fin. It is $^2/_3$ as high as the body below it. Its upper edge is concave and its last ray is prolonged. Caudal fin is deeply lobed, the lobes are more expanded in the immature fish than in the adult. Summit of head is dark olive and back is blush-green in colour in the adult, lighter in the immature fish. Their abdomen is silvery with bluish reflections.

Tarpon (*Megalops cyprinoides*)

Tarpon (*Megalops cyprinoides*) have large scales over the entire body. The fish is silvery in colour with a single dorsal fin with elongated last fin ray in a whip like filament. It has got a superior mouth and deeply forked tail. The margins of the scales are of brilliant silver as are also the lateral-line and sides of the head. The centre of the jaws are black. Dorsal and caudal fins are greyish, minutely dotted with black and the margins are blackish. Pectoral, ventral and anal fins are diaphanous with some black dots. The last anal ray is dark. Eyes are silvery, with the orbital margin having a dark tint. The pupils are oval pointing downwards.

Genus-*Chanos*

Fishes of this genus have moderately elongated and compressed body with rounded abdomen. They have four branchiostegals and pseudobranchiae is present. Gill-membranes are entirely united below and not attached to the isthmus. An accessory branchial organ is present behind the true gill-cavity. Mouth is small, anterior and transverse. Teeth are absent. Ventral fin arises opposite the dorsal fin and the dorsal fin has more rays than the anal fin. Caudal fin is deeply cleft, scales are small and lateral line is distinct.

Chanos chanos

Chanos chanos, an inhabitant of seas of India attain 90cm in length have elongated body. Upper surface of head is flat and upper jaw overhanging the lower jaw. The maxilla extends to under the anterior margin of the orbit. Opercle is nearly twice as high as wide. Small scales cover the body. There are some rows of enlarged scales over the nape. Gill-rakers are short. Dorsal fin is highest in front, its upper margin is concave and along its base are two rows of scales. It

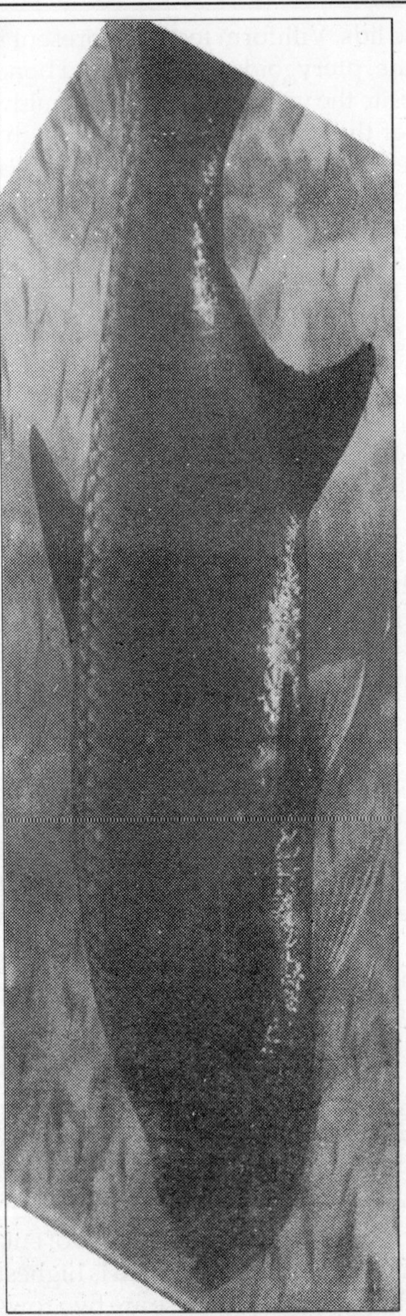

Megalops cyprinoides

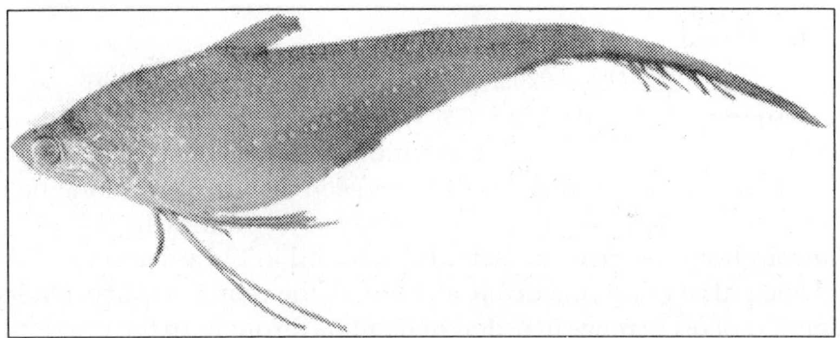

Coilia dussumieri

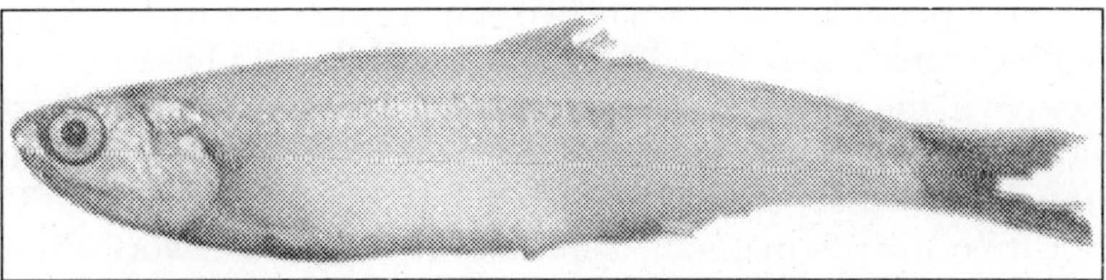

Anchoviella commersoni

arises midway between the front edge of the eye and the base of the caudal fin. Pectoral fin is pointed with an elongated scaly appendage at its base. Ventral fin is inserted under the middle of the dorsal fin, having a long basal scale. Anal fin is very small the lower margin of the fin is concave with two rows of scales along its base. Caudal fin is deeply lobed and its centre is covered by two semilunar laminae of scales. Summit of head of the fish and back are brilliant glossy blue in colour fading into silvery on the abdomen. Snout is light brown. Dorsal and caudal fin rays are greyish, which, as well as the membranes are dotted with light brown and both these fins are margined with black. The caudal semi-lunar laminae of scales are pale slate in colour. The pectoral and ventral fins are white with the anterior halves of their external surfaces are minutely dotted with dark brown colour. Their elongated appendages are bright silvery. The anal fin is white with the anterior half is dotted with black. Iris is silvery and orbital half is pale brownish.

The Hilsa

The hilsa, *Hilsa ilisha* (Ham) is a widely distributed species occuring in the Persian Gulf, Pakistan, India, Bangladesh, Mayanmar, and the Malay Archipelago. Though it is generally considered an anadromous fish, it is seen that the fish belonging to one size group or other are always to be found in the rivers throughout the year. In fact hilsa is found in the sea and in all the principal rivers at practically all times of the year. Such an extended period of occurrence in different habitats throughout the year is not normally expected of a homogeneous stocks of a migratory fish.

It is possible that there are two or more races or varieties of hilsa with different spawning grounds and habits. Old hilsa probably spawn in the higher reaches of the river, while younger ones more susceptible to the changes involved in the transition from the sea to the fresh water, probably spawn lower down the river. The occurrence of ripe males and females in the Pulta Water works during the rainy season indicate the probability of hilsa breeding in confined waters also. It was pointed out that there are differences between hilsa occurring in different rivers and also of the adjacent sea. Hilsa of the river Padma is thicker in structure and of bright silvery colour, while that of the Meghna river is a bit darker and the darkness becomes prominent in the flesh. Thus three types of hilsa has been recognized, namely, (1) those of the saline water of the seas, (2) the muddy freshwater rivers like Padma or Hooghly and (3) those of the clear freshwater like river Meghna, though some may migrate from place to place with the rise of water.

Hilsa fingerlings called "Jaitka" bear a series of dark, lateral, median transversely oval blotches even up to 6 or 7 on each side. These blotches were more distinct nearer the head and the first pair in all quite distinct in specimens up to 27 cm long in which there are 6 blotches on the right and 7 on the left.

The height of the body of hilsa can be used successfully for distinguishing Hilsa stocks. Two stocks co-extensive from Bangladesh to Orissa have been recognized. They migrate within this area. The assumption that the hilsa caught in Bangladesh, West Bengal and Orissa waters probably come from the same stock and that unregulated intensive fishing in any of these states is likely to affect the fisheries in the other states seems to find some support from biometric study of the stocks. A third stock of Hilsa recognisable is the Allahabad stock which appears to remain distinct from the rest. This, probably substantiates the assumption that the hilsa that spawn in the higher reaches of the rivers might be seperate stocks that spend their life cycle in fresh water.

Winter breeding of *Hilsa ilisha* in the Gangetic river system has for long been a controversial issue. Investigations in 1960s have shown that the species breeds in the fresh water sector of lower Ganga around Bhagalpur from late February or early March to the last week of April when the water warms up. It is shown that water temperature, current velocity and standing crop of plankton influence the spawning and that the resident slender variety is responsible for the post winter spawning.

The coastal hilsa fishery of Orissa, is small scale in nature consisting of mainly the traditional sector and small scale mechanized sector. Along the northern part of the Orissa coast, drift gill nets are employed in the night from fishing centres like, Kirtania, Chandipur, Adhuan, and Dhamra. The gear is deployed from 10m mechanized boats in the inshore waters within 10-15km from the foreshore. Some traditional fishermen use the conventional country boats for gill netting. The catch mainly consists of hilsa, pomfrets, seer fish, marine catfishes, silver-bars and sharks. The horse mackerel, long finned herring, tolishad, Indian salmon, serranids and sciaenids contribute to a minor part of the catch.

Off Chandipur (21°77'N and 78°2'E) hilsa (*Hilsa ilisha*) comprises nearly 22.4 per cent of the total gill net catch. The fishing season normally extends from September to March and is suspended thereafter owing to the roughness of the sea associated with the south-west monsoon. Occasionally a fishery develops earlier in August and continues briskly till April or May, depending on the weather conditions.

The hilsa fishery in the Orissa coast is closely correlated with the seasonal and annual rainfall. During the rainy season, which commences normally from June and persists till the end of October, the rivers of Orissa, like the Subarnarekha, the Kansbans, the Budhabalang, the Dhamra, the Rushikulya and the Bahuda carry huge quantities of nutrient-laden fresh water to the inshore seas. The riverine discharge and run-off coastal rain waters lowers the salinity and makes the inshore water more productive during the months of September October and November. This, in turn, ensures a rich crop of phyto and zoo plankton which attract the plankton feeders, like hilsa in the inshore seas, resulting in a lucrative fishery of *Hilsa ilisha* towards the latter part of the southwest monsoon. The species has also a preference for low-saline water. The scanty rain of the northeast monsoon in the coastal region of Orissa would have little effect on the coastal productivity, resulting in a drop in the hilsa catch from December onwards. The fall in catch rate may be due to their winter migration into the rivers, like the Ganga, the Dhamra, the Mahanadi and the Chilka lagoon. The years of heavy precipitation with intermittent showers cause the quick succession of phytoplankton blooms attract more pelagic fishes including hilsa in coastal waters resulting better harvest, which have been observed in 1971-72, 1973-74 and 1975-76. The rainfall in the catchment areas of the rivers and the prevailing current system are likely to affect the coastal hydrography and hilsa fishery during certain years.

Hilsa is fished in the Chilka lagoon throughout the year. The fish does not occur in appreciable quantities in the southern region of the lagoon, where the salinity is generally higher. Hisla is available in greater abundance in the northern section of the lagoon.

The abundance of hilsa correspond to two main waves of migration of fish into the lagoon, one at the close of the winter and the other at the commencement of the monsoons. In the Hoogly river also there are two periods of migration, one by about February, that is, at the close of winter and the other with the onset of the monsoons. The ascent of hilsa in Hoogly river during the floods is of considerable magnitude, compared to the earlier one. In Chilka lagoon, the late winter or the early spring migration is of equal, if not greater in intensity.

Rainfall in the hinterland which influences the flood levels in the rivers causes changes in the hydrological conditions and influence the hilsa fishery. Rainfall along the coast, when it is substantial, creates conditions which results congregation of shoals in the coastal waters and account for the heavy catch along the coast. When the flow of water from the lagoon is not sufficient due to insufficient rains in the catchment areas, into the sea, then the water current is not sufficient to stimulate large numbers of spawning hilsa to react against the water flow and thus ascend from the coastal waters to the lagoon. The hilsa in the Godavari river, owing to the fall of flood levels and the silting up of the river, did not enter the river Godavari but migrated towards north and contributed an unusual coastal fishery of considerable magnitude at Kakinada about 80 km away from Godavari river mouth.

The abundance of hilsa fishery thus seems to be dependent on a combination of factors for the migratory movements, such as, unusual turbidly of water and the direction and velocity of the wind, all are influential factors. Hisla react to the variations in the temperature, showing very restricted migratory movements during the cold season. In the coastal waters, however, fluctuations in temperature should he more frequent than in the open sea, and are to a very great extent influenced by currents, winds and tides.

Large shoals of hilsa generally come from the direction of the sea along the channel of the lagoon by February (January to March) when the fish are caught in appreciable numbers. An unusually heavy catch of hilsa from the sea at Puri in September, 1949, indicate the presence of hilsa in the coastal waters in the neighbourhood of Chilka lagoon.

Prompted by the spawning urge, hilsa breeders ascend the rivers and the spent fish drift with the flowing waters and the young also start on their gradual seaward movement.

Gonads of hilsa mature from August to October with September October as peak periods. Spent hilsa are collected from November onwards and sexually mature fish are rather rare thereafter till July-August of the next year and a regular extended breeding season do not appear to exist in Chilka hilsa as in Hoogly hilsa.

The young hilsa grows at an average rate of 25 mm, a month and within 10 months to a year old fish grows to about 300 mm in length. The breeding would have taken place by about August–September.

Chilka hilsa breeds in the lower reaches of Daya river and in its associated branches also. The fish negotiate, during the heavy floods, the Naraj anicut in the Kathjuri river (from which the Daya branches off) and reach the Mahanadi. The ascent of the hilsa in the main Mahanadi river, which is spanned by the Cuttack anicut, also takes place by September, depending on the intensity of the floods. During exceptionally heavy floods, hilsa is known to ascend as far as Sambalpur about 316 km from the sea.

Hilsa is essentially a plankton feeder. The oozing individuals do not appear to feed. There is no selective feeding as far as plankton is concerned. Copepods and diatoms have been found invariably to be the dominant food items depending on their availability. Spent fishes have more fine sand grains, showing more or less a sort of feeding habit at the bottom layers, while the young ones appear to be mid-water feeders.

The hilsa fishery in the Chilka lagoon cannot be regarded as a detached and isolated one itself as the crop is dependent on the stocks present in the adjacent coastal waters. Along Orissa coast, as at Chandipur and Talpada hilsa move about in shoals and these are caught regularly from the inshore waters from the month of November. The fish has been breeding in some of the tidal stretches of the rivers towards the end of summer or the beginning of monsoons, say May-June; and that the young ones are moving about in the coastal waters. Two to five months old hilsa feed in the estuaries and move about in shoals probably from estuaries upstream and along the coastal waters. In the river Ganges, the fish is found to move as high up as Goalundo from February to April.

The young hilsa after about two months stay in the sea, add considerably to their size and weight and towards the end of December they return to the estuaries in shoals. The fish at about this time are 17 to 22 cm. long and in search of new fishing grounds. Shoals of young hilsa (50 mm) appear to wander about along the Bengal-Balasore coasts where the hydrological conditions are similar. In the winter months most of the adult hilsa fall back from the rivers to the warmer waters of estuaries, but with the advent of the hot weather they begin to show greater activity. The comparatively low salinity favourable direction of currents, availability of food contribute towards the presence of hilsa in the coastal waters after the monsoons.

The currents in the sea indicate the net resultant movement of waters inclusive of the effect produced by the winds. Conditions are however, different in Chilka lagoon, where similar currents are absent, tidal influence is not much and the water is very shallow. The winds over the lagoon blow generally from south-west from March to September and from north-east in the remaining parts of the year, that is, opposite to the general direction of the currents in the sea. If the salinity, temperature and flow of currents have anything to do with the movements of fish, it will appear that the young fish in the tidal zone and the coastal waters is subject to a diversity of ecological conditions. In Chilka lagoon even moderate winds disturb the surface layer considerably and affect the movements of fish in the lagoon in view of its shallowness.

Oil sardine fishery along Indian Coasts

Sardine belonging to the family clupeidae and consisting of four species constitutes the single largest fishery of India. This incidentally is the biggest pelagic fishery too. *Sardinella longiceps*(Val) known as oil sardine is the most important among them and regularly contribute 20-30 per cent of the total marine landings in India.

Occurrence of *Sardinella longiceps* has been reported along the coast of Somalia, Seychelles, Gulf of Aden, Red sea, Gulf of Oman, Persian Gulf, Pakistan, Indian Peninsula, Srilanka, Malayan Archipelago, around Indonesian and Philippine waters as well as, off Vietnam, although their abundance varies with different regions. Around India, oil sardine is available from both coasts, west as well as east. Their occurrence in the east coast is reported as sporadic, the fishes are less fatty compared to their counterparts at the Malabar coast. Along the west coast the fishery is distributed from Ratnagiri in the north to Quilon in the south, Malpe and Calicut being the regions of maximum abundance. The other economically less important species of sardine found to occur along the Indian coasts and predominantly in the east coast are *Sardinella fimbriata*, *Sardinella gibbosa* and *Sardinella sirm*.

The appearance of oil sardine to the fishery exhibited wide fluctuations from as low 1 per cent to as high as 32 per cent. The reasons to this wide fluctuations in oil sardine fishery is attributed to the:

(*a*) Hydrological conditions such as temperature (27°-28° c), salinity (34-35ppt) and dissolved oxygen.

(*b*) Availability of certain species of diatoms which serve as food for the fish.

(*c*) Periodic migration to the off shore regions.

(*d*) Heavy natural morality or over fishing, and,

(*e*) Meteorological conditions, such as, average daily rain fall (30 mm) during the spawning fortnight and intensity of monsoon.

Though fishing season commences with the premonsoon shower, the fishery becomes regular only by July, the peak season being October to January, then falling and terminating by March. Seasonal variation in landing is more predominant along the Mysore coast than in the Kerala coast. However, of both the coasts, the largest catch comes from the last quarter of the year. The interquarter disparity is not so marked along the Kerala coast as on the Mysore coast.

The normal fishing area for sardine is confined to a narrow strip of coastal water extending to 12-15 km from the shore and at a depth of upto 15m. The craft engaged for fishing are mostly hand-rowed country crafts, a dug-out canoe of about 9-12 m, generally employed for operating boat seines and a similar one of smaller length of about 8-9 m used for operating drift-net and gill net. Along the Mysore and Konkan coasts in addition to the dug-out canoes out trigger boats of 5-15 m length are used for operating "*rampani*" There are three types of gears mostly in use for catching sardines. They are engulfing nets, boat seines, seine nets, beach seine and gill nets. Among these the most important gear is boat seines, used along the Kerala coast. Shore seine "*rampani*" is used along the Mysore coast to catch the sardine shoals which enter the shallow in shore waters. Gill nets are also operated with good results. With the introduction of monofilament synthetic twines for fabrication of the gear, the boat seines now operated have increased in length from the usual length of about 18m to about 30m.

Sardinella longiceps has a long and stout head, the body narrow and tapered towards the tail, covered with scales. In fresh condition the fish has an admixture of blue, green and brown colour on the back. Flanks are silvery with iridescent pink. During the peak season the catch comprises of medium sized fish ranging in size from 12 to 15 cm which belong to one year class, weighing 40 gram each. As the name suggests the fish has an oil content, the value reaching as high as 17 per cent on wet weight basis in November.

The proximate composition of oil sardine is:

Moisture: 65.28 per cent

Fat: 14.34 per cent

Crude protein: 18.10 per cent

Ash: 1.65 per cent

Inorganic phosphorus: 175mg per cent

Alpha amino nitrogen: 105mg per cent

Abundant in some years, occasionally forsake their haunts for several consecutive seasons, and returning again in enormous quantities, the oil sardine, *Sardinella longiceps (Cuv. and Val.)* ranks as one of the best known commercially, important fish of India and is extensively used as food in fresh and cured conditions. Its bye-products are of great value ; the oil is widely used in jute, leather, soap and other industries and also for painting canoes as it has a deterrent effect on ship worms and other timber boring organisms. The guano is an important fertilizer of tobacco, coffee, tea and other crops, besides being used as protein supplement of poultry and live stock feed.

The oil sardine is one of the principal fishes which contribute to the marine fish landings of the west coast. The oil sardine fishery always commences during the south-west monsoon (June-July) with the appearance of the spawners along the coast. They leave the inshore waters just prior to spawning which take place in July-August, once, in the year. The post-monsoon months show disappearance of the spawners and the entry of the juvenile oil sardines which practically form the bulk of the commercial catches during the peak period of the fishery. The spent oil sardines generally appear in the closing stages of the fishery.

The oil sardines reach the average size of 10, 15 and 19 cm at the end of the first, second and third year respectively and that the one-year old oil sardines are indeterminates, the two-year old, oil sardines are immature with developing gonads and the three-year old oil sardines are mature adults and active spawners.

Studies at Calicut on age and growth of oil sardine, *Sardinella longiceps* by means of length-frequency method indicated that the oil sardine attains the total length of 125-130 mm at the end of one year, 165-175 mm at the end of two years and 180-190 mm on completion of three years. From the monthly progression of modes during different years, it could be seen that the fishery was mainly supported by 0-year, 1-year and 2-year classes. The studies also revealed that the growth during the first year was more accelerated than during the rest of its life span.

The oil sardines show a definite preference for phytoplankton food. Phytoplankton production commences along the coast with the onset of south-west monsoon and paucity of phytoplankton production is generally observed during the summer months extending from March to May.

Growth rate of 10.0, 14.5, 18.3 and 20.5 cm during the first, second, third and fourth year of life of the oil sardines have been observed. The oil sardine attains sexual maturity and almost full adult size at the age of one year when it reaches 15 cm. Growth during the second year is very slow and amounts to only 1 cm. The life span of the oil sardine has been estimated at two and half years only.

The oil sardine shoals appear in inshore waters when the temperature is low and disappear from these waters because of higher temperature. The spawners enter the inshore waters when the temperature and salinity are at their lowest during the monsoon, but for the appearance of the juveniles which form the main stay of the oil sardine fishery every year, higher temperature and salinity, especially the latter, appear to be necessary. Immense shoals of juveniles have been encountered in the fishery during the peak period from November to December alone when the temperature and salinity vary from 27° to 28°C and 34 to 35 ppt respectively. These are the optimal hydrological conditions for the appearance of the juveniles in the coastal waters. The abundance of phytoplankton, particularly of *Fragilaria oceanica* appears to be another causative factor which influences the movements of the juvenile shoals, on the availability of which depends the success and failure of the oil sardine fishery every year.

The oil sardine and the mackerel are the two principal fishes which contribute to the marine fish production of the west coast. The fluctuation characteristic of oil sardine fishery does not normally affect that of the mackerel. An inverse relationship in the annual yields of these two fisheries have been observed earlier. The fluctuations in the oil sardine fishery seen within the season do not in any way influence the course of the mackerel fishery. The landings of these fishes have been uniformly good during the different months. The oil sardine fishery was poor during November and December, 1951, and a similar decline in the landings of mackerel was also observed during those months.

On the other hand the success or failure of the oil sardine fishery has been observed to exert a profound influence on the other sardine fisheries of the west coast, namely of, *Sardinella fimbriata, Sardinella brachysoma* and *Kowala coval*. The oil sardine being a more highly priced and esteemed fish than even the mackerel, its fishery is intensively exploited neglecting the other fisheries of the west coast. While in 1951-52 season showed prospects of a very good fishery of the other sardines and enormous shoals of these fishes frequently entered the very shallow inshore water, their fisheries were neglected by the commercial fishermen owing to the unusual improvement of the more lucrative oil sardine and mackerel fisheries at that time. Commercial exploitation of these lesser sardine fisheries was observed only on those days when the oil sardine and mackerel were scarce in the fishing grounds.

Oil Sardine Fishery in Saurashtra Coast

The Indian oil sardine, *Sardinella longiceps* is the most important pelagic fishery wealth of southwest coast of India and occasionally contributes as much as quarter of the marine fish landings. Normally the landings fluctuate between 10-18 per cent of the total marine fish catches. In the past several decades, the oil sardine fishery has shown remarkable fluctuations both spatially and seasonally. Since 1981-82 oil sardine fishery has been showing a declining trend.

Oil sardines though formed one of the major pelagic resources on the west coast, have never formed a fishery along the Saurashtra coast. They have been only once reported from east coast at Pondicherry during October to December, 1983. Part of west Saurashtra coast observed heavy landings of oil sardines for the first time during winter months of 1986-87. An estimated 352 tonnes of oil sardine were landed during mid November to February end. Peak landings were observed during 16th December to 31st January, particularly at Miani, Porbandar and Navibandar landing centres.

The oil sardine shoals appear at depths of 6 to 8 m and within 800 m range from the shore. During January, dense shoals, close to the shore were seen frequently by fishermen at Navibandar. These shoals were so dense that they changed the colour of water near the bank. The shoals were caught from dugout canoes and fibre glass boats using cast nets of mesh size 21-27 mm. The fish caught by the cast net ranged from 123 to 163 mm in total length with the dominant size around 138 mm belonging to 1 year class. The females constituted 96 per cent of the total oil sardine population and all the females were in immature stage. 98 per cent of stomachs of oil sardine were empty and the rest were partially filled. Food items mainly consisted of green algae and diatoms indicating its surface feeding planktivorous in nature.

Oil sardines do show their presence during winter months every year along the west Saurashtra coast. These shoals can either be migrating from south west coast of India or from the off shore. Several oceanographic as well as biological factors are responsible for oil sardine migrations. The wind driven surface currents of the west coast may be one of the main factors which influence the pelagic shoals of oil sardines. The sea water temperature and salinity also appear to significantly influence the oil sardine migrations.

Oil Sardine Fishery along East Coast of India

The oil sardine (*Sardinella longiceps*) is known to occur in the Indian waters from Maharashtra to Orissa. Large schools of fish are encountered in the inshore waters, sustaining a traditional commercial fishery along the coasts of Kerala and Karnataka. Its occurrence along the east coast is considered to be only sporadic and rare. In recent years the landings of oil sardine along the east coast have shown a marked increase. Over the 26 years (1961-86) period, Tamil Nadu and Pondicherry together accounted for 85 per cent of the oil sardine catch landed along the east coast followed by Andhra Pradesh 11 per cent and Orissa 4 per cent.

In Orissa about 99 per cent of the oil sardine catch was recorded from the southern district, namely, Ganjam and rest from Puri and Cuttack. In Andhra Pradesh oil sardine was recorded from Viziangram and Visakhapatnam in the north and from Prakasam and Nellore districts in the south. In Tamil Nadu including Pondicherry the oil sardine catches are mostly obtained from South Arcot district (37 per cent) and Pondicherry (27 per cent) followed by Chingelpattu (18 per cent), Kanyakumari (5 per cent), Tiruneveli, Ramanathapuram and Tanjore districts (each 4 per cent) and Tamil Nadu (1 per cent).

The best catches are witnessed during the third and fourth quarters along the Tamil Nadu–Pondicherry–Andhra Pradesh coasts and during the second quarter along the Orissa coast.

Gill nets, boat seines, bag nets, shore seines, drag nets and shrimp trawls landed the oil sardine along the east coast. In Orissa coast, bulk of the oil sardine catch is landed by gill net and rest by shore seine and drag net. In Andhra Pradesh bulk of the oil sardine

Moving Shoal of *Sardinella* sp.

catch is recorded from shore seines followed by boat seines, bag nets, gill nets and shrimp trawls. In Tamil Nadu and Pondicherry gill nets, boat seines and bag nets accounted for most of the oil sardine landings. At Tuticorin the oil sardine catch is mainly obtained by gill net though sporadically. Bulk of the catch is obtained during November–January accounting for 65.8 per cent of the annual catch followed by August-October (21.8 per cent) and February-May (12.43 per cent).

Oil sardine caught along the east coast ranged between 40 and 207 mm in total length. Around Gopalpur, in Orissa 120-175 mm size is caught in gill nets, boat seines and shore seines. Good quantities of fish measuring 50-100 mm is caught during May-July and December in drag nets, boat seines and shore seines.

In Andhra coast around Kakinada and adjacent centres towards north, size ranges of 170-195 mm in gill net, 125-198 mm in shore seine and 160-195 mm in trawl net are obtained during September-November. Dominant size ranges in different periods at Visakhapatnam are 100-160 mm during July, 40-145 mm during February, 60-135 mm during April-June. From the parental stock that start occurring in the area around July with the dominant sizes at 100-105 mm, the first progeny is recruited to the inshore catches at 50-85 mm size during May-July of the following year and the second progeny at 60-95 mm size during April-May in the third year indicating that the oil sardine has bred in the sea off Visakhapatnam and has established itself in the area.

In Tamil Nadu-Pondicherry area the oil sardines of 126-202 mm size range occur. At Pondicherry, during November-December 140-202 mm fishes are caught in the gill net, the bulk of the catch around 170 mm. At Madras, during September to March the size range of 126-195 mm with majority at 150-159 mm is met with in gill net catches.

The eggs and newly hatched larvae of oil sardine (Larvae up to 40 days after hatching) have been reported from Tamil Nadu coast. The occurrence of mature *Sardinella longiceps* during January-March were also reported off Tamil Nadu. Males predominate the catches. At Pondicherry fish caught during November-December have gonads in first and second stage of maturity and females out number the males. At Visakhapatnam fish with gonads in advanced stage of maturity (V-VII) are observed from January-September. There is evidence that fish spawn off Visakhapatnam during March-April. The adult moves away from the near shore waters for spawning.

In Orissa the fishermen have observed in many occasions large schools of oil sardine at the surface of the sea very close to Rushikulya river month. Around Kakinada, very good catches of oil sardine were obtained in artisanal gear and shrimp trawl soon after the severe cyclone in August, 1986. Similarly good catches of oil sardine are obtained off Penna river mouth in Nellore district.

Although the oil sardine has been occurring regularly in the catches of boat seines operated both within the break water area of the Visakhapatnam Outer Harbour and in the near shore area since July, the extent of its distribution towards north and south of the harbour is rather restricted to certain localities. In continuity with brackish waters oil sardines have been found to appear in large quantities.

It is interesting to note that the oil sardine catches along the east coast are obtained mostly in areas close to harbours, back waters and river mouths. This discontinuous distribution of the fish appears to indicate its affinity, particularly in juvenile phase, to areas that have a certain admixture of fresh or brackish water.

Oil Sardine in Mangalore Coast

The occurrence of large scale, small-sized oil sardine in the mangalore zone during 1960-66 is of special significance, since the fishery, though known for its erratic nature and annual fluctuations had become steadier during the period yielding an average annual catch of 135252 tonnes forming 17.89 per cent of the total marine fish landings in India. In the mangalore zone, young oil sardine (*Sardinella longiceps*) measuring 35 mm on wards were observed to occur on a large scale in the months of July, August, September and October during the year 1960-66. They were caught exclusively by, small meshed cast net, shore seine, boat seine operated in the inshore waters ranging in depth from ½ to 6 meters. The occurrence of such very small sized oil sardine in the shallow coastal water suggests that the spawning grounds of this commercially important species may not be far away. The size frequencies of young ones of oil-sardine during July-October period clearly indicate that the juvenile population is constituted by more than one group. Protracted spawning nature of the oil sardine and the extension of spawning period from June to November-December is subject to slight variations, depend upon the hydrological and ecological conditions. The modes at 47, 67, 77 and 82 mm in August, 52 and 77 mm in September and 92 mm in October can be considered as those representing the products of the current year's spawning. The fishery is supported by juveniles belonging to more than one "age class". While the size at which 1-year old enter the fishery during the commencement of the season in July is about 100 mm, the modal-size of 0-year class supporting the fishery during the months July-October appears to range from 47 to 92 mm.

Studies on the resolution of size distributions of oil sardine of the Mangalore zone revealed four components, (1) 87.9 mm, (2) 124.4 mm, (3) 156.9 mm and (4) 172.7 mm which appears to be the average length of fish at ages six, twelve, eighteen and twenty-four months respectively. In most of the years, the first three components support the fishery.

Lesser sardines

Among the clupeoids, the lesser sardines, namely, *Sardinella fimbriata* and *Sardinella gibbosa* occupy an important position in the inshore fisheries at Waltair, Andhra Pradesh as they contribute 20 to 35 per cent of the total clupeoid catches.

Sardinella fimbriata

The fishing season for *Sardinella fimbriata* starts in October-November and terminates in May-June.

The important feature of *S. fimbriata* fishery is that it is mostly sustained by length range of 50-120 mm, larger specimens being rare. This would indicate that spawning grounds might perhaps be away from the inshore fishing grounds. Certain size groups of *S. fimbriata* suddenly disappear from the fishing grounds during some months and certain other size groups showed up in the catches indicate that different broods enter the population. This might be due to multiple recruitment throughout the season.

The length-frequency analysis of *S. fimbriata* show that the fish less than 100 mm enter the fishery during the beginning of the fishing season, that is November to February and larger fish measuring more than 120 mm dominate the catches during March–June.

Fish measuring less than 50 mm feed mainly on diatoms, such as *Coscinodiscus* sp. Fish of 51-100 mm group also feed on diatoms particularly during November-December. During other months they feed on copepods, larvae of bivalves and gastropods. *S. fimbriata* measuring between 101-150 mm prefer copepods as food, the other items are larval forms of bivalves and gastropods, mysids, megalopa and alima larvae. Similar components could also be seen in the stomachs of fishes of 151-200 mm size group. The diatom, *Coscinodiscus* sp. *Calanus, Paracalanus, Macrosetella, Acartia, Microsetella, Temora, Oithona, Centropages, Oncaea, Pseudodiaptomus* and *Corycaeus* form the food of the fish.

S.fimbriata show a predominance of females over males as 2:1 of female to male and fecundity varies from 17974 to 34545. In fishes of 138 to 184 mm, the fecundity is from 5500 to 41700.

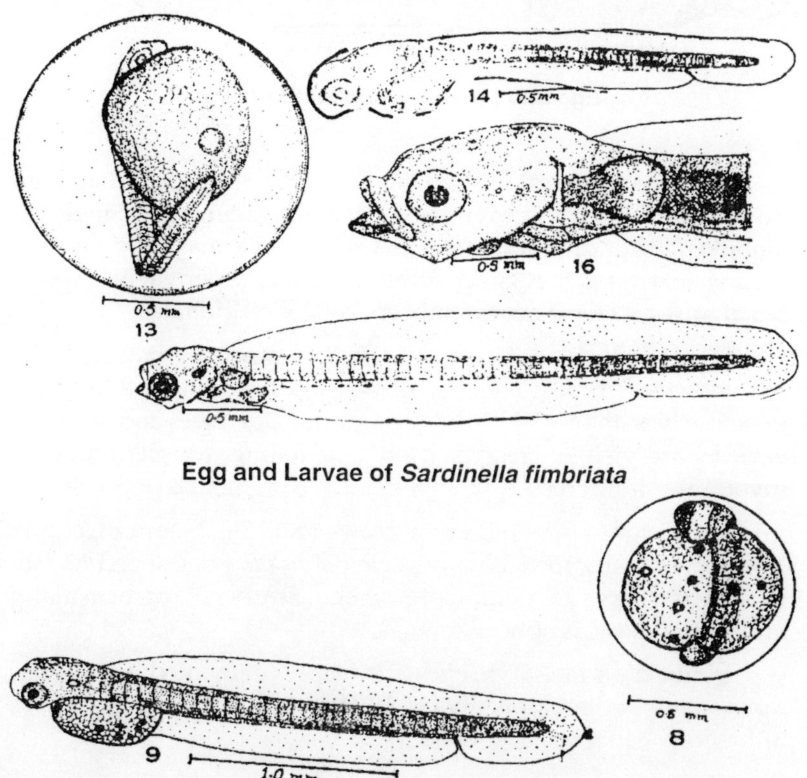

Egg and Larvae of *Sardinella fimbriata*

Egg and Larvae of *Kowala coval*

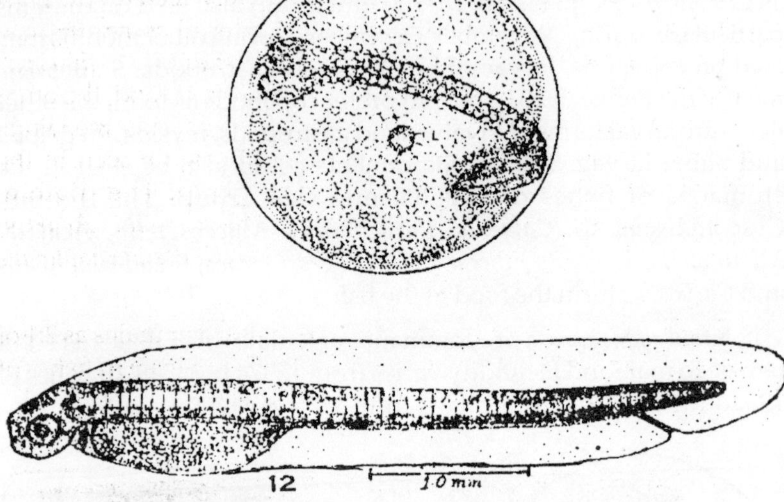

Egg and Larvae of *Dussumieria* sp.

Eggs are spherical in shape with 1.328-1.339 mm in diameter. Yolk colourless, transparent and coarsely vacuolated. Single oil globule, 0.099 mm in diameter is visible. Yolk mass and embryo occupy nearly half the egg capsule leaving a large perivitellir e space. No pigment is found on the embryo, yolk or oil globule.

Eighteen hours old larva is 3.08-3.14 mm long having 38-39 pre-anal and 7-8 post-anal myotomes, Yolk sac is found tapering posteriorly with oil globule on the anterior side. Eyes and auditory vesicles are visible. Heart started functioning. Muscle fibres on myotomes show crossed arrangement. No pigment is noticeable.

Forty eight hours old larva grows to 4.15-4.25 mm in length. Yolk is almost absorbed. Buds of pectoral fins have developed. Mouth has opened. Eyes are getting pigmented. Streaks of pigment along alimentary canal is noticed.

In five days old larva, eyes are deeply pigmented. Mouth has opened with developed jaws. Pigmentation along the entire alimentary canal has deepened.

Sardinella gibbosa

The fishing season of *Sardinella gibbosa* is shorter, from March to June-July. The important feature of *S.gibbosa* fishery is the dominance of adult fish in the population consisting of fish with mature and ripe gonads.

S. gibbosa fishery usually starts some times during the period February to March consisting of mature individuals measuring more than 120 mm in length and some times during October-December comprising immature fish measuring less than 100 mm in length.

Specimens of 50-100 mm size group mainly feed on copepods (*Paracalanns, Microsetella, Eucalanus, Oncae, Acartia* and *Undinula*) and other items, such as, alima larvae, larval forms of bivalves and gastropods. Fishes of 101-150 mm size feed on copepods, amphipods and diatoms as the main components of their food. Alima larvae and larvae of bivalves and gastropods which are included under other items are of secondary importance. In fishes of 151-200 mm size, besides copepods, bivalve and gastropod larvae, prawn larvae are encountered in the food.

In *S. gibbosa* the percentage of males is more than females in the population and the fecundity of the females varies from 12786 to 56378.

The maximum length recorded was 139 mm. The length-groups entering the fishery in April, May and June varied from 22 to 122 mm. The growth rate is about 55 mm in 8 months. The spawning season of *S. gibbosa* is from September to February and young fish appear in abundance during April and May. In *S. gibbosa* all fish below 85 mm were observed to be immature. The proportion of mature fish increases to 50 per cent in 92 mm group, 75 per cent in 97 mm group, 85 per cent in 102 mm group and 93 per cent in 107 mm group. Thus one-year-old fish are seen to mature in the case of *S. gibbosa*.

S. gibbosa fishery is confined mainly to the inshore areas of Palk Bay. Shore-seines, hand nets with torches and gill nets are the main types of gear employed to catch the species. The fishery operates mainly on 0-year class fishes, the proportion of older fishes in the populations is small. *S. gibbosa* spawn at about the end of the first year of their life. The annual replenishment of the 0-year class to the fishery is largely governed by hydrological and biological factors.

Indian sprat, *Sardinella gibbosa* feed mainly on copepods, Lucifer and diatoms. The fish changes from a predominantly crustacean diet to a mixed one of crustaceans and phyto plankton after attaining about 80 mm length. This is due to the effect of increasing number of gill rakers which attain almost the full complement as in adults in sizes above 70 mm. The bulk of the diet of *S. gibbosa* consists of crustaceans composed of zoea larvae, copepods and Lucifer. Though the blue green alga, *Trichodesmium erythraeum* has been reported as a favourite food item of this species the alga was found to be an occasional food item of this species. The juveniles of the species show a bottom feeding habit in the inshore regions.

Eggs have a diameter of 0.68-0.84 mm with frothy yellowish yolk and a small oil globule. The newly hatched larva has 29 pre-anal myotomes.

Sardinella albella

The maximum length recorded was 125 mm. The growth rate is about 40mm in 7 months. The rate of growth vary from year to year mainly due to changes in environmental conditions and density of population. The spawning season of the fish extends from February or March to June or even July. *S. albella* on the west coast spawns during May and June. Fishes less than 90 mm length is immature. Mature specimen is observed first in 90-95 mm size group of which 33 per cent were in stage III maturity or above. 55 per cent of fish is matured at 97 mm, 70 per cent at 102 mm and 87 per cent at 107 mm length, and the average size of 1-year old fish is 107 mm.

S. albella fishery is confined mainly to the inshore of water of Palk Bay. Shore–seines, hand–nets along with torches and gill nets are the main types of gear employed to catch the species. The fishery operates mainly on 0-year class fishes, the proportion of older fishes in the populations is small. *S. albella* spawn at about the end of the first year of their life. The annual replenishment of the 0-year class to the fishery is largely governed by hydrological and biological factors.

Sardinella melanura

Distributed in Arabian Sea, *Sardinella melanura* (Cuvier) has an elongate, oblong and moderately compressed body. Its upper jaw extend to the vertical through the anterior margin of pupil. The expanded portion of the second supra-maxillary bone more or less rounded ventrally but somewhat flattened dorsally. Gillrakers slender and is little longer than the longest gill filaments. Pseudobranch with out ventral ridge. Opeculum is narrow. Sub-operculum is rather rhombic. Minute teeth are present on jaws and middle of tongue. Eye covered with adipose tissue leaving a vertical slit at the pupil. Head bones covered with dermal sheath, without hollow venules under it.

Origin of dorsal fin is nearer to snout than to caudal base. Anal fin origin is nearer to caudal base than to pelvic fin base. Pectoral fins are nearly as long as head without snout. The dorsal, pectoral, ventral and anal fins bear 15-18, 14-15, 8 and 17-19 fin rays respectively.

The deciduous cycloid scales are found all over the body. Midlateral scales with one continuous transverse groove preceded by a set of two or five interrupted groves with a wide interspace between the discontinuous parts. Scale margin partly irregularly crenulated. Bases of dorsal and anal are covered by scale sheaths. Auxillary scales at the ventral are present. There is no pectoral axillary scale. An elongated enlarged scale on each caudal lobe near the fork is also present.

Belly of the fish is silvery but yellowish tinge present near the anal. Shoulder spot is pale dark. Pectorals, ventrals and anal are whitish. Minute dark spots are present on the dorsal rays. Tips of the caudal lobes are distinctly black.

White Sardine, *Kowala coval*

The white sardine, *Kowala coval* is one of the few economically important clupeoid fishes of the west coast of India, the fisheries of which assumed importance owing to the decline of the fishery of the valuable and much esteemed oil sardine, *Sardinella longiceps*. The highly fluctuating nature of the oil sardine fishery and its undependability, coupled with its failure during some years are the main factors responsible for the commercial exploitation of the fisheries of the white sardine and other sardines which were almost neglected earlier when the oil sardines occurred in plenty.

The white sardine supports an important fishery along the west coast and the fishing season extends approximately from July to March. Very heavy catches constituted mainly by fish ranging in length from 8 to 11 cm, frequently occurring during the September to November period of the season. The most important white sardine centres along the South Kanara and Malabar coasts are, Malpe, Kizhur, Puthiappa, Kadalundy, Parappanangadi, Tanur, Kootayi, Veliengode, Palapetty, Edakkazhiyur, Vadanapalli and Nathika.

The white sardine, *Kowala coval*, which is easily distinguished by its yellowish–white colour and silvery lateral band, is a widely–distributed species attaining a maximum size of 13 cm and is recorded from India, Pakistan, Srilanka, Mayanmar, Malaya, Malay Archipelago and China. It is a shoaling species, in–habiting the shallow coastal waters. Shoaling in vast numbers is known only from the west coast of India. Its occurrence in swarms have been even reported from the east coast. Species of *Leiognathus* and *Stolephorus* have frequently been observed to occur in large number in association with the shoals of white sardine.

The food of white sardine consists exclusively of plankton. The most favoured food of white sardine the Cladoceran, Evadne, which whenever it occurs in swarms in the foreshore waters has been noticed to constitute the bulk of the food of this fish. This tendency to feed on Evadne is particularly pronounced in the young. The penaeid larvae, larvae of other crustaceans, Lucifer and the common copepods of the genera *Acartia, Corycaeus, Temora, Oithona* and *Euterpina* are the other important zooplanktonic forms contributing to their food. Occasionally, fish eggs, mostly those of *Caranx* are consumed by the fish. Of the phytoplanktonic forms large numbers of *Thallassiothrix* occasionally constitute the main food. But other genera like, *Nitzschia, Fragilaria, Coscinodiscus, Chaetoceros* and *Rizosolenia* have also been observed to contribute regularly to the food of the white sardines.

It has been observed that intensive feeding takes place in the post-monsoon months, when there is rich production of plankton and this active feeding results in the accumulation of large quantities of fat with its characteristic sparkle in the abdominal cavity enveloping viscera. This fat is utilized during the spawning period and no trace of it is seen in the spawners and spent white sardines.

There is only one age group represented in the fishery. This year-class which dominates the fishery every successive year can only be the stock recruited during the previous spawning season. The rapid growth rate is evident from the progressive increase in the proportion of the different size-groups upto the 6.5 group after which a slight fall in next three groups is seen. The smaller fish usually enter fishery by May and a gradual increase in their number is seen upto August when the maximum number occurs in the fishing area. The complete disappearance from the foreshore waters in the succeeding months is mainly due to certain environmental factors caused by the southwest monsoon. The intensification of the monsoon in June and July results in the formation of "mud banks" on the foreshore areas, when the sea water becomes heavily laden with fine mud which gradually settles down after the cessation of the monsoon. This is followed immediately by the occurrence of swarms of *Noctiluca* imparting a red colour to the sea water (red water phenomenon) which is known to affect adversely the general fishery conditions. These factors particularly affect the post-larval and juvenile fishes which completely disappear from the foreshore waters during the August-October post monsoon period. Though the white sardine fishery commences immediately after monsoon, the juveniles take a much longer time to enter the foreshore waters. The average life span of the white sardine is about a year with only a relatively very small number surviving for a longer period. The new stock produced during the intensive spawning period extending from November to January apparently grows to a size of about 10 to 11 cms and dominates the best part of the white sardine season from September to November.

Sex differentiation takes place when the white sardines are small and the females become recognizable when they measure 7.5 cm in length, while the males could be distinguished only when they measure 8 cm. The females form a slightly higher percentage than the males in most of the size groups. The juveniles form about 19 per cent of the total of each sex, while the matured ones and spawners were found within a narrow range between 10 and 10.9 cm constituted about 61 and 64 per cent of the total males and females respectively. The higher groups were, represented by about 20 per cent in the case of males whereas an appreciable fall to about 17 per cent was seen with regard to females. The reduction in number of females is caused by a higher rate spawning mortality than in the males.

The complete atrophy of the right gonads is a characteristic feature of the white sardine. The left testis and ovary alone are functional and this is partly responsible for the very low fecundity. The estimated number of eggs in a mature female is about 8000.

Even though sex differentiation takes place at a very early size, the gonads attain full development only when the fish measure about 10 cm. In view of its short life span it is possible that each fish passes through the breeding season only once in its life time and the appearance of numerous small eggs in the ovaries of the spent and recovering individuals suggests the possibility of more than one spawning during the season.

The white sardine is a coastal breeder with the relatively shallow inshore waters as their breeding grounds. Spawning usually takes place at midnight or in the early hours of the morning. The eggs are pelagic, transparent and perfectly spherical with the diameter varying in the living condition from 0.7 to 0.8 mm. The yolk is transparent and spherical and shows distinctly the vacuolation characteristic of clupeoid eggs. Shining transparent spherical oil globules are present in the yolk mass and their number usually varies from 6 to 8. The oil globules show a tendency to crowd on the ventral periphery of the yolk mass.

Development of the fertilized egg is rapid and within six hours the embryo is well formed and occupies a little more than half the circumference of the yolk mass. The eyes, otocysts and the heart have formed, though the heart is not yet functional. Embryonic development is usually completed within 10 hours, when the egg is ready for hatching. Eight-hour old larva is 2.88 mm long. Head detached from yolk sac. Anus opens below the 30th myotome. Yolk sac is round with 4-6 oil globules at the posterior end. Auditory vesicles and eyes have appeared. Muscle fibres on the myotomes show a crossed arrangement. Melanophores are seen on the head and along the dorsal periphery of the myotome. The newly hatched larvae are very inactive and float on the surface of the water upside down due to the buoyancy of the yolk and its oil globules. They measure 1.5 mm in length. One day old larva are very active, the slightest disturbance making then dart from place to place. The usual stretching of the larvae has taken place and they measure 2.25 mm in length.

The white sardine fishery is confined to, narrow five to seven-miles strip of the coastal waters mainly due to the small size of the craft dug out canoe, commonly used by the fishermen along the coast. The most common net used for capture of these pelagic shoaling white sardines are gill net and boat seine.

The white sardine is mostly consumed by the local people in the fresh condition and during the glut period the excess is usually cured with salt in proportions ranging from 8:1 to 10:1.

Anchovies

The family Engraulidae (Anchovies) is represented by 5 genera in Indian oceans, namely *Thrissocles* (3 species), *Thrissina* (1 species), *Scutengraulis* (3 species), *Engraulis* (1 species) and *Stolephorus* Of all the five genera, the genus *Stolephorus* is of appreciable economic importance.

From the catch eight species of *Stolephorus* could be distinguished by counting branched and unbranched rays of the dorsal and anal fins. Pectoral fin ray and gill raker count are taken from the left side of the fish. The vertebral counts include the atlas and urostyle.

In *Stolephorus heterolobus* (Ruppell) anal fin originate under or behind the last dorsal ray. Muscular portion of isthmus is not reaching to hind border of branchiostegal membrane leaving portion of urohyal exposed. Alar scales are pigmented and readily discernible.

Maxilla tapering posteriorly, reaching at least to anterior border of preoperculum, lateral expansion of urohyal plate bony.

Total lower gill rakers 48 (43-53), head short and less deep; body sub-cylindrical and slender, posterior frontal fontanelles small; pre-operculum narrow, about half width of operculum; body depth 5.4-6.4 in S.L., maxilla with fine teeth of the same size, except 3 or 4 last ones in some juveniles.

In *Stolephorus purpureus*, maxilla 4-5 times in S.L., straight, just reaching to anterior border of pre–operculum. Lower gill rakers 23 24; pre-pelvic scutes poorly developed or may be absent; posterior frontal fontanelles not clearly discernible.

Stolephorus buccaneeri's maxilla 5-6 times in S.L., curved upwards, shorter than the lower lower jaw, not reaching to anterior border of pre-operculum. Lower gill rakers 23-27 in members, mostly 24-26.

Fishers with anal origin under dorsal base; muscular portion of isthmus extending forward beyond the hind border of branchiostegal membrane and hind border of pre–operculum indented near maxilla tip have been identified as *Stolephorus macrops.*

They have a deep body, depth equal to upper jaw, snout short and blunt and double pigment line on back behind dorsal bases. Posterior frontal fontanelles with lateral borders partly sigmoid; supra-orbitals projecting laterally from frontals and maxilla expanded above mandibular articulation. Pre-dorsal spine present but no spine on pelvic scute. Lower gill rakers are 20-27 in numbers.

In *Stolephorus holdon*, no double pigment line on the back, melanophores are absent or at most irregularly scattered. Their body is slender, its depth less than upper jaw; snout longer and pointed; posterior frontal fontanelles with lateral borders not sigmoid; supra-orbitals barely projecting laterally from frontals; maxilla not enlarged above mandibular articulation. Its head large and length 4.0-4.2 in S.L. 26-30 lower gill rakers; pre-pelvic scutes 6-9; posterior frontal fontanelles are broad anterior angles scute.

Head smaller in *Stolephorus andhraensis*, its length 4.2-4.5 in S.L. 19-21 lower gill rakers are present. Pre-pelvic scutes are 5-6 in members. Posterior frontal fontanelles narrower, anterior angles, more rounded.

In *Stolephorus tri*, hind border of pre-operculum evenly rounded near maxilla tip. Pre-dorsal spine present, a spine on pelvic scute in visible. Lower gill rakers count are 20-27. Posterior frontal fontanelles are very broad, their length is about one-third of eye diameter. Scales are adherent with reticulate striae.

Stolephorus indicus do not have any pre-dorsal spine and no spine on pelvic scute. Maxilla tip reaching to or just beyond anterior border of pre-operculum. Posterior frontal fontanelles are narrow, lateral borders straight with 4-5 pre-pelvic scutes.

Maxilla tips reaching to or beyond posterior pre-opercular in case of *Stolephorus commersonii*. Posterior frontal fontanelles are broad, lateral borders sigmoid. Supraorbital projecting laterally from frontals. Lower gill rakers more than 21 on whole of third arch.

Pre-pelvic scutes 3-5 (rarely 5 or more), head large, its length about 4 times in S.L. Two broad pigment lines on back from head to dorsal. Lower gill rakers 23-27; maxilla hardly expand beyond mandibular articulation.

Posterior frontal fontanelles are narrow in *Stolephorus bataviensis*. Lateral borders are straight. Supra-orbital not projecting from frontal; lower gill rakers not more than 23; less than 23 on whole of third arch.

Most species of *Stolephorus* are widely distributed through the Indian ocean. *Stolephorus indicus* being the largest of the species. All species of *Stolephorus* are caught throughout the year. The main catches of anchovies are made during the second half of the year (June to December), which is characterized by the southwest monsoon season. The observed predominant abundance of *S.commersonii* and *S.bataviensis* during the month from August to December has a strong bearing on the seasonal change in the anchovy fishery.

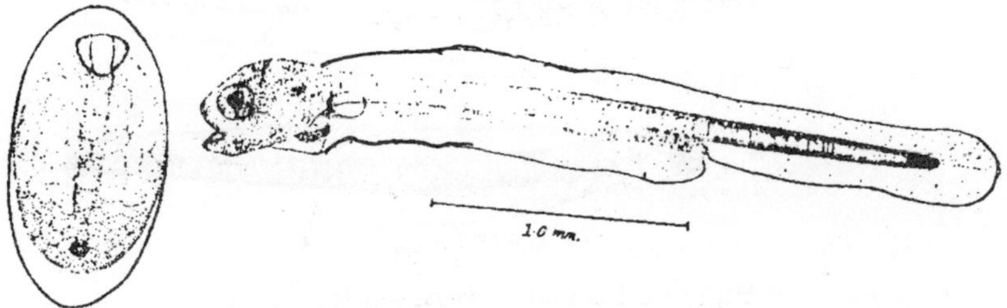

Egg and Larvae of *Anchoviella* sp.

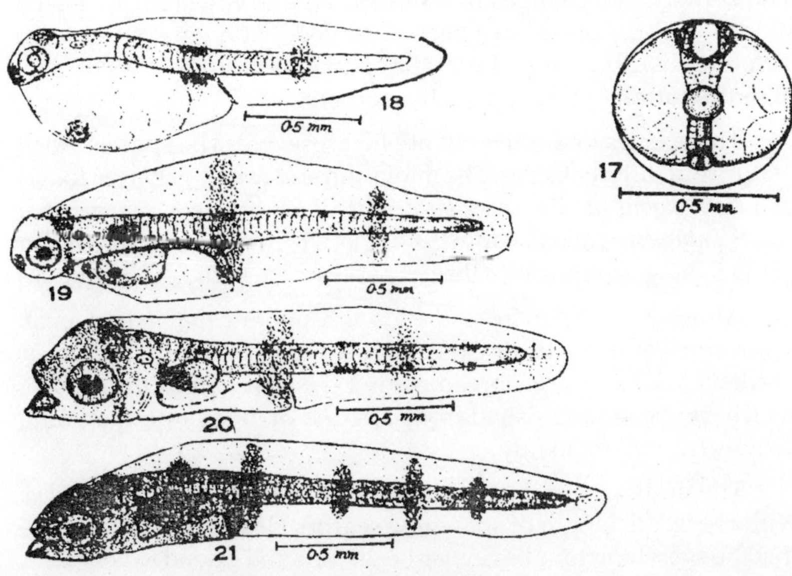

Egg and Larvae of *Carnax leptolepis*

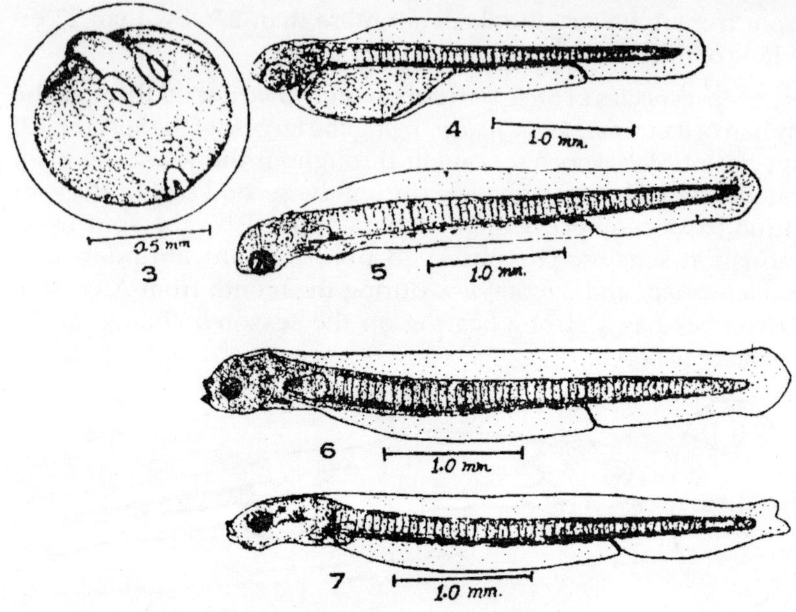

Egg and Larvae of *Thrissocles* sp.

Remarkable changes in the frequency of appearance of each species of *Stolephorus* have been observed. Different species have different distribution and that the special hydrological conditions of centres attract more species than others.

The species of *Stolephorus* are heterosexual. The species reach maturity at different sizes. The minimum size at maturity was found in *S. heterolobus* at 60 mm, *S. buccaneeri* at 65 mm, in *S. commersonii* and *S. bataviensis* at 65-70 mm. *S. indicus* which appears to move out of the fishing grounds into the deeper water to breed at 90 mm.

Although *S. heterolobus* breeds throughout the year, a peak spawning was noted during the northeast monsoon season (October to March) *S. heterolobus* breed during the first half of the season, which can be distinguished from a period of little or no spawning activity from April to July.

The findings in *S. buccaneeri* are similar to that of *S. heterolobus* with respect to length of spawning season. However it is probable that the spawning time begins earlier in June and extend to February. No spawning was observed in March, April and May.

S. indicus spawns over the longest part of the year. The species migrates into the deeper more saline water to breed.

Two type of eggs in the ovaries of different species of *Stolephorus* have been recognized. In the first type the chorion is with the knob at one end and in the other it is without knob. Pelagic *Stolephorus* eggs having oval or elliptical shape, without knob and with or without oil globules have been observed from the Indian seas. Bottle-shaped planktonic *Stolephorus* eggs with knob were collected from south west coast of India and also from surface plankton samples off Chennai at 40-45 m depth.

Eggs are elliptical in shape, ranging in length from 2.3-2.6 mm, including the knob, and in breadth from 0.53-0.72 mm. Yolk is colourless, segmented without oil globules. Chromatophores are absent. Embryo has been formed and yolk not absorbed.

Eggs of *Anchoviella (Stolephorus) bataviensis* is bottle shaped and the species forms the major component of *Stolephorus* catch in trawl net off Chennai waters. The specimens of *S. bataviensis* examined from the catches during the same period were in mature condition and the intra-ovarian eggs were also bottle-shaped like the pelagic eggs collected in plankton samples. The occurrence of mature eggs indicates the spawning of *Stolephorus bataviensis* in the off shore areas of Chennai.

The matured eggs of each species of *Stolephorus* can be distinguished by shape and size. While the eggs of *S.buccaneeri* and *S. heterolobus* are of oval shape, without a knob, the eggs of *S. indicus, S. commersonii* and *S. bataviensis* are characterized by the presence of a knob one pole of the oval eggs.

In *S. buccaneeri* the number of eggs varies from 7000 to 11000, in *S. commersonii* and *S. bataviensis* from 5000 to 10000 and in *S. indicus* from 9000 to 14000.

Species of *Stolephorus* spawn in the deeper part of the Bay. They have pelagic eggs. The planktonic egg catches by Hensen egg net indicate a peak of spawning period from June to August.

It appears that for most species of *Stolopherus,* the smallest fishes enter the fishery during the dry season, February to April and even until May which is the end of northeast monsoon, although some recruitment is noted also in September, at the end of the southwest monsoon. In *S. heterolobus,* the smallest fish were 18 mm; in *S.*

buccaneeri, S. commersonii and *S. bataviensis* 38 mm and in *S. indicus* *66mm*. The finding is directly related to the distance of the main spawning centres of the different species from the fishing grounds.

The largest specimen of *S. heterolobus*, 98 mm; of *S. buccaneeri*, 13C mm; of *S. commersonii* and *S. bataviensis*, 108 mm and of *S. indicus*, 146 mm have been recorded in the Indian oceans.

Stolephorus heterolobus grow to 33 mm at the end of its first year, about 60 mm at the end of second year and did not get much older than 3 years. Largest specimen found measured 98 mm. The species most likely spawn in the second year of life and may probably spawn at least twice during its life.

Fishes younger than one year are only caught occasionally in inshore areas during the dry months of February to April and fishes of less than one to one and half years old leave the inshore areas during the rainy season when the salinity decreases below 29 ppt. Younger fishes return to the inshore areas after the rainy season. There are very numerous discrete populations of varying age groups available to the fishery throughout the year. Although the fishery is seasonal due to the prevailing monsoon, unrelated populations are available to the fishery.

Thryssa mystax (Block and Schneider)

Anchovies make a significant contribution to the clupeiod landings on the northern Arabian Sea coasts throughout the year. *Thryssa mystax* and *T. dussumieri* are the common species of anchovies found in the Arabian sea.

The eggs of *T. mystax* occur in all seasons of the year and are common in December-March. Its spawning seasons continue for 8 months from September to May, but its maximum spawning is limited to November-March. *T. mystax* have two spawning seasons a year, each fish spawning only once in a season.

Different species of *Thryssa* apparently mature at different sizes. The average size at first maturity is 146-155 mm in *T. hamiltonii*, 170 mm in *T. purava*, 140-150 mm in *T. mystax*, 117 mm in *T. baelama* and 125 mm in *T. dussumieri*. In case of *T. mystax* about 52 per cent males and 47 per cent females are matured at 140 mm length. The maturation peaks at 140 mm in gonado-somatic index, relative condition and percentage occurrence of maturity at different size groups confirm the attainment of minimum size for maturation and breeding is at 140 mm.

Fecundity in the genus *Trissocles* (*Thryssa*) shows variations. *T. purava* produces 5342-23878 eggs at 170-237 mm; *T. hamiltonii*, 12495-23060 eggs at 150-173 mm; T. *baelama* 1171-3556 eggs at 110-127 mm; and *T. dussumieri* 1585-7943 eggs at 106-139 mm. The fecundity of *T. mystax* is 3581-24180 at 130-205 mm. The fecundity of *T. mystax* of the same length, fish body and gonad weight, varies considerably, suggesting that the ova are shed in batches.

The female-male ratio of *T. mystax* is found to be 1.26:1.

Eggs are spherical 0.896-1.248 mm diameter, without oil globule. Size of eggs ranged between 0.929 and 1.162 mm. Egg membrane thin without markings. Yolk is colourless and coarsely vacuolated. Developing embryo with yolk is 0.747-0.830 mm. Large perivitelline space is present. No pigment is visible on yolk or embryo.

Newly hatched larva measures 3.35 mm long. Anus is located below 31st myotome. Ten post-anal myotomes are visible. Muscle fibres show a crossed arrangement. Yolk sac is tapering posteriorly. Eyes developed with a few melanophores at the posterior end. Heart start functioning. Auditory capsules are visible. Larva is transparent.

First day larva is 3.75 mm long. Eyes brownish in colour. Buds of pectoral fins have appeared. Few melanophores along the ventral side of the alimentary canal and yolk sac is seen. On the second day the larva grows to 4.182 mm in length and become deep brown.

On the third day eyes turned completely black. Yolk completely absorbed and mouth opened. Pectoral fins have developed. Anus is located below the 30th myotome, 14 post-anal myotomes developed. Pigmentation along the entire length of the alimentary canal has increased.

On the fourth day pigmentation has increased further. Striations of the future caudal fin rays become distinct.

Bifurcation of the caudal fin take place on the fifth day. Striations of pectoral fin rays have appeared.

During February-March specimens of *Anchoviella indicus* with mature ova are common in plankton collections.

Ellipsoidal eggs 1.112-1.361 mm diameters, with transparent colourless yolk found to be coarsely vacuolated. Single transparent, colourless oil globule ranging from 0.099-0.116 mm is present. Perivitelline space is seen at the two poles. No pigment on embryo, yolk or oil globule has been observed.

Newly hatched larva of *Anchoviella indicus* is 2.20 mm long. Yolk tapers posteriorly with the oil globule near posterior extremity of yolk sac. Its eyes are not pigmented. Anus opens below 27th myotome. Auditory vesicles appeared and heart found to functioning. Muscle fibres on body show characteristic crossed arrangement, typical of clupeid larvae.

Two day old larva is 3.00 mm long. Position of anus not changed; 12-13 post-anal myotomes develop. Eyes become pigmented black. Mouth found open. Buds of pectoral fins are visible. Body found unpigmented.

The anchovy, *Satipinna taty* (Valenciennes) occupy a major place in the commercial catch of Hoogly estuary. The fishery of *S. taty* is mainly confined to the lower reaches of the Hoogly estuarine system, which lies in lower Sunderbans. It is particularly increased in magnitude in winter months, when maturing and matured individuals are landed from this region.

The fish attain maturity at 135 mm total length. Their fecundity varies between 805 to 3668 in fishes of 135 to 171 mm total length. The commercial catches of the species is dominated by the above size group.

Chapter 9

Indian Mackerel, *Rastrelliger kanagurta* (Cuv)

The genus *Rastrelliger* have four species dwelling in Indian Oceans. They are *Rastrelliger faughni*, recognized by short gill rakers, 21-25 present on lower part of first gill arch; *Rastrelliger brachysoma* identified by long gill rakers with more than 30 on lower part of first gill arch, head length distinctly shorter than body depth. Body depth 28.5 to about 34.0 per cent in fork length, depth ratio about 2.7-3.2 without dark stripes or regular rows of dusky spots along upper half of the body. In *Rastrelliger kanagurta*, however, head length is distinctly greater than body depth. Dark stripes or rows of dusky spots are present along upper half of the body. Body depth is 23-27 per cent in fork length and depth ratio about 3.1-4.0.

Indian mackerel inhabits both Tenassarim and Arakan coastal waters of Mayanmar. Two species of Indian mackerel, namely, *Rastrelliger kangurta* and *Rastrelliger neglectus* have been met in Mayanmar waters.

Indian mackerels are caught in Mayanmar waters by purse-seine net. The fishing operations are carried out only during moonless nights. The schools of the mackerel are located by their fluorescence.

The gizzard shad, *Anadontostoma chacunda* and fringe-scale sardine *Sardinella fimbriata* are caught sometimes by purse seines together with the Indian mackerel.

Sometimes smaller specimens of *Rastrelliger neglectus* come close to the shore and they are caught by lift nets in May and June. The species is also caught occasionally by drift nets with large mesh of 87 mm in November and March.

Most of the landings consisted only of one *Rastrelliger* species but part of them included both the species. Generally *R. neglectus* predominated in the purse-seine catches.

The distribution of species of *Rastrelliger* in Mayanmar waters confirms once again the conclusion that *R. kangurta* is apparently an open sea form and *R. neglectus* is a coastal or inshore form.

R. neglectus from Mayanmar waters has a bigger length than the fishes from Thailand. But *R. kanagurta* from Mayanmar waters are smaller in size than the specimen from Thailand.

In Mayanmar waters, the increase of mackerel size occur because of the growth of the fishes, but the decrease is caused by the appearance in the fishing area of smaller fishes evidently belonging to the new year class.

Among *R. kanagurta*, females constituted 58.2 per cent and males 41.8 per cent. The sex ratio was 1.4:1. But among *R. neglectus* the predominance of females were more pronounced as they constituted 62.4 per cent whereas males only 37.6 per cent, the sex ratio being 1.7:1. In Indian waters, the sex ratio of *Rastrelliger* spp. is 1:1 But in Mayanmar waters females predominate among both species. The same phenomenon is observed in Gulf of Thailand.

R. kanagurta have quite similar size compositions both of females and males, but in case of *R. neglectus* females are bigger than males. *R. neglectus* of same length size classes has higher average weights than *R. kanagurta* and this phenomenon is quite natural as *R. neglectus* has a higher body.

R. neglectus caught in purse-seines are of 14.1 to 25.0 cm in length with the predominance (79.4 per cent) of 18.1 to 21.0 cm size group. Fishes with 21.1-23.0 cm in length accounts for only 5 per cent. With drift nets a little bigger size of 19.8 to 20.21cm *R. neglectus* are caught. The smallest specimens (11.1 to 16.0 cm) of *R. neglectus* are caught with lift nets in Mayanmar coasts.

R. kanagurta caught with trawls are smaller than those caught with purse-seines.

Indian mackerel, *Rastrelliger kanagurta* (Cuvier), widespread in the Indo-West Pacific from South Africa, Seychelles and Red Sea east through Indonesia and off northern Australia to Melanesia, Micronesia, Samoa, China and the Ryukyu Islands. The species has entered the eastern Mediterranean Sea through the Suez canal.

The fish has elongated slightly compressed body with pointed snout. Its front and hind margins of eye is covered with an adipose eyelid. Its teeth in upper and lower jaws are small and conical, and teeth is absent from vomer and palatine bones, on roof of mouth. Body of the fish is moderately deep and its depth at margin of gill cover is 4.3 to 5.2 times of length. Its head is longer than depth of the body. Maxilla is partly concealed, covered by the lacrimal bone, but extending to about hind margin of eye. Its gillrakers are very long, visible when mouth is opened, 30 to 46 on lower limb of first arch. A moderate number of bristles are found on longest gillraker, 105 on one side in specimens of 12.7 cm, 140 in specimens of 16 cm, and 160 in specimens of 19 cm length. Two widely seperated dorsal fins, interspace atleast equal to length of first dorsal fin base, have 8 to 11 spines in the first dorsal. The second dorsal and anal fins have 12 rays. Anal spine is rudimentary. 5 dorsal and 5 anal finlets are present. Interpelvic process is small and single. Pectoral fins are short with 19 or 20 rays. Scales behind head and around pectoral fins are larger and more conspicuous than those covering rest of the body, but do not have well developed corselet. Two small keels on each side of caudal peduncle are present, but no central keel between them. Swimbladder is present. A total of 31 vertebrae is present, the first interhaemal bone anterior to haemal spine of 14th vertebra has been observed. Last branchiostegal ray form wide plate. Intestine of the fish is 1.4 to 1.8 times of body length. Back of the body is blue green in colour with narrow dark longitudinal bands on upper part of the body (golden in fresh specimens) and a black spot on body near lower margin of pectoral fin. Dorsal fin are yellowish with black tips, caudal and pectoral fins are also yellowish, but other fins are dusky.

Preopercular spines are absent in the larvae of Indian mackerel. Head and mouth are relatively small. Dorsal profile of head of the larva is rounded. Its snout blunt, black pigment dots appear along ventral edge of tail. The larvae develop 31 myomeres. Its abdominal cavity is elongated and ovoid. Anus located at or slightly posterior to middle of the body. Second dorsal fin develops before the first dorsal fin. Juveniles have a gap between the first and second dorsal fins.

Indian mackerel an epipelagic neritic species occur in areas where surface water temperatures are at least 17°C. Schooling of the species occurs by size. The spawning season around India seems to extend from March through September. Spawning take place in several batches.

Juveniles of the species feed on phytoplankton, diatoms, and small zooplankton such as cladocerans, ostracods, larval polychaetes etc. With growth they change their dietary habits, a process that is reflected in the relative shortening of their intestine. So adult Indian mackerel prey primarily on macroplankton, such as, larval shrimps and fish. Their longevity is believed to be at least four years.

The species attain first maturity at about 23 cm length. They grow to a maximum length of 35 cm, but common size is 25 cm.

Indian mackerel is a very important species in many parts of its range. The world catch of *R. kanagurta* alone fluctuated between 96000 metric tons (1975) and 128000 metric tons (1981) with a peak of 186000 metric tons in 1978. Indonesia, Thailand, India, Malaysia and the Philippines reported most of the landings. Indian mackerel is caught with purse seines, encircling gill nets, lift nets and bamboo stake traps and marketed fresh, frozen, canned, dry-salted and smoked.

The Indian mackerel, *Rastrelliger kanagurta* is a very important pelagic resource along the west coast of India, showing wide fluctuations from year to year. The yield per recruit indicates that mackerel population can withstand heavy fishing mortalities. The mackerel fishery in India experience wide annual fluctuations, the catch ranging from 21703 tonnes to 2,04,575 tonnes. Due to either heavy spawning mortality or due to some behavioural or ecological changes after spawning, there is quarterly decrease in recruitment and during some seasons the recruitment is not complete in the last quarter of the calendar year. Therefore fishing at very high intensities must be taken up with caution, especially because the age of mackerel at first spawning is higher than the age at entry into the exploited phase. It may affect the recruitment. The present age at exploitation is 6 months in the Mangalore area. In the mackerel fishery along the west coast the age at entry into the exploited is low in the southern centres and it increases as the fishery extends towards north. Increasing the age at entry into the exploited phase will result in limiting the mackerel fishery to the northern centres only and restricting the non-selective gears to certain periods.

The mackerel season in the Mangalore area usually extends from August to April. The relative abundance of juveniles and fluctuations in monsoon intensity seem to have great bearing on the fishery. There appears to be a direct relation between the amount of rainfall in a year and the total catch of mackerel in the following season. In 1970 there was a heavy rainfall and 1970-71 was the best season for mackerel. In the subsequent years the amount of rainfall gradually decreased and a downward trend in mackerel catches was also noticed.

The mackerel has a prolonged spawning season. Mackerel in advanced stages of maturity occurred almost throughout the year except during October to February. Their peak spawning periods are around July and January. During February more than 95 per cent of the catches comprised of spent fishes. Observations show that the mackerel measuring less than 200 mm length are immature. Individuals in the 200-209 mm group are mostly immature and maturing. Fishes measuring more than 210 mm comprised of all stages of maturity.

The normal modal size of one-year-old mackerel appears to be 12-15 cm and the fish measure 21-23 cm at the end of second year. The length of mackerel at the end of first year, second year, third year and fourth year of their life is found to be 11-15, 21-24, 25-27 and 29-28 cm respectively.

Juveniles of the Indian mackerel, *Rastrelliger kanagurta* of 86-142 mm size weighing 5 to 22 g were recorded for the first time on 24-12-1975 from Bombay waters. Their occurrence off Bombay (Lat. 18°-19°N and Long 72°-73°E) appears to be isolated. The food of these young mackerel consisted mainly of planktonic organisms, like *Coscinodiscus* and copepods.

Sarda orientalis (Temminck and Schiegel)

Striped bonito, Sarda orientalis (Temminck and Schiegel), are wide spread in tropical and sub-tropical waters of the Indo-Pacific region. In the western Pacific, it occurs northward to the northern end of

Honshu, Japan at 41°N and is rare in the Indo-Australian Archipelago, but is found in north western and south western Australia. There are recent records from off the west coast of Sumatra, south of Java and near Bali, from Ambon (Indonesia) and the Gulf of Papua. Further east, it occurs around the Hawaiian Islands and along the Pacific coast of America to Cabo San Lucas at the southern tip of Baja California and the Tres Marias Islands southward to the Galapagos Island and Gulf of Guayaquil.

The species is characterized by its elongate and slightly compressed body. Mouth is moderately large, 12 to 20 large conical teeth in the upper jaw and 10 to 17 teeth on the lower jaw. Vomerine teeth is absent and supra maxilla are narrow. 8 to 13 gillrakers are present on first arch. Laminae of olfactory rosette 21 to 39 and interorbital width is 21.3 to 30.2 per cent of head length. Dorsal fins are close together, the first dorsal with 17 to 19 spines. The length of first dorsal fin base is 28.2 to 32.7 per cent of body length. The second dorsal with 13 to 18 rays is followed by 8 finlets. There are 14 to 16 rays in anal fin. Anal finlets are usually 6. Pectoral fins are short with 23 to 26 fin rays. Interpelvic process is small and bifid. There is single lateral line gradually curving down toward caudal peduncle. Body completely covered with very small scales posterior to the corselet. Caudal peduncle is slender with a well developed lateral keel between two smaller keels on each side. A total of 44 or 45 vertebrae are present in the fish. Swimbladder is absent and the spleen is large and prominent in ventral view, located in the posterior half of the visceral cavity. Liver with elongate left and right lobes and a short middle lobe is present. Oblique stripes have been observed on dorsal side.

As an epipelagic, neritic species, they occur in waters of 13.5 to 23°C, schooling with small tunas. Off the south-west coast of India fully matured striped bonito are found from May to September, followed by juveniles from October to November. Off south and southwest Sri Lanka it occurs throughout the year, with mature fish prevailing between September and February. Juveniles are encountered off the west coast of Sri Lanka from June to August. Striped bonito prey upon clupeids and other fishes, cephalopods and decapod crustaceans.

In Indian Ocean, the species attain a maximum length of 101.6 cm, but the common size is 30 to 50 cm. In Japanese waters they grows to 80 cm in length weighing 3 kg. The all-tackle angling record is 10.65 kg fish with a length of 89.5 cm caught in Mahe, Seychelles in 1975.

Fisheries for striped bonito are not well developed in most parts of its range. In Japan it is caught often together with other scombrids by various types of gear including trolling lines, pole-and-line, purse seines and set nets. There is a minor trap fishery for the species in the Philippines and small-scale drift net operations around Sri Lanka and off southwestern India.

Larvae of striped bonito have weakly developed and rounded supraoccipital spine. Preopercular spines also are weakly developed. Supraorbital ridge is well developed and serrated with presence of spines. Snout is moderately long (snout length is one and half times of eye diameter). Its large mouth has well-developed teeth. Conspicuous black pigment on isthmus and just anterior to anus is noticed. Usually 6 to 10 large melanophores are embedded in skin on ventral edge of tail (may be as few as 4 in small specimens). Forebrain pigment is present in specimens larger than 3 mm standard length. Pelvic fins are pigmented in specimens larger than 5 mm standard length, with 42 to 46 myomeres.

Rastrelliger faughni (Matsui)

Known as Island mackerel, *Rastrelliger faughni* (Matsui) inhabit central part of the Indo-West Pacific from Taiwan Island, south through the Philippines and New Britain and east to Fiji, west through Indonesia, Thailand and Malaysia, to India at least as far as Tamil Nadu.

The fish has a slim body. Its depth at margin of gill cover is 4.9 to 6.0 per cent of body length. Head is longer than body depth. Maxilla covered by lacrimal bone extending only three-fourth of the length of the lacrimal. Gill rakers are shorter than snout. When mouth is opened wide, gillrakers do not extend far into the mouth. 21 to 26 rakers are present on lower limb of first gill arch. Few bristles are found on longest gill rakers, 30 to 55 on one side. Intestine is short, less or about equal to body length.

The fish belly is yellowish silver in colour with 2 to 6 large spots at the base of first dorsal fin, visible from above. Two faint stripes at level of lateral line are present in some specimens. A black blotch is present behind pectoral fin base. Outer margin of dorsal and pectoral fins are dark.

Island mackerel, like all species of the genus *Rastrelliger* is epipelagic, neritic, occuring in waters where surface temperature do not fall below 17°C. It feeds on the largest zooplankton organisms, thus complementing planktonic food spectrum of the other two *Rastrelliger* species.

The species attain a maximum size of at least 20 cm length and 0.75 kg in weight.

Seperate statistics are not reported for Island mackerel, but it is taken along with other species of *Rastrelliger* off Taiwan Island, the Philippines, Indonesia, and Malaysia. In the Philippines it is taken in commercial quantities by, bag nets and round seines.

Rastrelliger brachysoma (Bleeker)

Popularly known as short mackerel, *Rastrelliger brachysoma* (Bleeker) is distributed in central Indo-West Pacific from the Andaman Sea east to Thailand, Papua New Guines, Philippines, Solomon Islands, and Fiji.

An epipelagic, neritic species it tolerates slightly reduced salinities in estuarine habitats and occurs in areas where surface temperatures range between 20° and 30°C. It schools by size. Batch spawning is believed to extend from March through September. The short mackerel feeds chiefly on microplankton with a high phytoplankton component.

The species is known by its very deep body, the depth at margin of gill cover is 3.7 to 4.3 per cent of its length. Its head is equal to or less than body depth. Maxilla is covered by lacrimal bone but extending nearly to end of lacrimal. Gillrakers are very long, visible when mouth is opened, 30 to 48 are present on lower limb of first gill arch. Numerous bristles are visible on longest gill raker, about 150 on one side in specimens of 12.7 cm, 210 in specimens of 16 cm and 240 at 19 cm length. Intestine is very long, 3.2 to 3.6 times of length. Its spinous dorsal fin yellowish with a black edge, pectoral and pelvic fins dusky and other fins are yellowish.

The species attain its first maturity at about 16 cm length and attain a maximum length of 34.5 cm, but common size is 15 to 20 cm in length.

Catches of this speciess are usually recorded as *Rastrelliger* spp or are combined with *R. kanagurta.* It is the most important commercial species of mackerel in the Philippines, caught the years round with indegenous purse seines and fish corals in Manila Bay and by dynamiting. In India the species is fished with a variety of gear, such as, gill nets, seines and cast nets and drift nets operated from boats with out-triggers and from dugout canoes. The catch in the Philippines fluctuated between 25183 metric tons in 1978 and 18962 metric tons in 1981.

Gymnosarda unicolor (Ruppell)

Popularly known as dogtooth tuna, *Gymnosarda unicolor* (Ruppell), an epipelagic species, inhabits in tropical Indo-West Pacific from the Red Sea and east Africa east to Japan, the Philippines, Papua New Guines and Australia and out into the islands of Oceania, the Marquesas, Tahiti, Tuamotus, Pitcairn and Oeno Islands.

The fish with elongated and moderately compressed body have fairly large mouth, upper jaw reaching middle of the eye with 14 to 31 large conical teeth in upper jaw and 10 to 24 teeth in lower jaw. Two patches of teeth are seen on upper surface of tongue. 11 to 14 gillrakers are present on first gill arch. Laminae of olfactory rosette are 48 to 56 in numbers. Interorbital width is 32 to 40 per cent of head length. Dorsal fins are close together, the first dorsal with 13 to 15 spines, its margin almost straight. The second dorsal is followed by 6 or 7 finlets. The anal fin is with 12 or 13 rays followed by 6 finlets. Pectoral fins are with 25 to 28 rays and interpelvic process is large and single. Lateral line is strongly undulated. The fish body is naked, posterior to corselet except for the lateral line. Dorsal fin base, caudal keel and caudal peduncle are slender with a well developed lateral keel between two smaller keels on each side. Swimbladder is large and spleen is visible in ventral view on the right side of the body in the anterior half of the visceral cavity. Liver is present with elongated left and right lobes and a short middle lobe. A total of 38 vertebrae is present. The back and upper side of the fish is brilliant blue-black in colour with lower sides and belly silvery. No lines, spots and other markings are visible on the body. Anterior edge of first dorsal fin is dark, while other fins are greyish.

The species is encountered around coral reefs, at water temperatures ranging between 20° and 28°C. Dog-tooth tuna are generally solitary or occur in small schools of six or less individuals. They are voracious predators on small schooling fishes such as scads (*Decapterus*), *Caesio, Naso, Cirrhilabrus, Pterocaesio* and squids. In Fiji spawning occurs over the summer months, but little is known about their biology.

They attain a maximum size of about 150 cm in length weighing 80 kg in weight. The all-tackle angling record is 131 kg fish with a length of 206 cm caught at Kwan-Tall Island, Korea. At Ogosawara Islands sizes of fish commonly caught vary between 100 and 150 cm in length and 20 to 30 kg in weight. In Fiji the species are caught between 65 to 100 cm in length and 5 to 15 kg in weight. The fish attain maturity at about 65 cm length.

There are no fisheries directed specifically at dogtooth tuna. The species is regularly caught in small numbers during certain seasons at Port Blair, Andaman Islands, the Philippines, outside the reefs of Queensland and near the Ryukyu and Ogasawara Islands, Japan, near offshore reefs in Fiji, off. Tahiti, Western Samoa and the Marquesas. Hand lines, Pole-and-line fishing and surface trolling are the usual methods of capture. Initial high catches are usually not maintained.

A moderate ciguatoxic reaction was produced by feeding 6 to 13 large fish from Enewetok Atoll (55 to 135 cm length weight 3.2 to 35.4 kg) to mongooses.

Larval head of dogtooth tuna is large with very long snout (snout length is more than twice eye diameter). Well developed teeth is present and jaw tips bear canine teeth. Preopercular spines are well developed. Supraoccipital crest is absent, while supraorbital ridge is present. Body short with 39 myomeres. Triangular abdominal cavity is visible. Melanophores present on branchiostegal membrane, gill filaments and operculum. Pigment is absent from isthmus and anterior to anus. Black pigment spots are absent from tail.

Grammatorcynus bilineatus (Ruppell)

Double-lined mackerel, *Grammatorcynnus bilineatus* (Ruppell), an epipelagic species, is mostly found in shallow reef waters, where it forms large schools. The species inhabit in tropical and sub-tropical waters of the Indo-West Pacific from the Red Sea east to the Andaman

Sea, Java Sea, the Philippines, Ryukyu Islands, Irian Jaya, Papua New Guinea, northern coasts of Australia, Solomon Islands, New Caledonia, Caroline and Marshall Islands, and Fiji.

The species have elongated and slightly compressed body with a relatively small mouth, upper jaw reaching about to middle of eye. There are about 20 to 30 slender conical teeth in upper and lower jaws, patches of fine teeth on palatines and vomer and a rectangular patch of small, sharp teeth on tongue. Eyes are large, 7 to 9 per cent of fork length. There are many gill rakers, 19 to 24 on first gill arch. Doral fins are close together, the first with 11 to 13 spines, usually 12 and its margin is almost straight. The second dorsal with 10 to 12 rays followed by 6 or 7 finlets and anal fin with 11 to 13 rays followed 6 or 7 finlets are present. Pectoral fins are stout, with 22 to 26 rays and interpelvic process is short and single. There are two lateral lines, the first extending from the opercle to the lateral caudal keel, in the usual position for scombroid lateral lines. The second lateral line branches off from the first under the third spine of the first dorsal fin, descends below the level of the pectoral fin and runs posteriorly to join with the first lateral line at about the level of the last dorsal finlet. The body of the fish is covered with moderately small scales and no prominent anterior corselet is present. The caudal peduncle is slender with a well developed lateral keel between the two smaller ones on each side. The fish do have swimbladder and a total of 31 vertebrae as in the mackerel. The back and upper sides of fish is metalic blue-green in colour and belly silvery white with a golden tinge. There is no spot along the ventral surface of the body.

Reports on reproduction of *G. bilineatus* are available for the Andaman Sea, the Sulu Sea south of the Philippines and Fiji. The spawing season in Fiji extends from October through March. An ovary weighing 34.4 gram contained about 93000 eggs.

The food of the species includes adults and larvae of crustaceans, and fishes, particularly clupeoids like *Sardinella, Thrissocles,* but also other fishes, such as, *Sphyraena* and *Balistes.*

The fish attain a maximum size of about 60 cm in length weighing 3.5 kg. Maturity seems to be attained at a length of 40 to 43 cm.

Double-lined mackerel is taken incidentally with hand lines off Port Blair, Andaman Islands. It is common in the off shore zones of Fiji. The flesh is mild and pleasantly flavoured, but it is necessary to remove the kidney tissue before cooking to avoid the ammonia smell.

Dorsal profile of double-lined mackerel larvae's head are round, snout relatively blunt and supraorbital ridge are weakly developed. Their preopercular spines are also weakly developed. Thirty one to thirty two myomeres are present. Small specimens have a line of black pigment mid-laterally just anterior to the caudal peduncle. Large post-larvae and juveniles have large, saddle-shaped pigment blotches on dorsal side of the body.

Acanthocybium solandri (Cuvier)

Besides Indian Oceans, *Acanthocybium solandri* (Cuvier) inhabits the tropical and subtropical waters of the Atlantic, Pacific including the Caribbean and Mediterranean seas. The species, commonly known as Wahoo, have very elongated, fusiform body which is slightly compressed. Mouth is large with strong triangular, compressed and finely serrated teeth closely set in a single series. Their snout are as long as the rest of the head. Gillrakers are absent and posterior part of maxilla completely concealed under preorbital bone. Two dorsal fins, the first with 23 to 27 spines, the second with 12 to 16 rays followed by 8 or 9 finlets. The fish have iridescent bluish green colour on the back, sides silvery with 24 to 30 cobalt-blue vertical bars which extend to below lateral line, some doubled or y-shaped, becoming dusky grey after death. Their anal fin with 12 to 14 rays followed by 9 finlets have small and bifid interpelvic process. Single lateral line abruptly, curving downward under first doral fin have a total 62 to 64 vertebrae. Body is covered with small scales, without any anterior corselet, but with slender caudal peduncle and a well defined lateral keel between the two small ones on each side. The fishes have got the swimbladder.

The species roam in epipelagic regions of oceanic province frequently solitary or forming loose aggregations rather than compact schools. Spawing seems to extend over a long period; fish in differently maturity stages are frequently caught at the same time. Fecundity is quite high; some 6 million eggs per spawning were estimated for a 131 cm. long female.

Wahoo are known to prey on scombroids, porcupine fishes (Diodontidae), flying fishes (Exocoetidae), herrings and pilchards (Clupeidae), scads (Decapterus), lanternfishes (Myctophidae), other pelagic fishes and squids. They attain a maximum size at 210 cm and 83 kg or more. The size of fish in most surface fisheries ranges between 100 to 170 cm. Like other scombroids, wahoo shows size variations associated with changes in latitudes and hence water temperature Average weight tends to increase northwards and southwards of the equator.

There do not appear to be any organized fisheries for the wahoo, but it is highly appreciated when caught. In a number of areas, the species is primarily a game fish taken on light to heavy tackle, surface trolling with spoons, feather lures or strip bait like flying fish or half beak. There is a long line base in Samoa where the species is landed as by-catch and canned for local consumption. Catches were reported from Venezuela, St. Helena and Kiribati. Wahoo is marketed fresh, salted or spice-cured.

Giant digenetic trematodes, indentified as *Hirudinella ventricosa* were found in 80 per cent of the 885 stomachs of wahoo caught in the southeastern Atlantic and Gulf of Mexico.

The larvae of *Acanthocybium solandri* have very long slender body (62-64 myoneres). They develop long slender abdominal cavity. Anus appear posterior to middle of body. Preopercular spines are present. Very long jaws and large mouth (snout length more than 3 times of eye diameter) are formed. Well-developed pigment on snout is noticed. Peritoneum pigments developed. Second dorsal and anal fin bases have strong pigment patches directly opposite to one another.

Snake mackerel–*Gempylus serpans* (Cuvier)

Gempylus serpans, a bathypelagic species is circumtropical distribution. The species is known to occur in tropical and subtropical parts of the Atlantic and Pacific Oceans, East London, South Africa, Laccadives and in the Indian Oceans.

A male specimen of this species has been caught in the Gulf of Mannar at a depth of 12-15 m in drift gill net.

The body of the 37 mm long fish is very much elongated and compressed and weighs 520 gram. In fresh condition the entire body.

caudal fin, pectoral fin, pelvic fin and interspinous membrane of first dorsal, all jet black in colour. Lateral line is double, commencing from upper angle of the opercular region. The first lateral line run parallel to the base of the first dorsal and ending at 25th spine of the first dorsal fin. The second lateral line gently curves down above the base of pectoral fin, running midlateral and ending at the base of caudal peduncle. Upper jaw of the fish is shorter than the lower. Teeth are present in a single series on the sides of jaws. Three pairs of large compressed canines are present anteriorly in front of upper jaw. The upper jaw has 34 teeth (17+17); while the lower jaw has 36 teeth (18+18).

Bombay Duck *Harpodon nehereus*

Bombay Duck, *Harpodon nehereus* is semi-transparent like gelatine with minute black or brown dots along the head, back and sides. The abdomen is silvery white. The fins are transparent, but in some specimens there are black dots.

The length of fish head is one-sixth of the total length. Bombay Ducks are elongated and rather compressed fishes in which the head is thick and short and are provided with very short rounded snouts. The scales are small and transparent and can only be distinguished with great difficulty in fresh-fish. The cleft of the mouth is very wide. There are teeth in a band in both the jaws. The structural features, eg., the gelatinous consistency, large mouth and re-curved teeth on the lower jaw, indicate that the fish is a deep sea form, but it is not known definitely, whether it is a demersal or a pelagic fish.

The fish attains 40 cm in length. This is a shoaling fish swimming near the surface and is very common at Bombay, but decreases in numbers down the Malabar Coast. It is not common in the Coromandel Coast, but north of Tamil Nadu along the Andhra Coast it reappears in large numbers and finally is very abundant in the rivers and estuaries of Bengal. This migration has been shown to have nothing to do with reproduction of the species, but in some way connected with the marked changes in the physical conditions of the shore waters.

Bombay duck represented by a single species, *Harpodon nehereus* has a wide discontinuous distribution around Indian Ocean, along the coasts of East Africa, Indian subcontinent. Malaya, Indonesia and China. In India it is taken in large quantities on the south and

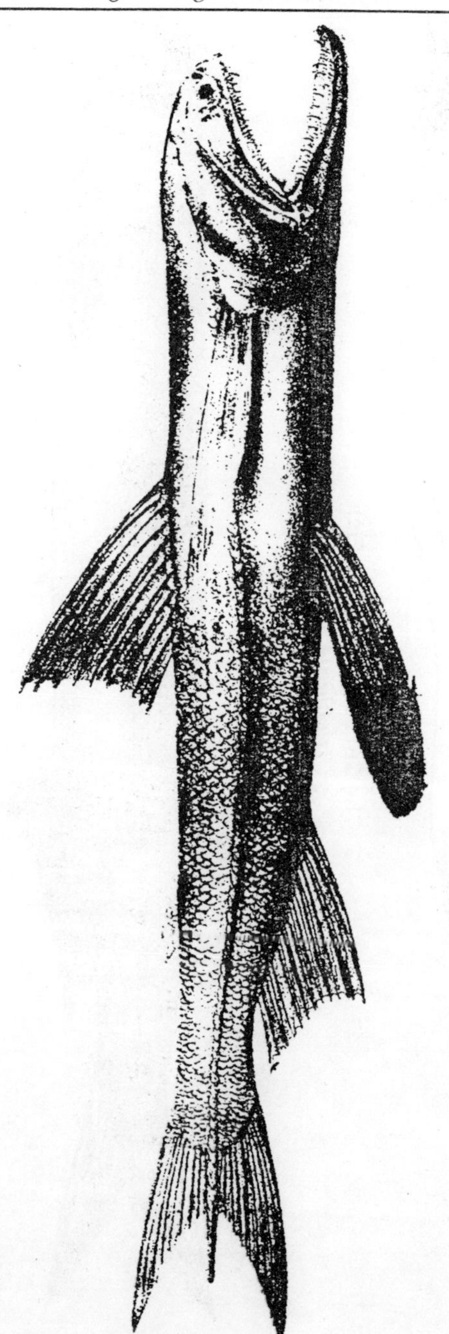

Bombay Duck *Harpodon nehereus*

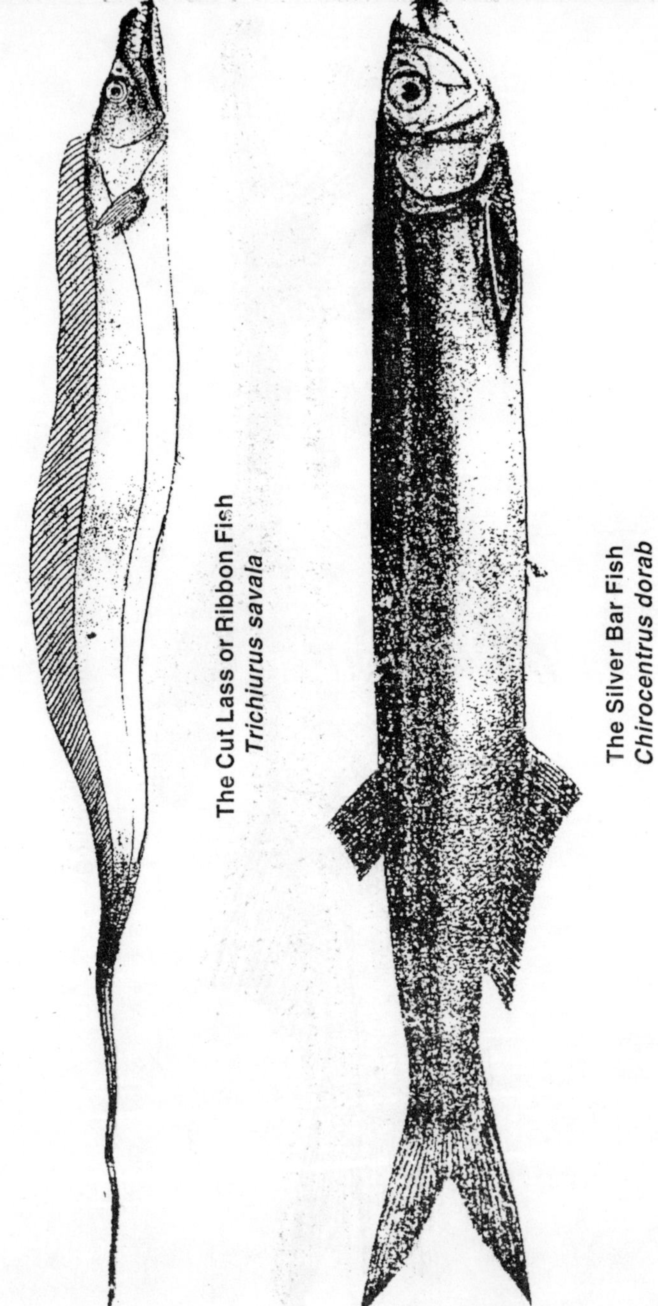

The Cut Lass or Ribbon Fish
Trichiurus savala

The Silver Bar Fish
Chirocentrus dorab

south-east coasts of Saurastra, on the Gujarat coast and Konkan coast of Maharashtra. It is also taken in appreciable quantities on the Andhra-Orissa coasts and from the estuaries of Bengal. The discontinuous distribution of Bombay duck along the coasts of India has been attributed to various factors, the principal ones being the distribution and movements of various food components, variation of salinity along the coast, the 70°F isotherm barrier of July etc. It is felt that the presence of low surface temperature in the area of occurrence is primarily responsible for peculiar distribution of the species.

Harpodon nehereus contributes to about 10 per cent of the estimated average annual marine landings of India. With a peculiar discontinuous distribution of the species, the fishery is of utmost importance in two maritime states of India, Gujarat and Maharashtra, where over 98 per cent of the all-India Bombay duck landings take place. In fact, from the bulk of the landings and the economic returns, the Bombay duck fishery in Gujarat and Maharashtra is equivalent to oil sardine and mackerel fisheries in Mysore and Kerala states.

Females of *H. nehereus* predominate the catches through out the year except in the months of July and August, when male predominates. In view of the continuous breeding habits, the recruitment to the stocks is also continuous and is not restricted to definite periods.

The fish appears to attain an average length of 127 mm at the end of the first year and 210 mm at the end of second year. The frequent appearance of the '0" year class shows that the recruitment is continuous and is more pronounced during the post monsoon period, that is, September to December.

The wanderings of the Bombay duck, which bring about fisheries of much regional importance along certain coasts seem to be influenced by two main factors, that is, the availability of food and the favourable condition of the water temperature. The low surface temperature in the areas of occurrence is probably responsible for the peculiar distribution of the species to a great extent than other factors.

In South and south east coast of Saurashtra, about 500 boats operate in the area during the fishing season, each boat carrying 7-

9 fishermen. They operate 'Dol' nets, 6-12 miles from the coast, in waters of 10-15 fathoms deep. The majority of fishermen operate two nets at a time from each boat, but some units operate three nets. The fishermen of Gujrat coast fish Bombay duck by gill nets along a stretch of 10 miles of coast line in inshore water during the months of June to September. Theses gill nets are 30 feet long and 3 feet high with a mesh of 1 inch from knot to knot.

Indian Lizard Fishes (*Saurida* spp.)

Of the four species of the genus *Saurida* namely, *S tumbil, S. undosquamis S. longimanus* and *S. gracilis* occurring in Indian seas, the first three species are available in the north-western part of the Bay of Bengal. *S. gracilis* are found in the south-west coast of Indian Ocean.

Fishes of genus *Saurida* have elongated and more or less rounded body with deciduous scales without photophores. Caudal region is sometimes depressed, with or without lateral keels. Snout obtusely pointed, rather short. Eye moderate with anterior and posterior adipose eye lids. Their head is depressed, naked above, and cheeks and opercular bones are scaled. Cleft of mouth is very wide more or less oblique bordered above by the long premaxillary with which the slender maxillary is connected. Teeth in jaws present in several rows, those of the innermost row is largest, slender, pointed and depressible. Similar teeth on the palate in a double band on each side, the inner part much shorter than the outer. Small teeth sometimes present on the head of the vomer; teeth is also present on the tongue and on the branchial arches. Gill openings are very wide, 13 to 16 branchiostegal rays, pseudobranchiae are well developed and gill rakers are rudimentary. Dorsal fin with 10-13 rays is situated nearly in the middle of length and the adipose fin is small situated above the short anal. Anal fin with 9-13 rays originate nearer to caudal base than to ventral base. Anal base is widely seperated from caudal. Pectoral fins with 11-16 rays are rather short, placed above the middle of height. Pelvic fins, 9-rayed is placed anterior, not far behind pectorals. The inner rays are not much longer than the outer ones. Pelvic bones is with short, laminar, posterior processes. Caudal fin is forked. Vent is situated posterior, nearer to base of caudal fin than to insertion of pelvics. Vertebrae are 45 to 62.

Saurida tumbil (Bloch)

The species characterized by elongated and more or less rounded body with deciduous scales, is distributed in east coast of Africa and Red Sea through the Indian Ocean and Archipelago to Australia, the coast of China, Japan and the Pacific. In India the species is available along both west and east coasts. They are common in offshore waters at 15-45 m depth.

Body of the fish is brownish above and silvery white below. Posterior part of caudal fin is blackish. Anal and pelvic fins are yellowish. Dusky brown spots may be present on the upper edge of the caudal and the front edge of dorsal fin. Fishes are caught in trawl nets at Waltair, Kakinada, and Tuticorin on the east coast and Mangalore and Bombay on the west coast.

The dorsal, pectoral, ventral and anal fins have 11-13, 14-16, 9 and 10-13 fin rays respectively. Its head is 3.77 to 4.19 and height is 5.25 to 7.58 part of standard length. Lateral line is covered with 53 to 58 scales. Caudal peduncle of the species is depressed with lateral keels. Teeth in jaws are slender, pointed, unequal, depressible and in several rows. Tongue of the fish is covered with fine teeth, gill rakers are rudimentary, numerous and arranged in 3 or 4 series.

Saurida undosquamis (Richardson)

The species is brownish or greyish above and lighter below. A series of 8 to 10 dark spots are visible along sides. The anterior edge of the dorsal and the upper edge of the caudal fins are with a row of dark dots.

The species is distributed in East African coast and Red Sea through the Indian Ocean and Archipelago to Australia, Japan and the Pacific. In India the species is available along both the west and east coasts. They occurs in offshore areas at a depth of 25 to 50 m. The species is found to be available in trawl catches off Puri, Waltair and Kakinada on the east coast.

The dorsal, pectoral, ventral and analfins have 11-13, 14-15, 9 and 10-13 fin rays respectively. Its head is 3.5 to 4.0 parts and height is 6.5 to 8.6 parts of standard length. Its eye diameter is 4.3 to 5.7 in length of head. Its scales are deciduous, 45-49 present in the lateral line. Caudal peduncle is depressed with lateral keels. Teeth in jaws are some what narrower than in *S. tumbil*, slender, pointed, unequal depressible and are arranged in several rows. Tongue is covered with fine teeth. Gill rakers are rudimentary, numerous and arranged in 3 or 4 series.

Saurida longimanus (Norman)

The species is distributed in Gulf of Oman and western side of Bay of Bengal. Dark brown above and silvery white below, the distal parts of dorsal, caudal and pectoral fins of the species are blackish. Sometimes faint traces of dark marks along upper edge of caudal fin may be present. In India the species are obtained from trawl catches in the east coast off Puri, False Point, Waltair and Kakinada from 75 to 100 m depth. *S. longimanus* was observed in the trawl catches during the upwelling period in the month of February to April along with other deep-water forms like *Psenes indicus, Nemipterus japonicus, Priacanthus* sp etc. This species has not so far been recorded from the west coast of India and also from the other regions of the Indo-Pacific.

The dorsal, pectoral, ventral and anal fins of the species bear 11-13, 14-16, 9 and 12-13 fin rays respectively. Its head length and body height is 3.4 to 3.9 and 6.5 to 8.3 parts of standard length. Caudal peduncle of the species is depressed with lateral keels. Teeth in jaws are slender, pointed, unequal, depressible and arranged in several rows. Its tongue is covered with fine teeth. Gill rakers are rudimentary, numerous and arranged in 3 or 4 series.

Saurida gracilis (Quoy and Gaimard)

Brownish above, silvery white below and a series of dark cross-bands or blotches are found along the middle of the sides of *Saurida gracilis*. Their fins are barred or with spots or patches. Black dots are present along the upper edges of caudal fin.

The species is distributed in East African coast and the Red Sea, through the Indian Ocean and Archipelagos to the Pacific. In India the species are available from south-west coast from the trawl catch at a depth of 100 fathoms off Cannanore (Lat. 12°04' N and Long. 74°27'E). The species was also collected from Port Blair, Andamans.

Head of the species is 3.17 to 3.45 and height 6.89 to 7.60 part of standard length. The dorsal, pectoral, ventral and anal fins of the species bear 10-11, 12-13, 9 and 10-11 fin rays. Teeth in jaws thin, pointed, unequal, depressible and in several rows. Outer bands of palatine teeth in 1 or 2 rows anteriorly, where the two bands are narrowly seperated. There is no teeth on vomer.

Caudal peduncle is depressed with lateral keels. Gill rakers are rudimentary and numerous.

Three species of *Saurida*, namely, *Saurida tumbil, Saurida undosquamis* and *Saurida longimanus* feed on organisms found on or near the bottom. They are mainly piscivores, the percentage of fishes in their stomach content being about 80. The other items of food are the crustaceans and the cephalopods. It has been reported that *S. tumbil* from the east China Sea and Yellow Sea feed mostly on fishes, clupeids and *Acropoma japonicum* is their favourite food item. 80 per cent of the food of *S. tumbil* from the Philippines sea consisted of fishes, the rest being composed of squids and prawns. Amongst fishes, *Stolephorus* spp, *Gobius* spp and *Leiognathus* spp occupied the first three places in the order of importance. A small amount of copepods, cirripede larvae, decapod larvae and sagitta were also found in the stomachs of adult fishes occassionally.

The feeding intensity of *S. tumbil* has been found to be low during spawing period, October to November. The intensity of feeding in *S. tumbil* is highest in the morning and thereafter gradually declined in the evening followed by slight rise in the night. The maturing fish has the highest feeding intensity.

Cannibalism is observed in all the size groups, but is more pronounced in the size group of 31 to 45 cm in case of *S. tumbil* and less than 16 cm in case of *S. undosquamis*.

All fish between 16 to 45 cm length, the general value of feeding index and feeding intensity gradually increase from February onwards reaching the peak in June or July and then drop to low level in August. The value shot up in September followed by a decline in October-November and another rise in December-January.

Besides the dominant items *Leiognathus bindus* and *Stolephorus* spp. other species of fishes, a considerable part of the food of *S. tumbil* shows that in the absence of these two items, *S. tumbil* feeds on other available fish not sparing even individuals of its own species. Cannibalism is observed frequently in *S. undosquamis* also.

Three populations of *Saurida tumbil* in Indian seas have been confined to the coasts of (*i*) Visakhapatnam-Kakinada, (*ii*) Bombay-

Mangalore and (*iii*) Tuticorin with reference to the analysis of morphometric characters, such as;

1. Total length-Distance from the tip of the snout to the tip of longest caudal ray of the upper lobe when the upper lobe is laid back parallel to the scale.
2. Length of the head-Distance from the tip of the snout to the outer edge of the operculum
3. Length of pectoral fin-Distance from the base of the pectoral fin to the tip of the longest ray.
4. Snout to origin of dorsal fin-Distance from the tip of the snout to the origin of the dorsal fin.
5. Snout to origin of adipose fin–Distance from the tip of the snout to the origin of adipose fin.
6. Tip of mandible to ventral fin–Distance from the tip of the mandible to the origin of ventral fin.
7. Tip of mandible to anal fin–Distance from the tip of mandible to the origin of anal fin
8. Height (Depth) of the body–Depth of the fish taken at the origin of dorsal fin.

It appears that:

(*a*) Visakhapatnam and Kakinada samples come from a single stock as they do not show significant difference (at 1 per cent level) in any of the six characters.

(*b*) Mangalore and Bombay samples are probably derived from a single stock or closely related stocks since significant differences at 1 per cent level between them are observed in respect of two characters only.

(*c*) Mangalore stock resembled more distant Visakhapatnam and Kakinada stock rather than the nearer Tuticorin stock.

(*d*) Bombay and Tuticorin stocks differ much from Visakhapatnam and Kakinada stocks.

(*e*) Bombay and Tuticorin stocks differ from each other to the maximum extent.

Analysing and considering the following meristic characters,

(*i*) Number of rays in the dorsal fin; (*ii*) Number of rays in the anal fin; (*iii*) Number of scales in the lateral line, (*iv*) Number of predorsal scales and (*v*) Total number of vertebrae including the urostyle, it can be seen that Mangalore and Bombay stocks do not show any significant differences between them in respect of all the five characters, there by suggesting that they are likely to be comming from a single stock. Visakhapatnam, Kakinada and Tuticorin samples differ from each other significantly at 1 per cent level in respect of two characters. Bombay as well as Mangalore samples differ significantly from those of Visakhapatnam, Kakinada and Tuticorin at 1 per cent level in respect of three or more characters. This suggests that the east coast samples are derived from stocks independent of the west-coast stock. At 5 per cent level, Tuticorin samples significantly differ from that of Kakinada as well as Mangalore in respect of three characters and, in view of the significant (1 per cent level) differences between Tuticorin and Visakhapatnam samples in two characters, it appears likely that Tuticorin sample may be comming from a seperate stock intermediate between the divergent west-coast and east-coast (Visakhapatnam-Kakinada) stocks.

The fact that no significant difference at 1 per cent level is observed between Visakhapatnam and Kakinada fishes (*S. tumbil*) in respect of any of the six morphometric characters and only one out of the five meristic counts shows that Visakhapatnam and Kakinada samples belong to one stock. Similarly, the fact that Bombay and Mangalore samples do not show any significant difference between them at 1 per cent level in any of the meristic characters and show significant difference in respect of only two morphometric characters indicates that Bombay and Mangalore fishes come from a single stock. The highly significant differences between Bombay and Visakhapatnam fishes in 5 out of 6 morphometric characters and in all the meristic characters substantiate the view that Visakhapatnam and Kakinada fishes on the east coast and Bombay and Mangalore fishes on the west coast belong to different stocks.

Tuticorin sample differ significantly from Bombay sample in all the morphometric characters and 4 out of 5 meristic characters. Tuticorin fishes differ significantly from Visakhapatnam and Kakinada fishes in 4 morphometric and 2 or 3 meristic characters. This shows that Tuticorin fish is probably derived from a seperate stock independent of those on the west-coast but closer to the latter.

The fish populations resemble each other more if the distribution is closer, and differ more if the distance becomes greater.

Ribbon Fishes, *Trichiurus haumela* (Foroskal)

Ribbon fishes of the genus *Trichiurus* occupy an important place among Indian marine fishes with respect to the magnitude of the fishery they support.

The family *Trichiuridae* consists only a single genus *Trichiurus* which can be easily identified by its elongated ribbon-shaped and

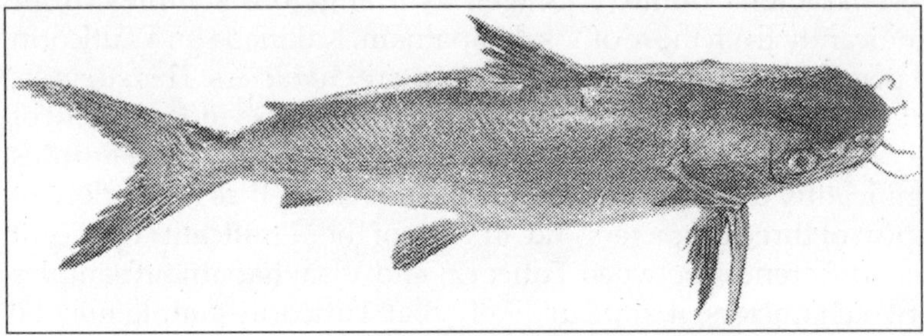

Arius maculatus

Arius thalassinus

Sphyraena jello

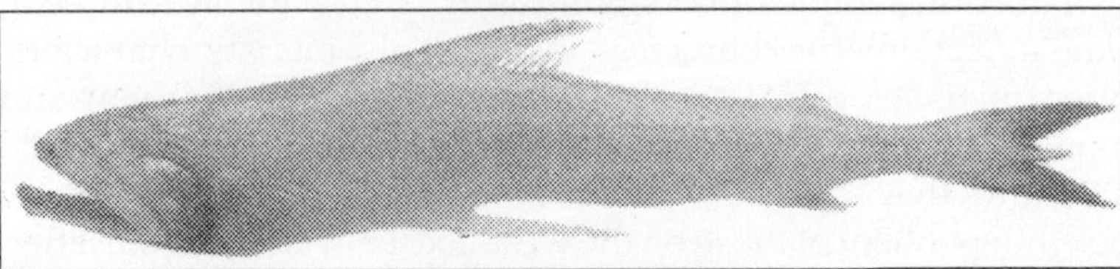

Harpadon nehereus

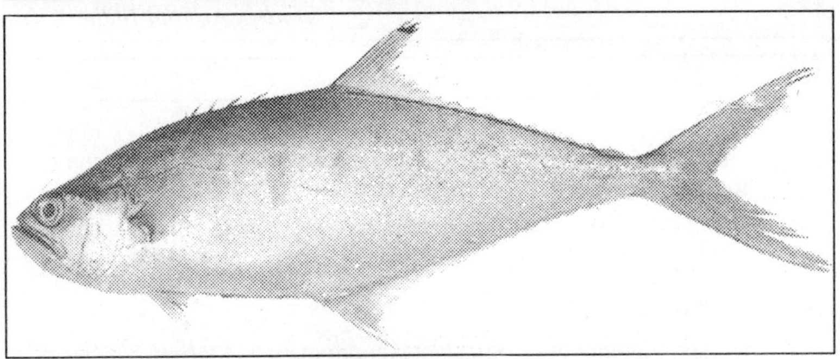

Scomberoides lysan

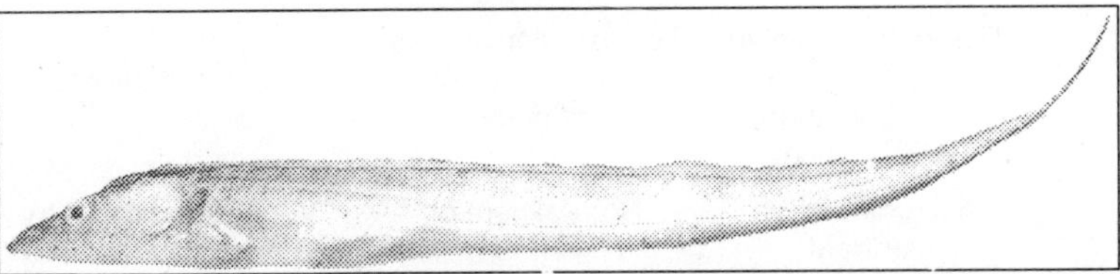

Lepturacanthus savala

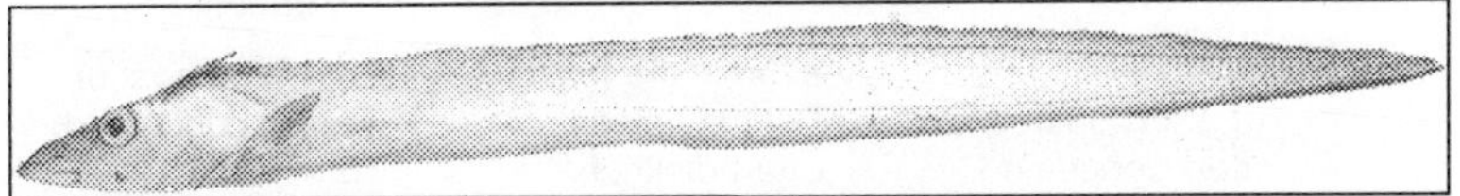

Trichiurus lepturus

laterally compressed body tapering towards the posterior end. There is no caudal fin. The dorsal fin extends from the hind edge of the preopercle along the whole of the back and is many-rayed. Ventrals are in the form of scales and anal spines are minute and sometimes concealed under the skin. The entire body is scaleless and is coated with a shining silvery substance. The cleft of the mouth is deep with teeth in jaws and palatines, those on the premaxillaries being arched and strong. The lateral teeth are lancet-shaped. Lateral line is just below the middle of the body and is very clear.

The genus *Trichiurus* is represented in the Indian seas by three species, *T. haumela, T. savala,* and *T. muticus.* The chief characters by which these three species can be identified in the field are;

Sl.No.	T. haumela	T. savala	T. muticus
1.	Lower jaw considerably longer than the upper jaw	Lower Jaw considerably longer than the upper jaw	Lower jaw more prolonged than the upper jaw
2.	Dorsal profile, between end of snout and eye, rather concave	Interorbital space nearly flat.	Inter orbital space with a keeled ridge
3.	Medium number of dorsal rays (127-33)	Smallest number of dorsal rays (112-20)	Largest number of dorsal.rays (140-50)
4.	The longest rays in the middle of the dorsal equal to height of body	Longest dorsal rays equal to height of body	Longest dorsal rays equal to only half height of body
5.	No rudiment of ventrals	No rudiment of ventrals	Ventrals indicated by two small rounded scale like projections on the lower surface of the abdomen
6.	Anal fin in the form of short spines often concealed or blunted at their extremities	Anal fin in the form of short spines, often distinct, especially the first which is twice as long as in *T. haumela*	Anal fin almost or entirely concealed under the skin
7.	Lateral line gradually descends until above the commencement of the anal fin where it is in the lower third of the body	Lateral line passes downwards to the lower third of the side	Lateral line almost straight and a little below the middle of the body especially in the last part of the course
8.	Upper half of the body rendered dark by numerous black dots. Fins pale yellow.	Silvery fins yellow-white	Burnished silver Fins yellowish

The ribbon fishes have a wide range of distribution throughout the warm seas. Besides being landed at several places along the eastern and western coasts of peninsular India, *T. haumela* has been known to be the most common species occuring in the Indian Ocean, around the Archipelago and in various parts of Pacific ocean.

It is believed that *T. haumela* of the Indo-Pacific region is identical with *T. lepturus* found in the tropical Atlantic, on the coast of Brazil, in the Gulf of California, the West Indies, the Gulf of Mexico and north of Woods Hole, Massachusetts. This species also occurs on the coast of Europe and also in Southern England where they occur abundantly, but they do not however enter the Mediterranean.

In the Indian waters the three species shoal seperately at different places. On the east coast *T. haumela* predominates in the catches at Tamil Nadu and in north Andhra, north of Visakhapatnam. *T. savala* has been reported to be more common on the coast of central Andhra, Visakhapattinam to Masulipattam; whereas in the south near Gulf of Mannar and Palk Bay, in addition to *T. haumela* and *T. savala* which occur in very small numbers, the third species *T. muticus* is also quite common. In the southern part of west coast consisting of Travancore-Cochin and Malabar and in the northern part comprising the Konkan, Bombay and Ratnagiri coasts, *T. haumela* and *T. savala* respectively are the common species caught.

The ovary in *T. haumela* is located just above the alimentary canal in the abdominal cavity and is bilobed. The two lobes are asymmetrical in length as well as in girth and they appear to be in the form of a U-shaped loop of which the left forms a part of one limb and is nearly two-thirds the length of the right one. During spawning season, the ovaries get distended and ova burst out from the follicles and lie in the lumen of the ovary, to be subsequently shed during the spawning season.

T. haumela around 42 cm have immature gonads, where testis is slender and transparent in males; while in females ovaries are also slender, transparent and eggs are not visible to the naked eye. Around 46 cm length, the gonads become more mature, when testis is slightly enlarged and opaque in case of males, while in females the ovary is slightly enlarged, eggs having granular appearance. In matured fishes of 48 cm length, the testis become greatly enlarged with pale white in colours. Ovaries in females also elarged greatly, with yellowish eggs visible to the naked eye. Average size of the largest eggs at this stage become 1.1 mm in diameter. The minimum size at which *T. haumela* mature is 47-48 cm in length.

The conspicuous absence of ribbon fishes with fully ripe gonads along the coastal waters and the reappearance of these fishes with spent gonads, suggest that they spawn in the off shore waters. So far neither developing eggs nor larvae of *T. haumela* have been collected from the coastal plankton samples suggest that the extreme rarity of the post-larval form in the inshore waters is due to their oceanic habitat. The occurrence of gravid specimens measuring 48 cm and above during the months of April and May and their reappearance in large numbers in spent condition in July, lead to the conclusion that the spawning of the species should have taken place towards the end of June and that the spawning is restricted to a definite and short period.

Young ones of this species measuring 7-9 cm in length have been observed in large numbers only once a year towards the end of July in coastal areas of Tamil Nadu and thus it is quite probable that the post-larvae of *T. haumela* grow to a size of 7 to 9 cm in length within a period of one and half month.

T. haumela grows an average length of 18 cm in the first year, 12 cm in the second year, 16 cm in the third year and 8 cm in the fourth year. The mature individuals of 52-56 cm size group and in all size groups ranging from 44-56 cm disappear from the coastal waters in the months of April and May respectively. Such an off-shore migration of mature individuals in April and May and their appearance in small numbers in June and in larger shoals in July in the inshore waters indicate that the mature individuals move away from the coastal waters for the purpose of spawning.

The production of eggs in *T. haumela* is the dominant function of the ovary and a close correlation is observed between the weight of ovary and the number of eggs produced. A 43–45 cm long matured *T. haumela* produce about 6000 to 6500 eggs; while 48 to 50 cm long fishes have 14900 to 15500 matured eggs.

T. haumela is a voracious carnivore and a selective feeder. Immature fishes of 17-45 cm length show a tendency to feed on prawns and small sized fishes, such as, *Stolephorus* spp and *Leognathus* spp. Adult forms (46-62 cm long) feed on fewer prawns and smaller fishes. The mature and spent fishes have been found to be voracious feeders, even resorting to cannibalism, where as individuals just prior to spawning show a definite slackness in feeding. The active feeding in *T. haumela* is evident soon after spawning in July and continues upto February.

The ribbon fish, *Trichiurus pantului* (Gupta) form a seasonal fishery in lower Sunderbans during winter months of October to

February. The species occupy a major place in commercial catch of the area.

The fish attain maturity at 425 mm total length. Their fecundity varies between 1672 to 9435 in fishes of 425 to 590 mm total length. The above size group dominate the commercial catch of the region.

Catfish (*Tachysurus thalassinus*)

The catfish forms a very important fishery along the east coast of India. Three species, namely, *Tachysurus thalassinus*, *T. tenuispinis* and *T. coelatus* contribute more than 20 per cent of the demersal catches by the trawlers. Along Andhra-Orissa coast, *T. thalassinus* alone contributes to 38.2 per cent of the catfish catch. *T. thalassinus* is found to attain an average length of 18 cm at the end of its first year, 35 cm at the end of second year and 42 cm at the end of third year.

The spawning season of *T. thalassinus* generally starts from around April each year and continues upto July/August. The resultant young ones start entering the fishery usually from November-December. It is this recruitment that is responsible for the well-marked first mode during the period. The growth of the '0'–year class could be estimated before the start of the next spawning season. The products of spawning of a particular fishery season grow to commercial size and enter the fishery of the very next season. The modal length of these recruits were between 18 to 21 cm. The minimum size group noticed was 16-17.9 cm while the maximum size was 22-23.9 cm. The difference in the growth rates of the '0' year class observed in the different years is seen also with respect to different year classes of the species. Again in the same year class there were more than one modal size group representing probably the products of the late spawning of the previous spawning season. Variations observed in the sizes of the fish constituting different year classes in different years may be on account of shifting of the spawning period and/or the post-larval recruitment.

Considering the composition of the different year classes annually, it is seen that the one-year-old fish contributes most to the fishery. In some years however, both 0-year and 1-year olds shared equally in the composition of the catch.

Catfish (*Tachysurus tenuispinis*)

Cat fishes form about 4 per cent of the annual demersal fish landings. *Tachysurus tenuispinis* alone forms 60.7 per cent of the total cat fish catches along the Andhra coast. Fecundity in *T. tenuispinis* of size range 28.5-42.4 cm varied from 29 to 82. It increased with increasing size. The linear relationship between weight and fecundity was more valid than that of length and fecundity. The fish breed once a year during May-September. The minimum size at first maturity was 27.5 cm total length.

Matured eggs of *T. tenuispinis* are large, attaining a size of about 13.5 mm diameter, with heavy yolk deposit. The matured ovaries contain ova of all stages of maturity.

In the state-I small, slender ovary occupy less than one third of the body cavity. Ova is not visible in naked eye, transparent with distinct nucleus and devoid of yolk deposition. They vary in size from 0.07 to 0.25 mm.

In the maturing II stage, ovary is slightly enlarged, but occupy less than one third of the body cavity. Majority of the ova are small, still transparent and not visible to the naked eye. Yolk deposition has commenced in larger ova at central semitransparent portion. Ova varies in size from 0.07 to 0.25 mm.

In the III stage ovary is further enlarged but still occupies less than one third of body cavity. Ovas are small, 0.25 to 0.60 mm in diameter, centre fully yolked and opaque, but periphery is transparent.

Enlarged ovary in the IV-stage occupy one third of the body cavity. Ovas are white and completely opaque and fully yolked, but still contained in the follicles. The size ranges from 0.6 to 3.5 mm.

Ovary become larger in the V-stage, yellowish in colour, occupying more than one third of the body cavity. Fully matured large, translucent ova burst free from the follicles, the diameter range being 4.0-10.5 mm.

In the VI-stage, free and large eggs with a clear transparent portion appearing at one side is ready for liberation. The size ranges from 10.5 to 13.5 mm.

The females dominated the commercial catch and the male : female ratio was 1 : 1.79.

Males of *T. tenuispinis* exhibit parental care. The fertilized eggs are kept in the buccal cavity for hatching. Some fertilized eggs were recorded from stomach also. These eggs might have been swallowed accidentally during the process of capturing the fish, as all the eggs recovered from the stomachs were found intact and fresh. Altogether 72 eggs (size range 10.0-13.5 mm) could be recorded from the mouth and stomachs during June to September from fish ranging 27.8 to 34.5 cm in length. The maximum number of eggs recorded from the mouth of fish of 27.8 cm size was 28. In all the eggs embryonic streak was clearly formed. In some eggs development had advanced further, although fins and eyes were yet to appear. The minimum size at which *T. tenuispinis* egg is fertilized is 10.0 mm.

Tachysurus tenuispinis is landed chiefly by trawl nets and bottom-set gill nets in north-east coast of India. A considerable quantity is caught by boat-seines and hook and line also. It has been found that the fishing mortality varied from 0.30 to 0.92 during the period from 1964 to 1976. Towards the latter part of the period (1974-76), the fishing mortality was higher (0.58 to 0.96) due to introduction of bottom-set gill nets and entry of higher number of small private trawlers, which has gone as high as 200 in numbers.

In the south west coast of India, most of the catfishes landed belonged to the family Tachysuridae and the important species identified as *Tachysurus thalassinus, T. dussumieri, T. sona, T. jella* and *Osteogeneosus militaris. T. thalassinus* and *T. dussumieri* among the catfishes are abundant in Kerala and make up an important group in commercial landings. *T. thalassinus,* the large catfish coming in the catch, were measuring 0.56 to 1.01 m in total length. The dominant size group of this species landed during December to January-March was 0.65 to 0.85 m. The species feeds on prawns, sciaenids, *Apogon* sp., carangids, leiognathids, crabs, squids and flat fishes in order of abundance.

Maximum landings of catfishes in the area were obtained in December accounting for 37.18 per cent of the total fish and the minimum landings were obtained in April forming only 22.16 per cent of total fish catch.

Flat Fishes

The annual average catch of the soles and other flat fishes in India is 10,045 tonnes forming 1.28 per cent of the marine fish landings. *P. erumei* occurs in some quantities in Bombay-Saurashatra and south eastern coast. Along other coasts it occurs in small quantities. At Porto Novo, the average annual catch was 8.6 tonnes.

Small mechanized boats operate trawl nets from February to June at a depth of 18-22 m depth upto a distance of 10 km from the shore. The fishery in general, forming a minor one, occurs throughout the year.

Psettodes erumei is a carnivorous fish subsisting mainly on fishes (81.73 per cent) crustaceans (11.93 per cent) and molluscs (6.34 per cent). Their spawning season is observed to be a prolonged one extending over a period of seven months from November to May as against the short and restricted spawning of this species in west coast of India. The animal being a bottom dweller is able to feed on both pelagic and demersal forms.

The fishery, in general, forming a minor one, occurs throughout the year. The first quarter register maximum and decline gradually from the second quanter onwards, with some improvement in the last quarter. The catches touched the lowest level during the third quarter. 41 to 78 mm size groups constituted the catch.

Eleven species of flat fishes (Cynoglossids) belonging to two genera, *Paraplagusia* and *Cynoglossus* were collected from commercial fishing trawlers in coastal waters of Indian Ocean.

Paraplagusia blochii (Bleeker)

Distributed in Orissa coast of India *Paraplagusia blochii* resembles closely to *Paraplaguisia bilineata*, but the dark conspicuous mark on the operculum in *P. blochii* is distinctly identifiable. The species occurs off Karachi and Pasni of Pakistan at a depth of 20 to 50 m.

The angle of the mouth of *P. blochii* is nearer to the gill opening than the snout. Lips are fringed. Snout is either obtusely pointed or obtuse. Two nasals are on either side of the body. Two lateral lines on the ocular side and a single indistinct lateral line on blind side is visible. Scales are ctenoid on the blind side. Caudal fin is acutely pointed and confluent with the dorsal and anal fins. The dorsal, anal, ventral fins have 99 to 105, 75-78 and 4 fin rays respectively. In adults eye diameter is equal to the inter-orbital width.

Adults are dark brown in colour with conspicuous dark mark on the operculum of the ocular side. The colour of younger fishes is light brown with distinct white spots spread all over the body. The marginal fins are dark at proximal end and light at the distal end.

Fishes of total length between 16 to 172 mm are available in inshore waters.

Paraplagusia bilineata (Bloch)

Lips of *Paraplagusia bilineata* are fringed. Angle of the mouth is nearer to gill opening. Rostral hook extend beyond the hinder borders of the eye. Snout is obtusely pointed, some times equal or little longer than head. Majority of fish have longer snout. Scales are ctenoid on both sides of the body. Two lateral lines are present on the ocular side. Single indistinct lateral line is found on the blind side. Left pelvic fin is united with anal fin.

Dorsal, anal and caudal fins have 104 to 108, 89 to 92 and 15 fin rays respectively.

Ocular side of the fish is dark brown in colour with pigments all over the body. 90 to 400 mm sized fish (Total length) are available at a depth of 20 to 50 metres. Besides Indian coasts, they also occur off Karachi, Pasni and Gawadar coast of Pakistan.

Cynoglossus puncticeps (Richardson)

Distributed in shallow waters at 10 to 20 metres depth off Bombay coast of Indian Ocean, the species have very small eyes with narrow interorbital space. Snout is rounded, rostral hook is small covering the anterior margin of mandibular symphysis. Their pelvic fin is short and seperate from anal fin. Dorsal, anal and caudal fins have 100 to 109, 79 to 84 and 10 fin rays respectively. Vent is situated on blind side. No distinct seperation of urinogenital papilla and the anus is visible. Lateral line is absent on the blind side.

Fresh specimens are brown in colour with vertical rows all over the body. Operculum with prominent spot. Vertical fins are with black stripes.

Fishes of 90 to 260 mm total length is seen in the population. Largest specimen of 175 mm total length were reported from Karachi waters.

Cynoglossus bilineatus (Lacepede)

The species occurs in Indian Ocean, usually at a depth of 30 to 50 metres. Their rostral hook is short, reaching the mandibular symphysis. Its upper eye is a little in advance of the lower. Two nostrils are far apart, the anterior tubular one is situated above maxilla and the posterior between the two eyes. Scales are ctenoid on ocular side and cycloid on the blind side. Snout is somewhat rounded. The species have three lateral lines on ocular side, while the third lateral line is present, but incomplete, a common phenomenon observed in cynoglossids. Their dorsal, anal, ventral and caudal fins bear 105 to 118, 81 to 91, 4 and 12 fin rays respectively. The population mainly consist of fishes of 190 to 214 mm total length.

Cynoglossus dispar (Day)

Available in Tamil Nadu coast of India, *Cynoglossus dispar* can easily be seperated from other species as its second lateral line terminates in the posterior one fourth of the body. The scales on lateral line is feebly ctenoid. The rostral hook of the species is little longer than the mandibular symphysis Two nostrils are present on the ocular side, the tubular one is anterior to the lower eye and the other in the interorbital space. Scales are strongly ctenoid on the outer margins and feebly ctenoid on the lateral line. Two lateral lines are present on the either side, the second lateral line ending in the posterior one-fourth of the body.

Their dorsal, anal and caudal fins have 106 to 114, 85 to 88 and 10 fin rays respectively.

The ocular side of the fish is dark brown in colour with black markings. Vertical fins are blackish or dark brown in colour. They occur in areas of depth between 20 to 40 metres. Their size range commonly consists of 20 to 280 mm in total length.

Cynoglossus semifaciatus (Day)

The species occur in Tamil Nadu and Malabar coast of India. They generally inhabit at a depth of 10 to 20 metres and their common size range is between 80 to 120 mm in Pakistan coastal waters.

The species can be identified by their dorsal, anal and caudal fins having 102 to 105, 70 to 82 and 12 fin rays respectively. The rectus of the mouth reaches posterior boundary of the eyes. Rostral hook is short touching the mandibular symphysis. Two nostrils are present on the ocular side close together. The anterior at the level of the maxilla, while the posterior lies between the eyes. Dorsal and anal fins of the fish are confluent with the caudal. Left pelvic fin is well developed and attached to the anal fin.

The ocular side of the fish is brown in colour with black pigments all over the body. The marginal fins are black. There is a prominent opercular mark. Vertical bands on ocular side is observed at times.

Cynoglossus borneensis (Bleeker)

The fish body is flat and pointed towards the tail. The posterior boundaries of maxilla extend to middle of the eye. The two nostrils on the ocular side are seperate, one tubular and the other simple. Scales are ctenoid on the ocular side and cycloid on the blind side. The interorbital depth is narrow and somewhat concave. The ocular side of the fish is with uniform brown in colour. Dark spots are present on the vertical fins. Fishes of 120 to 160 mm total length occur in Indian Ocean at a depth of 20 metre.

The dorsal, anal, ventral and caudal fins of the fish have 111 to 113, 80 to 91, 4 and 12 fin rays respectively. Caudal fin with 12 rays discriminates the species from others like, *C. arel* and *C. dubius*.

Cynoglossus cynoglossus (Hamilton-Buchanan)

The body of the fish is flat, tapering posteriorly. The ramus of the mouth extends behind the eyes. The rostral hook is short, reaching ventral margin of the body. Snout is short and obtusely pointed. Nasals are close together, the tubular one is close to maxilla and the simple one is placed anterior to the interorbital region. Eyes are small and close together. Distinct lateral line is visible on the ocular side. The dorsal, anal and caudal fins of the species have 102 to 107, 76 to 82 and 10 fin rays respectively.

Ocular side of the fish is brownish grey in colour with dark brown spots all over the body. Fishes of 13 to 140 mm total length are available in 10 to 30 metre deep waters. Available in Indian oceans along with Sumatra and river Ganges. The species form the food of cat fish, *Arius serratus*

Cynoglossus lida (Bleeker)

The species inhabiting at a depth of 10 to 20 meters of Indian Oceans has very small eyes. Their maxillary extend below the posterior edge of the eye. They have two nostrils, one tubular in front of lower eye and the other is simple between two eyes. Scales are ctenoid on both sides. Two lateral lines are present on ocular side. The dorsal, anal and caudal fins of the fish have 104 to 185, 80 to 91 and 10 fin rays respectively.

Ocular side of the fish is reddish, brown marginally and blackish along the lateral line. Vertical fins are blackish in colour.

They normally occur in size range between 50 to 180 mm in total length.

Cynoglossus arel (Schneider)

Eyes are prominent in *Cynoglossus arel*, upper in advance of the lower, and the lower reaches the margin of the maxillary. Rostral hook is short and touches the mandibular symphysis. Ramus of the jaw extends posterior to the eye. Tubular nostril is at the upper margin of maxilla in front of the lower eye and the other nostril is situated between the two eyes. Scales are feebly ctenoid on ocular side, cycloid on blind side. Two lateral lines are present on the ocular side. Gill opening is rather straight and nearer to the mouth than the snout.

Ocular side of the fish is grey in colour. Vertical fins with dark grey margins. Pelvic fin is colourless.

Their dorsal and anal fins have 119 to 126 and 92 to 106 finrays respectively. Fishes of 200 to 340 mm total length occurs in Indian Oceans from 10 to 50 metres depth.

Cynoglossus dubius (Day)

The species with a size range of 300 to 310 mm total length inhabit at a depth of 20 to 30 metres of Indian Oceans.

Their rostral hook extend little beyond the mandibular symphysis. Snout is obtusely pointed. Two nostrils present on the ocular side, the tubular one in front of the lower eye. Scales are cycloid on either side of the body. On the ocular side there are six rows of ctenoid scales at upper and lower margins of the body.

The dorsal, anal and caudal fins of the fish have 118 to 120, 89 to 91 and 12 fin rays respectively.

Ocular side of the fish is brown with grey vertical fins. Operculum is provided with prominent black spot.

Fishes of the genus *Cynoglossus* have spherical eggs of 0.835–1.035 mm in diameter. Yolk is colourless and transparent. 20-30 small light yellow oil globules with diameters 0.04-0.08 mm are present. Perivitelline space is very narrow.

Yolk mass is rounded in newly hatched larva. Anus opens below 12th myotome. Ten to eleven post-anal myotomes are present.

Eyes are not pigmented. Auditory vesicles are visible. Xanthophores are present on head, finfold. Yolk sac and a small vertical band is seen near the tail.

On the fourth day mouth opens. Eyes are black. Yolk is absorbed. Fan-like pectoral fins developed with striations of rays. Large tentacle is present on head. Head tentacle and body are pigmented with xanthophores.

Large-scaled Tongue Sole (*Cynoglossus macrolepidotus* (Bleeker)

Cynoglossus macrolepidotus, the large-scaled tongue sole forms a good fishery along the Porto Novo coast during January to June. Their population is supported mainly by the size range of 200 to 334 mm with IV and V stages of maturity. The food of *C. macrolepidotus* consists of crustaceans bivalves, polychaetes fish larvae, phytoplankton and algal matter of which the crustaceans appear to be highly favoured. Ova-diameter measurements show that spawning takes place from August to October, pre-monsoon period. The fecundity of the fish which increases with its size, shows a range of 25,663 to 3,17577. The size of the female at first maturity is 191 mm.

Sole

The fishery for soles is important only along the Malabar coast, although these fish occur in some quantities along the adjoining coast of South Kanara as well. Many species of soles including *Cynoglossus semifasciatus*, are, however represented on the east of India also. *C. semifasciatus* is the chief among the species that are marketed together under the name of sole. While other species of the genus, such as, *C. dubius*, *C. puncticeps* and *C. bilineatus* do occur in the region, but they are always so few that they are never considered as forming part of the sole fishery. *C. semifasciatus* is the only flat fish that occurs in large shoals. The best season for the fishery is immediately after the south-west monsoon, September being always the month of peak commercial catches. The species has, however, been known to occur in shoals from August of one year to February of the following year.

Cynoglossus semifasciatus is a fish of small size, growing commonly to about 15 cm in length, the maximum size so far recorded being 17.5 cm. The body is flattened and leaf-like, one side lying on the substratum and being white in colour. Both the eyes are found on the side away from the substratum and this side is pigmented with irregular brown half-bands across the body. The mouth is narrow and symmetrical. The dorsal and anal fins are long, extending along the margins of the body. There are two lateral lines on the ocular side and none on the blind side, which is the right side of the animal.

Length of head and height of the body are $4^2/_3$ and $3^2/_3$ in the total length of the fish respectively. Eyes are close together, the upper scarcely in advance of the lower. Angle of mouth one diameter behind posterior edge of lower eye and slightly nearer to snout than to gill opening.

A single ventral fin is attached to the anal. The dorsal, ventral and anal fins bear 98-103; 4 and 75-80 fin rays respectively.

Scales are ctenoid on both sides. Two lateral lines are present on the coloured side, seperated at furthest part by 12 or 13 rows of scales. A single lateral line is present on the blind side.

Deep brown in colour with vertical incomplete or half bands irregularly disposed. They extend on to the dorsal and anal fins.

The bulk of the commercial catches of soles consisted of individuals of one year age group, the older individuals being negligible in proportion. The products of spawning of a particular fishery season grow up to the commercial size and directly enter the fishery in the next fishery season.

During September-October fishery season, the size group with the highest frequency was 10.0-10.9 cm. But in the fishery season of next year higher growth was recorded (12.0-12.9 cm) with a decrease in the total catches of the area.

As a bottom feeder C. *semifasciatus* seems particularly to favour polychaetes, usually dominated along the coast by *Prionospio pinnata* during the post monsoon months. During the south-west monsoon season every year, the inshore sea bottom is highly agitated and most of the bottom fauna dispersed. During this period, soles move into deeper waters. In August or September, as soon as the monsoon condition cease, the soles return to the inshore fishing grounds and form shoals. This is the main fishery season along the entire coast for this fish. A coincidence has been noticed between the sole fishery and the commencement of polychaete settlement in the area. It is likely that the food factors play an important role in the large scale appearance of the fish in the inshore waters in this period, in view of the fact that there seem to be very few organisms of value as sole food available on the sea bottom in the deeper regions at this period.

The species is a bottom feeder and a carnivore. It favours a diet of polychaetes, amphipods and small lamellibranchs. *P. pinnata*, a polychaete, is its chief food. The fish does not usually occur in aggregations in the inshore fishing grounds after October, but large shoals occur at all depths in the inshore waters during September and October, which form the chief commercial fishery season for this species. During the monsoon months of June to August the soles migrate to deeper waters and return to inshore immediately after the disappearance of monsoon, conditions, coincident with rising salinity and the commencement of bottom animal settlement. In October they begin to migrate to deeper regions again for breeding as the gonads are in an advanced stage of maturity in September and October, and juvenile soles are recruited to the inshore fish stocks from December onwards. Spawning starts in October and continues till about the commencement of the south-west monsoon though with variable intensity at different periods of the long-drawn out-spawning season. Spawning takes place in deeper waters and the post-larvae and young move inshore to shallow waters. From January onwards in any year the juveniles of the same season predominate the stocks.

The sexes are recognizable in the Malabar sole of 6 to 7 cm. total length. There is no external marks to distinguish between two sexes, but in the case of females with gonads in advanced stages of maturity the ovaries can be easily seen through the body wall when the fish is held against light.

The two sexes of Malabar soles do not occur in the same proportions during the entire period, but showed considerable fluctuations. During the months of September to December, there are more males than females in the population. But in the subsequent period, except during a few months the females are more than the males. The main spawning grounds of the species lie somewhere in deeper waters and the migration out of and into inshore grounds is taking place in batches. Spawning migration commence sometime in October with the weakening of the commercial fishery.

The smallest specimen with ovary developed to fifth stage was noticed when the fish grows to 10 cm in total length. Normally the ovary is not ripe below a size of about 12.0 cm total length.

The ripe ovarian egg is spherical. Several oil globules are present in each egg. The ripe ovary is filled with transparent eggs which are all shed in one or more installments in the same season. There are no different batches of eggs maturing at different periods in the same spawning season. But different individuals may mature at different times and thus prolong the breeding season. Ripe and full ovaries of specimens measuring 15.6 and 15.9 cm were found to have 42,200 and 65900 ripe and ripening ova. Spawning commences in the species in October and continues upto May with varying intensities in between. Eggs have been found in offshore plankton samples in November and the post-larvae were collected in April.

The 2.8 mm larva has a symmetrical and laterally compressed body. There is a very prominent bulge on the ventral side containing the intestinal coil. On a prominent protuberance on the posterodextral aspect of the abdominal bulge is the anus. The head is large and there is a clear operculum. The mouth is anterior and slit-like. The eyes are large and prominent. There is a small slit-like nasal depression in front of each eye. The anterior margin of the head is straight and slopes upwards and backwards without any sinuosities. There is a continuous median finfold extending from the anterodorsal summit of the head to the posterior margin of the bulging belly. Except in the anterior most region of the dorsal fin no clear fin rays are yet visible in the finfold. The first two rays are very long and form a characteristic tentacular structure above the head. These rays are free along their distal portions. Behind these two rays the dorsal fin is narrow, but it widens out over the rest of its length. The anal fin is similar in structure to the dorsal fin, but has no clear fin rays in any region. Pelvic fins appear to be represented by two very small processes in front of the belly ventrally. A pair of large pectoral fins are present. Each of these is a pedunculated fan-like structure. The peduncular portion being narrow and the rest of the fin widen out fanwise distally. The distal part of the fin is thin and marked out abruptly from the thick proximal portion.

There are groups of dark pigment spots characteristically distributed over the body. The positions of brain, auditory vesicle, nerve cord and vertebral column can be easily visible. The caudal end of the vertebral column shows an upward kink. The larvae are transparent when alive and are seen to swim about actively at all levels.

In the post-larva, measuring 4.5 mm in total length, the body become more elongated and lanceolate. The bulge of the intestinal mass on the ventral side is less pronounced than the earlier stage. The anal protuberance is more median in position. The anterior end of the fleshy base of the dorsal fin has grown forward from below the tentacular process to form a rostral projection which is laterally compressed and slopes downwards and backwards so as to make an obtuse angle with the rest of the front of the head. The gill rakers are visible. The mouth is prominent and anterior, with very minute teeth on the jaws. The nareal depression is more prominent than in the earlier stage. There are distinct fin rays in the dorsal and anal fins, but all the rays are not yet distinctly formed in the caudal fin. There are about 98 rays in the dorsals fin excluding the first two long rays and about 78 rays in the anal fin. The pelvic fins are clearly seen, but their fin rays are faint. The pectoral fin continues to be without any fin rays. An air bladder is distinclty visible through the skin and there is a dense group of pigment spots in the region of the air bladder.

Post-larvae is slighly longer than the above stage and also show a similar structure except that the rostral projection grow further and produced downwards into a beak-like process.

The first sign of approaching metamorphosis is the further development of the rostral projection into a prominent parrot-beak like structure, rostral beak. In the most advanced symmetrical stage, this is projecting downwards and inwards, enclosing a small space between itself and the interorbital region of the head where a slights depression is developed. The distal end of the beak touching the head. The post-larva is found on the bottom, still in the vertical position with the head touching the substratum and the rest of the body disposed at an angle to the substratum. It look as if the larva is rubbing its head on the bottom and jerking the tentacle sometimes. Ocassionally it is lying on its right side and is showing occassional jerky movements. It do not assume the vertical position again after this. The lower jaw is moving constantly and rhythmically and the tentacle is lashing slightly now and then. The larva show occasional movements of the entirebody also, but slowly and still lying on the right side only. The first signs of asymmetry is noticed when the right eye has slightly shifted its position. The subsequent stages in shifting of the eye are reached rapidly, and in about two hours time the eye is completely on the left side. The right eye gradually moves towards the dorsal edge of the head where the rostral beak comes more and more in apposition with the front of the head and finally, when the eye has reached the lefts side, the entire beak presses against the head and finally fuses with it just leaving the eye clear of the left side. Within 12 hours the fusion of the beak is complete and the appearance of the head is very much as in adult. The eye does not shift further towards the original left eye, once it has completely shifted on to the left side of the body. The shifting of the eye is invariably taking place during nights.

The main phases of metamorphosis is over with the shifting of the eye and the correlated adjustments in the structure of the skull. The abdominal bulge disappear following the metamorphosis. Clear scale centres are noticed in the skin on the ninth day of metamorphosis. The light brown pigment spots of the body arrange themselves into irregular transverse bands across the body. Both the lateral lines are well formed in larvae of 23 mm total length.

Chapter 10
Scombroid Fishes

Five species of *Cybium* were recorded as occurring in Indian seas of which *Scomberomorus guttatus* (Bl. Schn.) (*Cybium guttatum*) is the commonest in Tamil Nadu coast. Seer fishes have a high market value in Tamil Nadu and are caught with hand lines and boat seines in offshore waters between February and June, when they do not occur inshore. From July to January, especially during the last five months, mature adults and spawners are landed from the coastal waters with the shore seine. The species breed in inshore waters approximately 16 km off Tamil Nadu coast from where eggs and larvae were collected. Several transparent spherical eggs of average diameter of 1.17 mm was obtained between September and January. The habit of *S. guttatus* of being in the surface waters near the shore from July to February, when they breed and descend to midwater during summer months of March–June may be associated with spawning migration. Feeding begins invariably with the transformation of the prolarva into post larva. The fish begins, on a diet of copepod nauplii during larval stages, but gradually change over to larger pelagic organisms as it grows and finally subsists on planktons as well as nectonic organisms. Analysis of the stomach contents of adult fish indicates a change from a pelagic habitat during July–February to life in deep seas during March-June. The breeding season occurs during the pelagic phase.

Length-frequency analysis indicates that the king seer (*Scomberomorus commerson*) attains a size of 402 mm, 726 mm 995 mm and 1186 mm total length at first, second, third and fourth years of age respectively. The streaked seer (*Scomberomorus lineolatus*) reaches a size of 350mm, 713 mm, 835 mm and 965 mm total length at one, two, three and four years respectively. The spotted seer (*Scomberomorus guttatus*) grows to a length of 369 mm, 532 mm and 640 mm at first, second and third years respectively.

A 1936 mm long and 33 kg king seer caught from Palk Bay, Tamil Nadu, in October 1975 appears to be the record for the Indian seas. The maximum size of spotted seer (705 mm length and 2070 gram weight), male streaked seer (940 mm length and 4250 gram weight), and female streaked seer (980 mm length and 4553 gram weight) were obtained from the commercial catch from Gulf of Mannar.

The theoretical maximum length computed by Raifail method from the von Bertalanffy growth equation are found to be 2081 mm (46.7 kg) for the king seer, 1683 mm (15.7 kg) for the male streaked seer, 1447 mm (24.3 kg) for female streaked seer and 1278 mm (9.6kg) for the spotted seer.

All the three species develop two rings a year in their otoliths (sagitta) at a regular interval of six months.

Scomberomorus guttatus (Bloch and Schncider)

Popularly known as Indo-Pacific king mackerel, *Scomberomorus guttatus* (Bloch and Schneider) inhabits along the shores of continental Indo-West Pacific from Wakasa Bay, Sea of Japan and Hong Kong south to the Gulf of Thailand and west to the Gulf lying between the Arabian peninsula and Iran.

An epipelagic, neritic species it is believed to be less migratory than *S. commerson* that may be encountered in turbid waters with reduced salinity. Its movements in the Gulf of Thailand might be deduced from seasonal changes in peak fishing months along the coast of Thailand. These peaks are November/December in eastern Thailand late December/January in the northern part of the Gulf and January to March in its western part.

Based on occurance of ripe females and size of maturing eggs, spawning probably occurs from April to July around Rameswaram

Island between India and Srilanka. Ripe females of 32.5 to 46.5 cm length are taken in May in Thai waters.

Depth of fish body is less than in *S. koreanus* (22.8 to 25.2 per cent vs 24.4 to 26.7 per cent fork length). Its head is larger than in *S. koreanus* (20.2 to 21.5 per cent vs 19.7 to 20.4 per cent fork length). Gillrakers on first arch is moderate, 1 or 2 on upper limb, 7 to 12 on lower limb totaling 8 to 14. The first dorsal fin with 15 to 18 spines and second dorsal with 18 to 24 rays are followed by 7 to 10 finlets. Anal fin with 19 to 23 rays is followed by 7 to 10 finlets usually 8. Pectoral finrays are few, 20 to 23 in numbers. Lateral line with many fine auxillary branches extending dorsally and ventrally in the anterior third, gradually curving down toward caudal peduncle. A total of 50 or 51 vertebrae are present. Intestine with two folds and three limbs. Sides silvery white in colour with several longitudinal rows of round dark brownish spots, smaller than eye diameter, are scattered in about 3 irregular rows along lateral line. First dorsal fin membrane is black, upto the 8th spine, white posteriorly, with distal black margin. Pectoral, second dorsal and caudal fins are dark brown, while pelvic and anal fins are silvery white.

As with other species of *Scomberomorus*, the food of Indo-Pacific king mackerel consists primarily of fishes. Juveniles in India feed mainly on teleosts, particularly clupeoids, such as *Anchoviella*, Adults also prey mainly on fishes with small quantities of crustaceans and squids. Anchovies are particularly important as food of the species; *Stolephorus* in Singapore Straits and *Anchoviella* in Waltair in India.

The species atttain a maximum length of 76cm. Size at first maturity ranges between 48 and 52 cm total length in southern India and about 40 cm total length in Thailand.

There are commercial and artisanal fisheries for *S. guttatus* in Kampuchea, Thailand, Malaysia and India, particularly in the lower Sundarbans, West Bengal, around Tamil Nadu, the Gulf of Mannar–Palk Bay area and Malwan, south of Bombay. These fisheries may be operational throughout the year, but with peaks that differ from region to region in correlation with differential abundance of Indo–Pacific king mackerel.

S.guttatus is one of the principal species in the drift net seerfish fishery in India, but the catch is not identified to species in the statistics. Indonesia reported between 4047 and 4639 metric tons per year in the period from 1978 to 1981 in Fishing Areas 57/71. At the same time vessels from Taiwan caught between 10838 and 14699 tonnes in Area 61.

The primary gear in most areas appears to be the drift gillnet, but the species is also caught in bamboo stake traps and with hand lines in Thailand, and by trolling or with hook-and-line in India and Malaysia. It is utilized fresh or salted in most areas (Kampuchea, Thailand, India). Although less abundant than the Indian mackerels (*Rastrelliger* spp),*S guttatus* is highly esteemed for food and commands a higher price in Thailand and India.

Scomberomorus commerson (Lacepede)

Narrow-barred Spanish mackerel, *Scomberomorus commerson* (Lacepede) are wide spread through out the Indo-West Pacific from South Africa and the Red Sea east through the Indo-Australian Archipelago to Australia and Fiji and north to China and Japan. A recent immigrant to the eastern Mediterranean Sea by way of the Suez Canal has been observed.

Elongate fish body is strongly compressed. Snout is much shorter than rest of the head. Posterior part of maxilla is exposed reaching to a vertical from hind margin of eye. 5 to 38 sharp, compressed triangular teeth is present in upper and lower jaws. Patches of fine teeth on palatines and vomer is present but no teeth on tongue. Gillrakers on first arch is few, 0 to2 on upper limb; and 1 to 8 on lower limb. The first dorsal fin with 15 to 18 spines, usually 16 or 17 followed by second dorsal with 15 to 20 rays but usually 17 or 18, followed by 8 to 10 finlets. The anal fin is provided with 16 to 21 rays, usually 18 or 19 followed by 7 to 12 finlets, usually 9 or 10. Pectoral fin have 21 to 24 finrays. Lateral line abruptly bent downward below end of second dorsal fin. 42 to 46 vertebrae are present. Intestine is with 2 folds and 3 limbs. Sides of the fish is silvery grey marked with transverse vertical bars of a darker grey colour. Bars are narrow and slightly wavy, sometimes breaking up into spots ventrally. In adults there are 40 to 50 bars but are usually fewer than 20 in juveniles upto 45 cm length. Cheeks and lower jaw are silvery white. The first dorsal fin is bright blue rapidly fading to blackish blue. Pectoral fins are light grey turning to blackish blue. Caudal fin lobes, second dorsal, anal and dorsal and anal finlets are pale greyish white turning to dark grey. Juveniles have the anterior membranes of the first dorsal jet black contrasting with pure white posteriorly.

An epipelagic, neritic species known to undertake lengthy long shore migrations, but permanently resident populations also are known to exist. Migrations extend along the entire east coast of Queensland. The migration route in the Gulf of Thailand has been mapped.

Depending on temperature régime, the spawning season may be more or less extended. In east Africa it extends from October to July, off Madagascar from December to February, in the coastal waters off Tamil Nadu from May to July, off Taiwan Island in spring, off Papua New Guines from July to December, on the Great Barrier Reef from October to December and around Fiji from October to February with peaks in December and January.

Like other species of the genus, *S. commerson* feeds primarily on small fish, particularly anchovies, such as *Anchoviella* and *Stolephorus* and clupeids such as, *Sardinella*. Other prey include small carangids, slipmouths (*Leiognathus*), squids (*Loligo*) and penaeid shrimps. Feeding apparently takes place in day and night.

The fish attain a maximum length of 220 cm, but 90 cm size are more common. The all-tackle angling record is 44.9 kg fish caught off Scottburgh, Natal, South Africa. The smallest mature males and females had length between 65 and 70 cm respectively.

This species is taken throughout its range by commercial, artisanal and recreational fisheries. There are important fisheries in three Fishing Areas 51,57 and 71. The world catch increased from 55452 metric tons in 1978 to 72281 metric tons in 1981. The five countries with the largest reported catch in this period were Indonesia, Philippines, SriLanka, Democratic Yemen and Pakistan. Approximately 1000 tons a year are landed in Queensland, Australia (1981) while the 1982 catch in Fiji probably exceeded 300 tons. There is also an important drift gill net fishery in India, but the catch is not identified to species. In Thailand and Malaysia drift nets also seem to be the most important gear deployed to catch the species. Other gear include shore seines in Taiwan and India, trolling lines on Taiwan Island, in Malayasia and in east Africa, where it is a priced market fish and hand lines baited with mackerel or squid and trolling lines with spoons in the Gulf of Thailand. In Samoa it is sold fresh and canned.

The fishing seasons change according to differential availability of fish as a function of variation in hydrographical conditions and weather conditions for fishing. They peak from August to September on the Great Barrier Reef, in spring off the Island of Taiwan, in the dry season between October to April/May off Kampuchea and in the Gulf of Thailand, in March/April, June/July and December in northeastern India, from September to April in southeastern India and in February/March and October to December off the southwestern coast of India, south of Bombay. It is marketed fresh, on ice or salted and dried.

Larvae of Spanish mackerel have well developed and pointed supraoccipital spine. Supraorbital ridge is finely serrated but without spines. Preopercular spines are well developed. They have long pointed snout, snout length is twice that of eye diameter. Mouth is very large with well-developed teeth. 41-56 myomeres are present. Large melanophores on larval fin fold just anterior to anus, on isthmus and on nasal area is noticed. Large stellate melanophores on ventral edge of tail are present.

Scomberomorus plurilineatus (Fourmanoir)

Kanadi kingfish, *Scomberomorus plurilineatus* (Fourmanoir) is confined to the western Indian Ocean along the coast of East Africa from Kenya to Zanzibar to Algoa Bay, Natal and along the west coast of Madagascar. The species is also found in Seychelles.

The species having an elongated and strongly compressed body has moderate gillrakers on first arch, 2 or 3 on upper limb, 9 to 13 on lower limb usually a total of 12 to 15. The first dorsal fin with 15 to 17 spines and second dorsal fin with 19 to 21 rays is followed by 8 to 10 finlets. The anal fin with 19 to 22 rays is followed by 7 to 10 finlets. In the pectoral fin, rays are many, 21 to 26. Lateral line without auxillary branches anteriorly, gradually curving down toward caudal peduncle. A total of 46 vertebrae are present. Intestine is with 2 folds and 3 limbs. Sides of the fish are silvery in colour with a series of about 6 to 80 interrupted horizontal black lines, much narrower than interspaces. Anteriorly, usually only one of these lines is above lateral line; replaced posteriorly by a number of short oblique black lines becoming somewhat confused and only 2 or 3 continue through caudal peduncle. Horizontal black lines on body are interrupted to varying degrees, beginning almost intact in places, but broken up into a series of small rectangular "spots" in others. Juveniles have spots but develop adult pattern of interrupted lines by the time they reach a length of 40 cm. Upper areas of caudal peduncle and median keel are black and lower areas are dusky. First dorsal fin is black except lower areas of membrane that may be pale posteriorly. Second dorsal fin with leading edges and tips of rays are dusky, rest silver to pale. Finlets are dusky with a silver area at centre. Anal fin, leading edges and tips of rays are dusky, rest are silvery. Finlets are white with a dusky central area. Pectoral fins are black inside and dusky outside with edges black in colour. Pelvic fins are pale whitish with outside of mid-rays dusky. Groove on body is little dusky.

As an epipelagic, neritic species, it is present in large schools in the Zanzibar Channel from March or April until August or September (average weights between 3.2 and 3.5 kg). Angling statistic point to a peak abundance in Natal, South Africa during May. Spawning probably takes place in August and September in the Zamzibar Channel. Kanadi Kingfish feed mainly on anchovies (*Anchoviella* sp.) clupeids (*Amblygaster* sp. *Sardinella fimbriata, S. perforata*), other small fishes, squids and mantis shrimps.

They attain a maximum fork length of 120 cm. The South African angling record is 10 kg fish. Sexual maturity is attained at about 80 cm. fork length.

Kanadi King fish is one of the Tanzanian staple food fishes during the fishing season extending from March or April until August or September in the Zanzibar Channel, where large schools are present. It is caught mostly in gill or set nets, but also on handlines baited with live sardines and by trolling. In the Malindi area of Kenya, Kanadi Kingfish continues to be taken mainly by trolling lines and handlines. The species is also important on both sides of the Mozambique Channel south to Durban and is a popular gamefish of the ski-boat fishermen and spear fishermen in Natal.

Larvae of *Scomberomorus* species have well developed and pointed supraoccipital spine. Their supraorbital ridge is finely serrated but without spines. Preopercular spines are well developed. Snout is long and pointed, snout length is twice eye diameter. Mouth is very large with well-developed teeth. 41-56 myomeres are present. Large melanophores are noticed on larval finfold just anterior to anus, on isthmus and on nasal area. Large stellate melanophores are also present on ventral edge of tail.

Scomberomorus lineolatus (Cuvier)

Popularly known as streaked seerfish, *Scomberomorus lineolatus* (Cuvier) is distributed in Indo-West Pacific from the Gulf of Thailand and Java west to Bombay, India.

The species have moderate gillrakers on first arch, 1 or 2 on upper limb, 6 to 11 on lower limb, totalling 7 to 13. The first dorsal fin with 15 to 18 spines and second dorsal fin with 15 to 19 rays are followed by 7 to 10 finlets. The anal fin with 17 to 22 rays is followed by 7 to 10 finlets. Pectoral fins are covered with scales having 20 to 24 pectoral fin rays. Lateral line without auxillary branches anteriorly, running almost straight below second dorsal finlet, then slightly bent downward toward keel of caudal peduncle, which is very wide. A total 44 to 46 vertebrae are present. Intestine with 2 folds and 3 limbs. Sides of the fish is silvery in colour and marked with series of irregular, horizontal, narrow black lines and few if any spots. First dorsal fin is black anteriorly, white posteriorly.

The biology of this epipelagic neritic species is poorly known. Unlike *S. commerson* and *S. guttatus* it is not encountered in very turbid waters or much reduced salinity. From single records, the reproductive season is deduced to be in fall including October off southern India and in winter including January in the south eastern Gulf of Bengal. Juveniles in India are found to feed on teleosts.

The species attain a maximum size of 80 cm.

There are small fisheries for *S. lineolatus* in the waters around Thailand, Malaysia and India. It is taken from October to November along the Thai coast of the Indian Ocean, but is less abundant than either *S. commerson* or *S. guttatus* in both this area and the Gulf of Thailand. Streaked seerfish is fished along both coasts of Peninsular Malaysia, along the west coast from November to February in the north and March to July in the south, and on the east coast from February to March and August to November. Fishing is mainly by gillnets, but on the east coast hand lines and trolling lines are also important. In India, there is an important coastal fishery for the three species of seerfishes that are much is demand, both fresh and salt cured, although they form a smaller proportion of the catch in

India than mackerels, *Rastrelliger* spp. *S. lineolatus* is the least common of these seerfishes. Small individuals, upto 50 cm, are taken together with *S. commerson* and *S. guttatus* from May to September in gillnets 5 to 12 miles off Tuticorin in the Gulf of Mannar, India. Gillnets, hook-and-line, and trolling lines are the most important gear types in India. *Scomberomorus* spp are highly esteemed food fishes in Thailand and are consumed as spicy fish balls or high quality salted fish. A monthly average of 100 metric tons, fresh or salted is consumed in Bangkok alone.

Scomberomorus koreanus (Kishinouye)

Known as Korean seer fish, *Scomberomorus koreanus* (Kishinouye) are inhabitant of continental Indo-West Pacific from Wakasa Bay, Sea of Japan and China south to Singapore and Sumatra and West to Bombay, India.

Their depth of body is greater than in *S. guttatus* (24.4 to 26.7 per cent vs 22.8 to 25.2 per cent of fork length). Head is shorter than in *S. guttatus* (19.7 to 20.4 per cent vs 20.2 to 21.5 per cent of fork length). Gillrakers on first arch are moderate, 1 or 2 on upper limb; 9 to 12 on lower limb totaling 11 to 15. First dorsal fin with 14 to 17 spines and second dorsal fin with 20 to 24 rays are followed by 7 to 9 finlets. Anal fin is provided with 20 to 24 rays followed by 7 to 9 finlets. Pectoral fin rays are usually 20 to 24. Lateral line with many fine auxillary branches extending dorsally and ventrally in anterior third, gradually curving down toward caudal peduncle. 46 or 47 vertebrae are present. Intestine with 4 folds and 5 limbs. Sides with silvery white colour with several longitudinal rows of round dark brownish spots, smaller than eye diameter are sparsely scattered along lateral median line. The first dorsal fin membrane is black in colour, while pectoral, second dorsal and caudal fins are dark brown and pelvic and anal fins are silvery white.

Little is known on the biology of this epipelagic, neritic species. Kishinouye (1923) reported that it spawns in July at the mouth of the Daidoko, Korea and that it feeds on sardines, anchovies and shrimps.

The species attain a maximum size of 150 cm in length and 15 kg in weight. But usually fishes of 60 cm is common. Sexual maturity is reached at about 75 cm fork length and a weight of 2.25 kg.

Korean seerfish is usually not distinguished from other species of seerfishes, but makes up an important part of the drift net fishery in Palk Bay and the Gulf of Mannar between southwestern India and Sri Lanka.

Scomber japonicus (Houttuyn)

Commonly known as Chub mackerel, *Scomber japonicus* (Houttuyn) is cosmopolitan, inhabiting the warm and temperate transition waters of the Atlantic, Indian and Pacific Oceans and adjacent seas.

The fish is known to have elongated and rounded body with pointed snout. The front and hind margins of eye is covered by an adipose eyelid. Teeth in upper and lower jaws are small and conical. Palatine bone, on roof of mouth is narrow with palatine teeth in single or double rows, when double rows are close and running into each other. Gill rakers are shorter than gill filaments, barely visible through open mouth, 25 to 35 on lower limb of first arch. Two widely seperated dorsal fins. The first dorsal with 9 or 10 spines. The space between first and second dorsal fin is less than first dorsal fin base. Anal fin originate opposite that of second dorsal fin or somewhat more posterior. The second dorsal and anal fins with 12 rays. Anal fin spine is conspicuous, clearly seperated from anal rays but joined to them by a membrane. Five dorsal and five anal finlets are present. Interpelvic process is small and single. Pectoral fin is short with 18 to 21 rays. Entire body of the fish is covered with rather small scales. Scales behind head and around pectoral fins are larger and more conspicuous than those covering rest of the body; but no well developed corselet. Two small keels on each side of caudal peduncle are located, but no central keel between them. Swim bladder is present. Vertebrae 31, first haemal spine is posterior to first interneural process and 12 to 15 interneural bones are found under first dorsal fin. Last banchiostegal ray is slightly flattened but not forming a wide plate. The body of the fish is steel blue colour on the back with oblique lines which are zigzag and undulated. Belly unmarked or marked by spotting or wavy broken lines, having silvery-yellow colour.

A primarily coastal pelagic species, to a lesser extent epipelagic or mesopelagic over the continental slope, occuring from the surface to about 250 or 300 m depth. Seasonal migrations may be very extended, the fish in the northern hemisphere moving further northward with increased summer temperatures and southwards for overwintering and spawning. The reverse pattern generally applies to populations in the southern hemisphere. Schooling by size is well developed and initiates at approximately 3 cm size. Schools of adults are the most compact and structured. Multispecies schooling in the Northeastern Pacific may occur with eastern Pacific bonito, *Sarda chiliensis,* jack mackerel, *Trachurus symmetricus* and Pacific sardine, *Sardinops sagax.*

Spawning most often occurs at water temperatures of 15° to 20°C, which results in different spawning seasons by regions. For example, off Peru from January through May and in September. Off northeastern Japan, the spawning season is from April to August with a peak in May, but initiating in March further south. Off California and Baja California, the season is from March through October with peaks between April and August. Spawning occurs in several batches of about 250 to 300 eggs per gram of fish with the total number of eggs per female ranging approximately 100000 to 400000.

The chub mackerel is believed to be in food competition with the species it schools with, such as the eastern Pacific bonito, jack mackerel and others. Its feeding is opportunistic and non-selective, the diet of adults ranging from copepods and other crustaceans to fish and squid. Its predators include tunas, billfishes, white seabass (*Cynoscion nobilis*), yellow tail (*Seriola lalandi*) and other fishes, as well as sea lions, sharks and pelicians.

They attain a maximum length of 50 cm, but common size is 30 cm. A fish of 47.6 cm length weigh 1.1 kg.

Chub mackerel supports important commercial fisheries in 40 countries, reporting catches of *S. juponicus.* The most prominent catches are generated in Fishing. Area 61, where they fluctuate between 1.3 and 2.2 million metric tons and most of which are caught by Japanese vessels. The nominal world catch decreased steadily after the record year 1978 when 286126 metric tons were reported. Among sharing this catch, the USSR ranked second behind Japan, Chile and Peru occupied third and fourth places respectively.

At present the principal method of fishing chub mackerel is purse seining, even though other types of gear are still being used. For example, lampara nets, set nets, trap nets, gill nets, large lift nets, spoon nets, trolling gear, balance nets, stake lines, long lines and even trawl nets are used to catch the species. Such gear is mostly used in small scale fisheries. Between 1971 and 1974 more chub mackerel were caught in the sports fishery than in commercial operations off California.

Round Scads, *Decapterus* spp.

The round scads are small to moderate sized fishes, with a taste and texture of flesh almost similar to that of Chub mackerel *Rastrelliger*. The fish is popular with poor people because it is cheap and is of good flavour.

Roundscads are widely distributed in Indian Oceans upto Philippine waters. They can be defined as coastal fishes which avoid salinities below 30 ppt and the phytoplankton concentrations. Their main area of concentrations are a little off-shore, characterized by water depths from 20-25 fathoms upto 100 fathoms occasionally.

Four species of roundscads have been reported to inhabit Indian Oceans. They are;

1. *Decapterus kurroides* (Bleeker), a rare fish only known from 10 specimens, 4 from East Indies and 6 from Philippines.
2. *Decapterus lajang* (Bleeker), found in East Indies and Philippines, north to Japan and in the Indian Ocean to South Africa.
3. *Decapterus macrosoma* (Blecker), found from Natal to the Philippines, north to Japan and the Bonin Islands, south through the East Indies to Australia.
4. *Decapteruss resselli* (Ruppell), inhabit in Red Sea and East coast of Africa to the Philippines, Japan and Australia.

Comparison of the species reveals that *Decapterus russelli* has a much longer pectoral fin, a larger body depth and head length, the larger number of scutes and a smaller number of scales along the lateral line and of fin rays in its second dorsal fin. The species has 51-66 scales along lateral line (region from head to the first scute). Its pectoral fin length in standard length is 3.7-4.7 and has 32-39 scutes. 30 to 33 finrays are present in the second dorsal fin. Its body depth and head length in standard length are 4.0-5.1 and 3.0-3.7 respectively.

In *Decapterus lajang* and *Decapterus macrosoma*, however, 74-93 scales are present along the lateral line. Their pectoral fin length in standard length have been calculated as 4.9-6.3. They have 26-32 scutes and 34-38 fin rays in their second dorsal. Their body depth and head length in standard length have been found as 4.7-6.5 and 3.4-4.3 respectively.

Spawning of both the species extends from November to March in some areas and seems to be delayed by 1-2 months in other areas.

Fecundity studies showed that *D. macrosoma* and *D. lajang* females bear 67900 and 106200 eggs, while in *D. russelli* much lesser number of eggs have been found.

D. macrosoma and *D. lajang* are typical zooplankton feeder, while *D. russelli* preys to a much larger degree on fishes. Both the species frequently eat *Stolephorus* eggs.

The sex ratio differs slightly with area and species. An annual growth of about 60 to 70 mm have been observed in both the species. The fish enter the fishery for the first time roughly at the beginning of their second year of life and are fully exploited in the same year. They get mature and spawn for the first time at the size of about 180-200 mm at the beginning of their third year of life, when they leave the fishery for spawning. Only a few fish return after spawning.

The carangid fish, *Decapterus dayi* forms an important fishery at Vizhinjam and also along the south west coast of India. The species attains a length of 150mm in the first year and 184 mm in 19 months. Maximum growth of 30 mm and 20 mm was noticed during the first and second months. Though many broods enter into the fishery in the same year, those born in April manifest strongly in the fishery than the earlier or later ones. The younger broods tend to catch up the older ones by faster growth and form an unimodal group at the size of 120-129 mm and are found as distinct unimodal size class then onwards. The fishery of *D. dayi* is mainly dependent upon fish of 0-year class.

Megalaspis cordyla (Linnaeus)

Megalaspis cordyla forms on an average about 40 per cent of the carangid catch at Vizhinjam.

The fishery of *M. cordyla* at Vizhingam extends for a period of 10 to 11 months in a year with two peak seasons, namely, premonsoon peak, from March to May and the post-monsoon peak from September to December. The latter being most important. The post-monsoon fishery is composed mainly of spent-recovering fishes and this period is observed to be the active feeding season for *M. cordyla* and that is why fish move to inshore waters, where good concentration of fish food is available.

Different stages of maturity of male and female gonads of *M. cordyla* have been described below;

Stage I–Immature

Ovaries are thin, small buttan shaped and pink coloured. Small, transparent ova measuring 0.0196 mm or less, visible under higher magnification only. Testis is thin, pale pinkish in colour.

Stage II–Immature

Ovaries are translucent, occupying $^1/_4$ to $^1/_3$ of the body cavity. Pinkish in colour. Ova size range from 0.0196 mm to 0.118 mm. Eggs transparent with visible nucleus and cytoplasmic layer but without yolk deposition, oval in shape and not seen with naked eye. Testis at this stage is pinkish in colour occupying about $^1/_3$ of the body cavity.

Stage III–Maturing

Ovaries are yellow in colour and occupying half of the body cavity. Walls are thick, as such eggs are not visible through the walls. Eggs are transparent as well as yellow, spherical, opaque with yolk present. Size of the eggs ranges from 0.0196 mm to 0.431 mm. Testis is pinkish in colour occupying about $^1/_3$ of the body cavity.

Stage IV–Mature

Ovaries are bright yellow in colour occupying $^3/_4$ of the body cavity. Large eggs are seen through the walls with naked eye. Both transparent and opaque eggs are present, size of eggs range from 0.0196 mm to 0.510mm. Testis is whitish in colour and extend to $^3/_4$ of the body cavity.

Stage V–Ripe

Bright yellow ovaries occupy the entire body cavity. Eggs measuring from 0.0196 mm to 0.627 mm are present. A few large eggs with oil globules are also observed. Whitish testis fill the entire body cavity.

Stage VI–Running

Ovaries occupy the entire body cavity, pushing aside the digestive organs. Large translucent ova with single oil globule is present. Size of the ova range from 0.137 mm to 0.176mm. Opaque yolked ova and small transparent ova are also present. Maximum size of the eggs noticed are 0.882 mm. Ovaries are pale reddish yellow in colour giving "plum pudding" appearance, because of the visibility of both translucent and opaque eggs through thin tunica.

The spawning of *M. cordyla* is of prolonged nature commencing from December and extending to July. The spawning take place during night hours.

Females upto 239 mm length are mainly virgin immatures. The minimum size at maturity is 240-259 mm. In the case of males the minimum size of first maturity is at 240-250 mm length group. The females of *M. cordyla* dominate the catch in almost all the months. The females occur in maximum numbers in all the size groups except in the 240-259 mm groups where more or less equal proportion of both sexes are observed. Equal number of females and males in 240-259 mm length group, the minimum size of first maturity suggests the congregation of fishes of both sexes for spawning.

M. cordyla grows to 250 mm in the first year and 290 mm in the second year, 320 mm in the third year and to 330 mm in 40 months.

The fishery is composed of 0-year and 1-year old fishes, both of them accounting to 90 per cent of the catches and generally former being the dominant.

Among various crafts and gears employed, boat seines, shore seines, hook and lines and drift nets are operated from dug-out or plank built canoes and raft catamarans to catch *M. cordyla*. Drift nets are operated at night and hooks and lines and shore seines during day, whereas boat seines are operated both day and night. The drift nets, which are employed from September to middle of May, while boat seine operation is limited to nine months in a year, March to November. Among the four gears the boat seines catch only individuals less than 200 mm, the hooks and lines and drift nets catch fish above 200 mm and the shore seines netted all size groups. Hooks and lines and drift nets fetch on an average of 95.8 per cent of the catch, while boat seines and shore seines land the rest. Both hooks and lines and drift nets are operated away from the near shore regions, where good concentrations of fish take place. Operation of drift nets in surface layers during night and the hooks and lines in slightly deeper waters during day suitably utilize the vertical migratory habits of the fish.

Carax leptolepis

Caranx leptolepis is the most carangid fish with ripe ova found in Rameswaram Dhanushkodi area in January-March. The species spawn in February and in August. Eggs are spherical 0.581-0.67 mm in diameter. Yolk coarsely segmented, colourless and transparent. Single, colourless, transparent oil globlue of 0.116-0.149 mm diameter is present. Perivitelline space is narrow. No pigment is found on yolk, oil globule or developing embryo.

Newly hatched larva is 1.85 mm long. Anus opens just behind the yolk sac below the 12th myotome. Sixteen post-anal myotomes have formed. Oil globule is located in the middle of the yolk sac on ventral side. Eye and auditory vesicles appeared. Heart started pulsating. Melanophores and xanthophores are seen on head in front of eye and in two vertical patches on the body. Melanophores unevenly distributed on the body and oil globule.

On the first day the larva grows to 2.15 mm. Eyes become deep brown. Most of the yolk is absorbed and oil globule become small. Buds of pectoral fin appeared. Post anal myotomes are 18 in numbers. Pigmentation in three vertical bands are present on the head, trunk and tail.

On the second day the larva become 2.33 mm long. Eyes turned black. Mouth opened. Lower jaw become slightly longer than the upper jaw. Pectoral fins found to be more developed. Gill slits are visible. Two more small pigment patches on the posterior side of the larva appeared.

On the third day, pectoral fins with striations of the rays are visible. Anus shifted slightly forward, opening below the 11th myotome. 20 post anal myotomes are visible. Pigmentation increased showing six vertical bands beginning from the head.

Caranx Eggs Available in Inshore Waters

(*a*) Eggs are spherical. Yolk colourless and coarsely segmented. Large yellow, oil globule 0.116-0.215 mm in diameter, profusely pigmented with melanophores are present. Narrow perivitelline space is visible. Few melanophores are found on ventral side of developing embryo.

Newly hatched larva measures 1.54 mm in length. Anus opens just behind the yolk sac, below 12th myotome. Ten to twelve post-anal myotomes are visible. Yolk sac is round and the oil globule is located at the posterior side of the yolk sac. Auditory capsules are visible. Eyes are not pigmented. Stellate pigment spots on head, oil globule and a vertical band comprising of xanthophores and melanophores are found on the body behind anus.

(*b*) Eggs are spherical 0.747-0.990 mm in diameter without oil globule. Yolk is colourless, transparent and granular. Perivitelline space is very narrow. Melanophores are present on the dorsal side of the developing embryo.

Newly hatched larva measures 1.4 mm. Anus opens just behind the yolk sac, below 13th myotome. Eight to ten post-anal myotomes are present. Eyes and auditory vesicles have appeared. Light brown pigment spots are found on the head.

In twelve-hour old larva, head is detached from the yolk sac. Heart started pulsating. Pigmentation on the head deepened. Pigments has also appeared on the dorsal periphery of the yolk sac and ventral part of the body.

Two days-old larva grows to 2.00 mm. Eyes turned black. Yolk completely absorbed. Mouth opens. Buds of pectoral fins developed. Thirteen post-anal myotomes formed. Position of the anus not changed. Large branched brown pigment spots along alimentary canal and ventral part of the body has appeared.

(*c*) Eggs are spherical with 0.630-0.818 mm in diameter. Yolk colourless, transparent and granular. A large oil globule 0.215-0.231 mm in diameter is found near the posterior end of the developing embryo. Perivitelline space is large.

Eighteen myotomes appeared on the developing embroyo. Melanophores have been found on the ventral side of embryo and on the oil globule.

Newly hatched larva measures 1.4 mm. Anus opens just behind the yolk sac, below 13th myotome. Eight to ten post-anal myotomes are present. Eyes and auditory vesicles have appeared. Light brown pigment spots are found on the head.

In 12-hour old larva, head is detached from the yolk sac. Heart started pulsating. Pigmentation on the head deepened. Pigment has also appeared on the dorsal periphery of the yolk sac and ventral part of the body.

Two days old larva grows to 2.00 mm. Eyes turned black. Yolk completely absorbed. Opening of mouth take place. Buds of pectoral fins developed. 13 post anal myotomes have formed. Position of the anus remain unchanged. Large branched brown pigment spots is visible along alimentary canal and ventral part of the body.

Chapter 11
Sciaenid Fishes

Sciaenid fishes form one of the important constituents of the demersal fish landings of India, forming about 6 per cent of the total marine fish landings. While the larger sciaenids like *Protonibea diacanthes* and *Otolithoides biaurites*, which attain more than one meter in length, are commercially important, occuring in large shoals in the north-west coast, smaller sciaenids form a seasonal fishery of varying magnitude along the entire coast of India.

Umbrina sinuata (Day)

Head of the fish is 32 to 39 per cent of standard length; and anal spine is 11 to 14 per cent of standard length. Soft dorsal fin bear 26 to 28 fin rays. Its gas bladder is simple without arborescent tubules. Mental barbel is provided with a median pore at its tip. Sagitta with the tail bent sharply. Teeth are uniform band situated on lower jaw. Body surface is provided with dark stripes. The species is available from east coast of Africa to Karachi.

Bahaba chaptis (Hamilton)

Head and anal spine of the fish are 27 per cent and 10 per cent of standard length of the fish respectively. Soft dorsal fin bears 24 to 25 fin rays. Its gas bladder with an anterior unbranched tubule extend backward. Sagitta is provided with enlarged anterior depression and posterior curved groove. Lower jaw of the fish is provided with an outer row of minute teeth and an inner row of villiform teeth. The first dorsal fin is distally black. Second dorsal and caudal fins is coloured with narrow black streak. Found in estuaries of Sittang and Ganges.

Macrospinosa cuja (Hamilton)

Head of the fish is 30 to 31 per cent of standard length. Second anal spine is very strong and 16 to 19 per cent of standard length. Second dorsal fin has 28 to 29 fin rays. Gas bladder is provided with two pairs of minute tubules at the anterior end. Mental pores are five in numbers. Sagitta is broad and thick with well-curved tail. Teeth are well differentiated in both jaws. Large black spots are visible on the dorsal fins. Body dorsally is with rows of spots. The species forms a fishery in the estuaries of Ganges.

Kathala axillaris (Cuvier)

Head length is 31 to 36 per cent and second spine is 11 to 14 per cent of standard length of the fish. Gas bladder is provided with a short curved horn-like tubule on each side anteriorly. There are five mental pores. Sagitta pointed anteriorly with a shallow anterior depression. Teeth are slightly differentiated in lower jaw. A large blotch is present on axilla. The first dorsal fin is provided with a dark grey blotch. The species is available in seas around India, Sri Lanka, Indo-China, South China, Java and New Guinea.

Otolithoides biaurites (Cantor)

Head and second anal spine are 27 to 29 per cent and 6 to 5 per cent of standard length respectively. Second dorsal fin is provided with 28 to 29 fin rays. Gas bladder with a pair of long tubular appendages originate from posterior end and extend into the head giving short branches. Sagitta is thin and elongated. Lower jaw is provided with three mental pores. Teeth are well differentiated in both jaws. The first dorsal fin is grey in colour. Opercle is provided with a grey blotch. The species inhabit in seas around India, Mayanmar, Malay Peninsula, Somalia, Borneo, Indo-China.

Otolithoides biauritus (Cantor) is one of the larger sciaenids found in Indian Oceans, growing to a size of over 160 cm. *O. biauritus* contributed substantially to the catch of the north west coast of India, along with other bigger varieties of fish, namely, *Pseudosciaena diacanthus*, *Congresox telabonoides*, *Polynemus indicus* and *Pomadysis hasta* till early seventies. Except *P. hasta*, all of them grow to more than a metre. The average length attained by *O. biauritus* at the end of one to five years of its life span has been worked out to 78.42, 122.33 146.92, 160.68 and 168.39 cms respectively. The species take about 2 years to mature at an average length of 122 cm. The catch of the species has shown wide fluctuation over 1993-94 to 2001-02, a nine years period. The species in Gulf of Kutch region, attain a growth of 41.8, 60.0, 74.1, 83.1, 107.9, 127.1, 134.7, 136.1, 142.0, 148.2, 149.0 and 152.0 cms in one to 12 years period as evidenced from their scales and otoliths studies. The largest size reported is 162 cm. The fish may survive for more than 15 years–a concept very difficult to accept for a fish from tropical waters where the growth rate is usually fast and consequently life span is short. The species being bigger in size and carnivore in nature, the chances of its natural death, particularly by predation would be rather less. The other bigger species in the north-west coast of India are sharks, rock cods and species such as *P. indicus, P. diacanthus* and *C. telabonoides*. The chance of *O. biauritus* being predated by them is less. The mortality due to fishing is on the higher side and nearly 62 per cent of the catch is below the length at first maturity. This is because most of the catch is comming from shrimp trawlers with cod end mesh size at times as low as 15 mm. With the introduction of shrimp trawlers in the north west coast of India, the catches of almost all the bigger varieties of fish including *O. biauritus* declined. It has been observed that the catches of fish increase if the biomass is reduced to half the biomass infinity. This allows the remaining stock to grow at a faster rate as the competition for food and space is reduced. A fish which is under a tremendous pressure tends to make up by faster rate of growth. The catches of all the bigger species on the north west coast of India have declined. But among the five bigger species whose catches have gone down, *Otolithoides* appears to be relatively better placed showing some recovery in the catches also. The chief reason for this is its faster growth rate in relation to other bigger species in the same area, higher fecundity and their offshore breeding ground.

Otolithoides pama (Hamilton)

Head and second anal spine are 28 to 34 per cent and 3 to 4 per cent of standard length respectively. Second dorsal fin is provided with 40 to 44 fin rays. Gas bladder is carrot-shaped with a pair of tubules originating near the posterior end and extend forward into the head ramifying as branches below ear ossicles. Sagitta is elongated with shallow anterior and posterior depressions. Lower jaw is provided with two pairs of mental pores. Teeth on the lower jaw are well differentiated. Tip of the first dorsal fin is grey in colour. Opercle is provided with a grey blotch. The species is available in seas around Karachi, north-east coast of India, Malay Peninsula, Sumatra, New Guinea and in estuaries of Ganges and Brahmaputra.

Panna microdon (Bleeker)

Head and second anal spine are 25 to 36 per cent and 6 to 8 per cent of standard length of the fish respectively. Second dorsal is provided with 31 to 35 fin rays. Gas bladder with an anterior tubule on each side, which bifurcates into a branched anterior appendage and an unbranched posterior tube. Sagitta is narrow, elongated with anterior depression pear shaped and posterior depression is deep. Lower jaw is provided with three pairs of mental pores. Upper jaw is provided with canine teeth and lower jaw with enlarged teeth. Two-third of spinous dorsal is black in colour. Opercle is provided with a blue spot. Found in seas around India, Mayanmar, Malay Peninsula, Sumatra, Borneo and Vietnam.

Pennahia macrophthalmus (Bleeker)

Head and second anal spine are 30 to 36 per cent and 6 to 10 per cent of standard length of the fish respectively. Mouth of the fish is terminal. Second dorsal is provided with 21 to 26 fin rays. The gas bladder is round anteriorly with 15 to 20 well-developed arborescent tubules without ventral limb. Sagitta is oblong with indistinct anterior depression with marginal deep groove. Lower jaw with two pairs of minute mental pores. Teeth on both jaws are well differentiated. Inner rows of lower-jaw teeth are enlarged. Caudal fin is truncate. Opercle is provided with a steel blue blotch. Pectoral axilla is with a black blotch. Found in Karachi and Chinese coast.

Argyrosomus hololepidotus (Lacepede)

Head and second anal spine are 28 to 34 per cent and 6.9 per cent of standard length of the fish respectively. Second dorsal is provided with 26 to 29 fin rays. Gas bladder is with 25 to 35 arborescent tubules. Sagitta is elongated with pouched head and J-shaped tail. There are three pairs of mental pores. Teeth are differentiated in both jaws. Spinous dorsal fin is dusky in colour and the caudal fin with dark margin. Found in seas around West Africa, Cape of good Hope to Natal, Madagascar, Kathiawar (India) and Australia.

Argyrosomus amoyensis (Bleeker)

Head and second anal spine are 29 per cent and 7 per cent of standard length of the fish respectively. Second dorsal fin is provided with 27 fin rays. Gas bladder is with 22 pairs of arborescent tubules. There are three pairs of mental pores. Mouth of the fish is terminal. Faint oblique stripes are present on the dorsal part of the body. Spinous dorsal fin is dark in colour distally. Found in Bombay and Gwadur coast.

Atrobucca nibe (Jordan and Thompson)

Head and second anal spine are 32 to 35 per cent and 11 to 14 per cent of standard length of the fish. Gas bladder is anteriorly rounded with 25 to 28 well developed lateral tubules with dorsal and ventral limbs ramifying profusely. Sagitta is oblong with indistinct anterior depression and distinct posterior groove. Lower jaw is with three pairs of mental pores. Teeth are well differentiated in both jaws. Mouth of the fish is terminal. Length of gill filament is 36 to 42 per cent of eye diameter. Caudal fin is lunate. The upper edge of first and second dorsal fin is grey. Present in East China Sea, Korea, Taiwan and Bay of Bengal.

Atrobucca trewavasae (Talwar and Sathiarajan)

Head length is 38 to 40 per cent of standard length of the fish. Second dorsal fin is placed anteriorly. Gas bladder round with 26 pairs of highly ramifying tubules with dorsal and ventral limbs. Sagitta is oblong with indistinct anterior depression and deep posterior groove. Teeth are well differentiated in both jaws. Mouth of the fish is terminal. Length of gill filament is 71 to 90 per cent of eye length. Caudal fin is lunate. Inhabit in deep sea off Tamil Nadu at a depth of 250 meters.

Chrysochir aureus (Richardson)

Head and second anal spine length are 27 to 33 per cent and 5 to 8 per cent of standard length of the fish respectively. Dorsal fin is provided with 25 to 27 fin rays. Gas bladder is round, anteriorly with 24 to 26 arborescent tubules which are not divided into dorsal and ventral limb. Sagitta is elongated with distinct anterior and posterior depressions. Lower jaw is provided with three pairs of mental pores. Teeth are well differentiated in both jaws. Upper jaw is with two pairs of canine teeth and overlaps the lower jaw. A grey blotch is visible on the dorsal fin. Opercle is provided with a blue blotch. Found in east coast of India, Malay Peninsula, Borneo and China.

Otolithus ruber (Schneider)

Head and second anal spine length are 29 to 34 per cent and 4 to 6 per cent of standard length of the fish. Second dorsal fin has 27 to 31 finrays. Gas bladder is round anteriorly with 30 to 38 pairs of branched tubules. Sagitta is elongated with indistinct anterior depression, but the posterior depression is expanded and distinct. Lower jaw of the fish bear two pairs of mental pores. The upper and lower jaw of the fish is provided with well-developed strong canines and differentiated enlarged teeth posteriorly. Mouth of the fish is terminal. It has 8 to 11 gill rakers. The body of the fish is silvery colour. Upper two-third of dorsal fin is grey coloured. The opercle bear a dark blotch. The fish is found in east coast of Africa, India, Indo-Australian Archipelago, Queensland, Philippines and Chinese seas.

Otolithus cuvieri (Trewavas)

Head and second anal spine length are 32 to 33 per cent and 5 to 7 per cent of standard length of the fish. Second dorsal fin has 29 to 32 fin rays. Gas bladder is round anteriorly with 25 to 28 arborescent tubules. Sagitta is elongated, with shallow anterior and posterior depressions. Lower jaw is provided with two pairs of mental pores. Upper and lower jaws bear canine teeth and teeth are well differentiated. The fish has a terminal mouth. 12 to 17 gill rakers are present on lower limb of first arch. Body is silvery coloured. Axilla with a blue spot. Opercle with a grey blotch. Fish is available in the coasts of India and Pakistan.

Pterotolithus maculatus (Kuhl and Van Hasselt)

Head and second anal spine length are 28 to 33 per cent and 4 to 6 per cent of the fish standard length. Second dorsal fin is provided with 30 to 34 fin rays. Anal fin contain 10 to 12 fin rays. Gas bladder is anteriorly round with 37 to 53 arborescent tubules with many branches covering the bladder dorsally. Sagitta is elongate, twice as long as wide, thin with a very faint anterior depression. Lower jaw is provided with a pair of mental pores. Upper and lower jaws are with well-developed canine teeth. Teeth are well differentiated in jaws. Mouth of the fish is terminal. Upper part of the body is provided with 3 to 4 rows of dark blotches. Blotches are present on the dorsal fin. Available in north-east coast of India, Mayanmar, Malay Peninsula and Borneo.

Protonibea diacanthus (Lacepede)

Head and second anal spine length are 30 to 33 per cent and 9 to 11 per cent of standard length of the fish. Second dorsal fin is provided with 22 to 25 fin rays. Gas bladder is anteriorly round with 19 to 20 pairs of arborescent tubules of which anterior branches are short. Sagitta is oblong with distinct anterior depressions. Three pairs of mental pores are present on lower jaw. Teeth on upper and lower jaw are well differentiated. Fishes below 28 cm length have broad vertical bands and small blotches. In fishes above 40 cm length, body is hyaline. The species is found in Gulf of Oman, India, Sri Lanka, Mayanmar, Malay Peninsula, Thailand, Indonesia Western Australia, Philippines, Chinese and Japanese waters.

Dendrophysa russelli (Cuvier)

Head and second anal spine length are 28 to 33 per cent and 12 to 17 per cent of standard length of the fish respectively. Second dorsal fin is provided with 25 to 28 fin rays. Gas bladder is anteriorly round with 15 to 17 short arborescent tubules. Sagitta is broad anteriorly, with indistinct anterior depression. Lower jaw is provided with five mental pores. Mental barbel is filiform with a median mental pore at its base. Mouth is inferior. Lower jaw with a band of villiform teeth. Available in India, Malay Peninsula, East Indies and Philippines to Chinese waters.

Nibea semiluctuosa (Cuvier)

Head and second anal spine length are 31 to 33 per cent and 11 to 14 per cent of standard length of the fish. Second dorsal fin is provided with 28 to 29 fin rays. Gas bladder is round anteriorly with 15 to 21 pairs of arborescent tubules. Sagitta is elongated with a shallow anterior depression and a distinct posterior groove. There are five mental pores. Teeth of the lower jaw is well differentiated.

Rows of dark narrow stripes found on the body. Fins are black in colour. Present in north-west coasts of India and Mekaran coast.

Nibea albida (Cuvier)

Head and second anal spine length are 28 to 31 per cent and 13 to 15 per cent of standard length of the fish. Second dorsal fin is provided with 23 to 24 fin rays. Gas bladder of the fish is round anteriorly with 18 to 19 arborescent tubules, of which first tubule is short. Sagitta is thick with distinct anterior and posterior depression. Five mental pores are present. A pair of minute mental barbels are situated on lower jaw. Lower-jaw teeth are differentiated. Spinous dorsal is black in colour. Fish body is greyish. Available in Indian coasts and ascend estuaries and backwaters.

Nibea maculata (Schneider)

Head and second anal spine length are 28 to 35 per cent and 9 to 12 per cent of standard length of the fish. Second dorsal fin is provided with 22 to 26 fin rays. Gas bladder is round anteriorly with 19 to 21 arborescent tubules on each side, the first one extending to the base of cranium. Sagitta is broad with posterior groove sharply curved. Five mental pores are present on the lower jaw. Teeth are differentiated in both jaws. Mouth of the fish is inferior. Nape is provided with a grey blotch expanding over opercle. Body of the fish is provided with four dark bands. Upper spinous dorsal is black in colour. Available in east and west coast of India and Sri Lanka.

Nibea chui (Trewavas)

Head and second anal spine length are 31 per cent and 13 per cent of standard length of the fish respectively. Second dorsal fin is provided with 24 fin rays. Gas bladder is round anteriorly with 18 pairs of arborescent tubules, the first tubule extend to the head. Sagitta is broad with curved posterior groove. Lower jaw is provided with five mental pores. Teeth are well differentiated in both jaws. Upper portion of spinous dorsal is grey in colour. Available in Bombay, Hongkong and Japanese waters.

Nibea soldado (Lacepede)

Head and second anal spine length are 26 to 32 per cent and 14 to 18 per cent of standard length of the fish respectively. Second dorsal fin bear 28 to 30 fin rays. Gas bladder is anteriorly round with 20 to 22 arborescent tubules, of which first tubule is longer. Sagitta is elongated with anterior distinct depression and posterior curved with deep groove. Lower jaw is provided with five mental pores. Teeth are differentiated in both jaws. A dark blotch on first dorsal and a blue blotch on the opercle are present. Base of each dorsal spine and rays are provided with a spot. The species is available in India, Sri Lanka, Malay Peninsula, Gulf of Siam, Indonesia North Australia and Queensland waters.

Johnius carutta (Bloch)

Head and second anal spine length are 28 to 33 per cent and 7 to 12 per cent of standard length of the fish respectively. Second dorsal has 27 to 30 fin rays. Gas bladder is with anterio-lateral expansion with 15 to 16 arborescent tubules. Inner branch of first tubule extends to head, whereas its lateral branch extends to opercular opening. Sagitta is very thick, of which anterior depression is distinct and posterior groove is deep. Three pairs of mental pores are present. Teeth are villiform in lower jaw. Lateral line is silvery. Caudal fin is nearly truncate. Body deep grey in colour Available in Indian and Thailand waters.

Johnius elongatus (Mohan)

Head and anal spine length are 25 to 32 per cent and 6 to 9 per cent of standard length of the fish respectively. Second dorsal is provided with 25 to 29 fin rays. Gas bladder is with anterio-lateral expansion with 15 tubules. The first tubule is well developed and extends to head and exterior. Sagitta is thick, anterior half is enlarged and posterior groove is deep. Five mental pores are present. A pair of minute barbel-like knobs at inner mental pores. Teeth on lower jaw are uniformly villiform. Fish possess an inferior mouth. Dorsal portion of fish body is grey in colour. Upper portion of the spinous dorsal is grey in colour. Found in west coast of India.

Johnius belangerii (Cuvier)

Head and second anal spine length are 27-33 per cent and 7 to 12 per cent of standard length of the fish. Second dorsal fin is provided with 27 to 31 finrays. Gas bladder and sagitta are "Johnius type". Five mental pores are present. Lower jaw is provided with uniform villiform teeth. Mouth of the fish is inferior, lips rather thick. Ventral fin are black in colour. Body of fish is grey to dark grey in colour. Found in Natal, India, East Indies and Chinese waters.

Johnius dussumieri (Valenciennes)

Head and second anal spine length are 29 to 33 per cent and 7 to 11 per cent of the standard length of the fish. Second dorsal fin has got 23 to 26 fin rays. Gas bladder with anterio-lateral expansion is provided with 14 to 15 arborescent tubules. Sagitta is "Johnius type". There are five mental pores. The median mental pore is at the base of a solid barbel. Scales on the body and head are cycloid. Lower jaw with uniform villiform teeth. Second and third dorsal spines are elongate, filiform (56 to 86 per cent of head). The first dorsal is dark grey in colour. Available in Natal, Mekaran coast, India, East Indies, Philippines, China and south-western Japan.

Johnius mannarensis (Mohan)

Head length is 29 to 31 per cent of standard length of the fish, while the second anal spine length is 29 to 34 per cent of head length of the fish. The second dorsal is provided with 27 fin rays. Gas bladder is "Johnius type" with lateral expansion and 13 to 15 arborescent tubules. Sagitta is "Johnius type". Five mental pores are present. Mental barbel is stout, median mental pore is at its base. Lower jaw with villiform teeth. Scales on the body are ctenoid. Second and third dorsal spines are not elongated. First dorsal fin is grey in colour. Available in Gulf of Manner.

Johnius coitor (Hamilton)

Head and second anal spine length are 25 to 28 per cent and 11 to 14 per cent of the standard length of the fish respectively. Soft dorsal fin is provided with 25 to 29 fin rays. Gas bladder is "Johnius type" with anterio-lateral expansion and 10 to 13 pairs of lateral arborescent tubules. Sagitta is "Johnius type". Five mental pores are present on lower jaw. Median mental pore is with a fleshy pad. Lower-jaw teeth are villiform, uniform or slightly enlarged. Mouth of the fish is inferior. Anal spine is robust pointed (43 to 56 per cent of head). Eyes are small (23 to 31 per cent of head). Opercle is provided with a blue blotch. Dorsal side of the body is light grey in colour. Available in estuaries and coastal seas of India, Indo-Australian Archipelago, east coast of Australia. Forms a fishery in Sunderbans.

Johnius glaucus (Day)

Head and second anal spine length are 25 to 32 per cent and 8 to 10 per cent of standard length of the fish. Soft dorsal is provided with 28 to 30 fin rays. Gas bladder is "Johnius type" with anterio-lateral expansion and 14 to 15 lateral arborescent tubules. Sagitta is "Johnius type". Five mental pores are situated on the lower jaw. Lower jaw is provided with villiform teeth. Eyes are large (24 to 30 per cent of head). Mouth is slightly oblique. The upper two third of spinous dorsal is black in colour. Branchial cavity is black. Available in north west coast of India. Form a fishery in Bombay and Veraval.

Johnius macropterus (Bleeker)

Head and second anal spine length are 24 to 29 per cent and 8 to 11 per cent of the standard length of the fish. Soft dorsal is provided with 30 to 34 fin rays. Gas bladder is "Johnius type" with anterio-lateral expansions and 15 to 16 arborescent tubules. Sagitta is stout "Johnius type". Lower jaw is provided with five mental pores. The median mental pore is situated at the base of a minute filiform mental barbel. Lower jaw is with uniform villiform teeth. Fish body is covered with deciduous ctenoid scales. Spinous dorsal ray is not elongated. Body dark grey in colour. Spinous dorsal is grey in colour. Found in Natal, India, Sri Lanka, Indo-Australian Archipelago, New Guinea.

Johnieops macrorhynus (Mohan)

Head and second anal spine length are 24 to 29 per cent and 6 to 10 per cent of standard length of the fish respectively. Soft dorsal bear 26 to 30 fin rays. Gas bladder with anterio-lateral expansion, "Johnius type" with 13 to 14 pairs of arborescent tubule. Inner branch of first tubule extends to head and outer branch extends to exterior through the opercular wall near the junction of cleithrum and supracleithrum. Sagitta is "Johnius type" with faint anterior depression and a deep posterior groove. Sagitta is thick. Lower jaw is inferior with inner row of slightly enlarged or molariformed teeth. Body is pale brown dorsally and yellowish ventrally. Found in India, Andamans and Singapore waters.

Johnieops aneus (Block)

Head and second anal spine length are 28 to 33 per cent and 8 to 12 per cent of standard length of the fish. Soft dorsal bears 25 to 30 fin rays. Gas bladder is "Johnius type" with lateral expansion and 13 to 14 arborescent tubules. Sagitta is "Johinius type". Lower jaw is provided with five mental pores. Inner row of lower-jaw teeth is enlarged and the outer row is villiform. Snout slightly projects beyond upper jaw. Dorsal fins are black in colour. Opercle is with a steel blue blotch. Found in Persian Gulf, India, Sri Lanka. Forms a fishery in Gulf of Mannar.

Johnieops dussumieri (Cuvier)

Head and second anal spine length are 29 to 33 per cent and 7 to 11 per cent of the standard length of the fish respectively. Soft dorsal is provided with 25 to 29 fin rays. Gas bladder is 'Johnius type" with anterio-lateral expansion and 14 arborescent tubules. Sagitta is "Johnius type". Lower jaw is provided with six mental pores and enlarged teeth. Caudal fin is lunate. Mouth of the fish is subterminal. Snout slightly projects beyond upper jaw. First dorsal is black in colour. Opercle is provided with steel blue blotch. The species is found in east coast of Africa, and India. Forms a fishery in south-west and south-east coasts of India.

Johnieops sina (Cuvier)

Head and second anal spine length are 29 to 35 per cent and 4 to 9 per cent of standard length of the fish respectively. Soft dorsal bears 26 to 31 fin rays. Gas bladder is "Johnius type" with anterio-lateral expansion and 15 to 16 arborescent tubules. Sagitta is "Johnius type". Three pairs of mental pores present without fleshy pad. Lower jaw with differential teeth. Mouth is terminal and cleft of the mouth is oblique. Upper two-third of spinous dorsal and opercle bear a steel blue blotch. Available from Natal to Malay Peninsula and Indo-Australian Archipelago.

Johnieops sina is one of the most common species among the small sciaenids contributing to the scianid fishery of the Malabar coast of India. Commercial catches of the species are landed by trawlnets, mackerel gill nets, boat seine throughout the year with peak catches between October to March. The percentage occurrence of matured fish and ova-diameter frequency in the species indicated prolonged spawning of the species. The minimum size at first maturity has been found to be at an average size of 115mm for females. The condition factor had the minimum value at a length of 115mm for females and 125mm for males, when they are considered to be mature for the first time. Fishess of both sexes appear in equal proportion in most of the months. Dominance of females in the population is noticed at 115mm total length and upwards. Fecundity has been found to vary from 12,774 to 1,51,697 in specimens of the size range 124-174mm. The population of *J. sina* breeds all the year round, with the peak spawning periods corresponding to the months of high percentage of occurrence of running and spent individuals.

Johnieops vogleri (Bleeker)

Head and second anal spine length are 31 to 34 per cent and 7 to 10 per cent of standard length of the fish respectively. Soft dorsal is provided with 26 to 31 fin rays. Gas bladder is "Johnius type" with anterio-lateral expansion and 14 arborescent tubules. Sagitta is "Johnius type". Six mental pores are present. No fleshy pad around the pores are present. Mouth of the fish is terminal. Lower jaw is provided with well-differentiated teeth with enlarged inner row. Upper jaw is with a pair of enlarged teeth. Mouth is oblique. Upper two-third of first dorsal is grey is colour. Found in India, Indo-Australian Archipelago, Australia, Philippines and Thailand. Forms a fishery in north west coast of India.

Johnieops vogleri, one of the major species constituting Dhoma fishery of Maharashtra coast accounts for an average of 10,684 tonnes during 1966-73 period. The fishery is supported by larger species like, *Nibea dicanthus, Otolithoides biaurites* and smaller species like *Otolithus cuvieri, O. ruber, Johnieops sina, J. vogleri, J osseus, Johnius glaucus, J. belengerii, J. dussumieri, J. carutta* and *Pennahia aneus*

The recruitment of *J. vogleri* in two batches namely, 15 to 35 mm and 95 to 115mm are observed in June-August period of every year. The former brood would have resulted from the spawning during June-July and the latter from that in October-November as spawning occurs in these period in the area.

Mature fish are available throughout the year. Fish in oozing condition are observed in June and July and again in October-November. This suggests that *J. vogleri* spawns twice a year, that is, June-July and October-November. As a result the juveniles of the species of 15-35 mm and 95-115mm length occur during July-August in the fishery.

Fully ripe eggs have a few oil globules. In the mature egg stock, one batch of ova grows faster and gets ready for spawning during the ensuing season. The rest of the mature eggs spawn subsequently taking some more time to attain the ultimate stage, since these eggs are already in mature condition. As such the spawning of this species is restricted to a short period with an indication of a second spawning, the duration in between being short.

Sexes can be differentiated from 80-89 mm size group onwards. The male to female ratios are 1.5 : 1; 1 : 1 and 1 : 1.2 in different years. Males are matured at 130-139 to 170-179 mm size groups, where as females attain sexual maturity in 220-229 to 240-249 mm size groups. Total stock of eggs varied from 26028 to 5,81295.

J. vogleri is a carnivorous like other sciaenids found in Indian waters. When young, it feeds primarily on crustaceans, especially prawns, and with growth it shows piscivorous tendency. The bottom feeding habit of this fish is evident from the presence of mud particles and organisms like *Eupagurus* spp. and *Squilla* spp and gastropod shells. The species has no preference or selection of food, but feeds, mainly on the available food.

Sudden change of pressure and shock, when they are hauled up from bottom waters, cause everted and disgorged condition of stomachs in *J. vogleri*. The percentage of everted condition is low among juveniles but is fairly common in size groups 140-149 mm onwards.

Sciaenid, *Pennahia aneus*

Sciaenid fishes form a considerable proportion of the catches all along east and west coasts of India. Among them are *Johnius aneus, Pseudosciaena diacnthus Sciaenoides brunneus, Otolithoides brunneus, Otolithus argenteus, Pseudosciaena aneus, Pseudosciaena bleekari, Johnius carutta, Otolithus ruber Johnius dussumieri* and *Pennahia aneus*.

The total length of *Pennahia aneus* has the fastest growth rate followed by the pre-anal length. Snout to ventral origin grows faster than snout to dorsal origin. A comparison of the relative growth of the fin rays shows that the ventral finray shows faster growth than the dorsal and the anal fin ray. The rate of growth for head length is more than the head depth. The interorbital space, snout length and depth at caudal peduncle have the same growth rate. The slowest growth rate has been observed in the eye diameter.

P. aneus feed on the fishes, juvenile prawns, mysis, *Platycephalus* spp, *Lucifer* spp, *Apogon* spp, Amphipod, *Squilla* spp, *Sepia* spp, *Cynoglossus* spp, gobids, crab juveniles and hermit crab. The juveniles of *P. aneus* feed only or prawn mysis and prawn juveniles.

The ripe ovary contains fully mature ova with oil globules and yolk and the modal range is 0.356 to 0.411 mm. The ovaries with ripe ova is obtained during the months of September and October. The spawning period of the species is short, from September to early November. The fecundity of the species increases in relation to the size of the fish. The maximum number of ova was 79835 for the fish of total length 252 mm and the minimum number 11423 for the fish of total length 189 mm.

The female and the male attain their sexual maturity at a total length of 134-148 mm.

Ghol, *Pseudosciaua diacanthus*, though present in the waters off Goa, never form a fishery comparable to that in Maharashtra or Gujarat. Being demersal in habit, the trawlers operating off Goa may bring a few fish during certain trips. Large shoals of 'ghol' are occasionally noted in the shallow regions off Goa. The largest single haul of 'ghol' by purse seine comprised of 349 fishes with an estimated weight of 4.1 tonnes were reported in 1917. In 19th September, 1973 an unusual catch of 2309 'ghol' weighting 30.24 tonnes were obtained from off Aguada light house Goa (Lat 15°29'N and Long. 73°46'E). A total number of 2309 *Pseudosciana diacanthus* was caught in a purse seine net. The size range of fish was 95 to 130 cm with an average length of 109 cm. The fishes were of five to seven years old. They mainly fed Malabar sole, *Cynoglossus macrostomus*. The number of soles in the stomach of the fish varied from 8 to 22 and their size ranged from 6 to 12 cm. Their stomach were fully packed with soles. The habit of extroverting the stomach and disgorging the contents have been observed in the species. It has been suggested that the shock sustained during the capture and sudden change of pressure experienced during hauling is the probable reasons for regurgitating the food and extroverting the stomach respectively. In the purse seine haul only regurgitation of food was observed but not with extroverted stomach which may be due to the fact that they came from the surface layers of water column and as such no pressure change was experienced by the fish while hauling.

The availability of food in the inshore waters off Goa is the prime factor for the inshore migration of 'ghol' during the monsoon months.

Chapter 12
Elasmobranch Fishes

Sharks constitute an important component in off shore catches of Sri Lanka accounting for 78 per cent and 14 per cent respectively of the catch in long line and gill net fisheries. The silky shark *Carcharius falciformis* made up 70 per cent and 88 per cent of the shark catches in gillnet and long line fisheries respectively.

The oceanic white tip shark is more prominent in the south-west and the south, while the hammerhead shark, *S. levini* and milk shark *R. acutus* catches are high in the west and north-west, compared to the southern areas.

One school of skipjack and yellowfin was observed in association with sharks. *C. falciformis* of 61-293 cm total length was caught in gill net; while sharks of 63-145 cm total length was caught in long line.

Sex ratios for two species of sharks, namely, *C. falciformis* and *C. longimanus were female is to male were 1 : 0.82 and 1 : 0.55 respectively.*

C. falciformis prey on squids (21 per cent); frigate tuna (13 per cent); crabs (7 per cent) and skipjack (5 per cent). Other unidentified items (due to advanced stage of digestion) account for the rest 50 per cent.

Sharks belonging to genus *Chiloscyllium* have elongated body and trunk shorter than tail. Their eyes are without nictitating membrane. Nasoral grooves and cirri are present. Spiracles are situated below eyes. Five pairs of gill-openings are visible. Two spineless dorsal fin, first dorsal behind the pelvics and anal fin are present. 1-3 dermal ridges are seen on the back. Teeth are small triangular, with or without lateral cusps.

Fishes below to this group are distributed in the Indian Ocean along the coast of South Africa, Red Sea, India, Sri Lanka, Singapore, East Indies, Indo-China, Formosa, Japan, Philippines, Australia and Melanesia.

In *Chiloscyllium ocellatum* mouth nearer to snout than the vertical from front eye-edge. A clear large black ocellus is present just above pectoral fin.

In.*Chiloscyllium indicum* three dermal ridges on the back is present.

In *Chiloscyllium plagiosum* mouth is nearer the vertical from front eye-edge than the snout end. No clear large black ocellus present just above pectoral fin. One dermal ridge on the back is present. Origin of first dorsal above middle of ventral bases. Body is with white spots.

In *Chiloscyllium griseum*, origin of first dorsal above ends of ventral bases and body with dark spots.

Fishes of genus *Nebrius* are distributed in Madagascar, Red Sea, India, Sri Lanka, Malay Peninsula, Malay Archipelago, Indo-China, Melanesia and Polynesia.

Their elongated body with spindle-shaped trunk region have short snout. Eyes are small without nictitating membrane. Nasoral grooves and cirri are present. Spiracles are minute located behind eyes and five pairs of gill openings are present. Two spineless dorsal fins, the first dorsal fin situated opposite the pelvics. Anal fin is present. Their teeth are multicuspid.

In *Nebrius concolour*, teeth are situated in three rows and the second dorsal fin is longer than the anal fin. But in *Nebrius ferrugineum* more than three rows of teeth are present and second dorsal fin is smaller than the anal fin.

Sharks of genus *Stegostoma* has got the only species, *Stegostoma varium* are found in Indian and western Pacific Oceans. They are found in India and Sri Lanka. Its body is slender and elongate, tail longer than trunk and snout obtuse. Eyes are small without nictitating membrane. Nasoral grooves and cirri are present. Labial folds are well developed. Spiracles are situated behind eyes. It has got five pairs of gill openings. Two spineless dorsal and anal fin are present. Caudal fin is very elongate. Teeth are trilobed.

Only species of the genus, *Rhincodon typus* is found in India and Sri Lanka. It is the largest shark in the world growing upto 45 feet in length and are also available in tropical Atlantic, South Africa, Seychelles, India, Sri Lanka, East Indies, Philippines, Japan, Australia and East Pacific Ocean.

Its body is fusiform and massive. Snout is broad, flat and short. Eyes are very small without nictitating membrane and located just behind the mouth. Nasoral grooves and cirri are absent. Spiracle small and 5 pairs of gill-openings are present. Two spineless dorsal fins and anal fin are present. Sub-caudal lobe is well developed. Several keels along the sides and caudal pit are present. Teeth are very small, pointed and numerous.

The only species of the genus, *Carcharias tricuspidatus* is recorded from Indian Ocean. It is also found in Atlantic Ocean, South Africa, India, East Indies, Into-China, Japan, Australia, and Tasmania.

With fusiform body, trunk is about more than twice the tail. Eyes are small without nictitating membrane. Nasoral grooves and cirri are absent. Spiracles are small and located behind eyes. There are 5 pairs of gill-openings. Two spineless dorsal fins and anal fin are present. Caudal pit is present. Teeth very large awl-shaped, are smooth except at base where there exists a basal cusp on either side.

Alopias vulpinus, the only species of the genus *Alopias,* found in Atlantic Ocean, Arabia, South Africa, India, Sri Lanka, China, Korea, Japan, Australia, New Zealand, Hawaii and Eastern Pacific Ocean.

With fusiform body its trunk is equal to the extraordinarily elongated upper lobe of the caudal fin. Eyes are large without nictiting membrane. Nasoral grooves and cirri are absent. Minute spiracles are located behind the eyes. Five pairs of gill-openings are present. There are two spineless dorsal fins of which the second dorsal is very small and equal to the anal fin. Caudal pit is present and lateral keel on tail is absent. Teeth are simple, smooth and sharp-edged.

Scyliorhinus capensis the only species of the genus founds in India. The species is also distributed in South and East Africa, Gulf of Aden, Arabia, Gulf of Oman, Andamans, Malay Archipelago, Philippines, Korea, Japan, Australia and South-East Pacific.

The species has got an elongated body, depressed head and slightly shorter trunk than the tail. Snout is obtuse, short or elongated. Large eyes with nictitating membrane. Nasoral grooves are absent or rudimentary. Mouth is wide and nasal cirri are absent or present. Labial folds are present on both jaws or on lower jaw only. Spiracles are present and 5 pairs of gill-openings are narrow not so wide as the orbit. Two spineless dorsal fins; first dorsal fin behind or above the pelvics. Base of anal fin is distinctly longer than the base of the second dorsal. Caudal pit is absent. Teeth in numerous rows, tri-pentacuspid.

In *Scyliorhinus hispidum* first dorsal is larger than the second dorsal.

In *Scyliorhinus indicus* first dorsal not larger than the second dorsal. First dorsal is smaller than the second dorsal. Nasal cirri is well developed.

In *Scyliorhinus burgeri* first dorsal and second dorsal are subequal. Nasal cirri is absent or rudimentary. Origin of first dorsal above the middle of the ventral bases. Transverse bands are present spotted with black.

In *Scyliorhinus quagga* origin of first dorsal is a little in advance of the hind ends of ventral bases. Transverse bands are 20 or more, not spotted with black.

Atelomycterus marmoratum, the only species of the genus is found in India and Sri Lanka. The fish is also found in Malay Peninsula, Malay Archipelago, Siam, Indo-China, Philippines.

Characterized by the slender and elongated body, the trunk of the species is shorter than tail. Its eyes are large, oblong orbit with nictitating membrane. Nasoral grooves are present, but cirri absent. Spiracles are small and located close behind the eyes. Labial folds are well-developed. Five pairs of gill-openings are present. Two spineless dorsal fins, first dorsal fin is behind the ventrals. Base of the anal fin is equal to the base of second dorsal. Caudal pit is absent. Teeth are small, tricuspid, median cusp is the longest.

Proscyllium alcocki is the only species of the genus found in India. It is also available in Formosa and Japan.

The body of the species is slender and elongated. Trunk is shorter than tail. Snout is rounded. Eyes are large with rudimentary nictitating membrane. Nasoral grooves and cirri are absent. Labial folds are rudimentary. Spiracles are small and located close behind eyes. Five pairs of gill-openings are found. Two spineless dorsal fins, first dorsal fin is situated before the pelvics. Anal fin is long. Caudal pit is absent. Teeth are dimorphous on the upper jaw and tetramorphous on the lower jaw.

Physodon mulleri, is the only species of the genus found in India. It is also found in China and Australia.

Characterized by elongated and slender body, the trunk of the species is equal to tail. Snout is elongated and pointed. Eyes are small with nictitating membrane. Labial folds are present only on the lower jaw. Nasoral grooves and cirri are absent. Spiracles are also absent. There are 5 pairs of gill-openings. Two spineless dorsal fins are present along with anal fin. Caudal pit is present. Teeth are smooth, the central ones are smaller than those at the side, which bear swollen bases with oblique and narrow cusps.

Scoliodon sorrokowah has elongated and slender body with trunk equal to tail. Snout is elongated and pointed. Moderate eyes with nictitating membrane. Nasoral grooves and cirri are absent. Labial folds present on both the jaws and spiracles are absent. There are 5 pairs of gill-openings. Two spineless dorsal fins and anal fin are present. Caudal pit is present. Teeth are with smooth edges, all oblique and without swollen bases. Its second dorsal fin is posterior to the base of anal fin.

All the three species of the genus are found to inhabit in the oceans of South Africa, Red Sea, Arabia, Mekran, India, Sri Lanka, Mayanmar, Malay Peninsula, Java, Indo-China, Formosa, Japan and Philippines.

In *Scoliodon palasorrah* the second dorsal fin over end of base of anal fin.

In *Scoliodon walbeehmi* labial fold is extending to the upper jaw.

Aprionodon acutidens, the species of the genus is found in Indian Ocean. It is also found in Western Atlantic, Red Sea, Seychelles, Arabia, Indo-China, Japan, Australia and Micronesia.

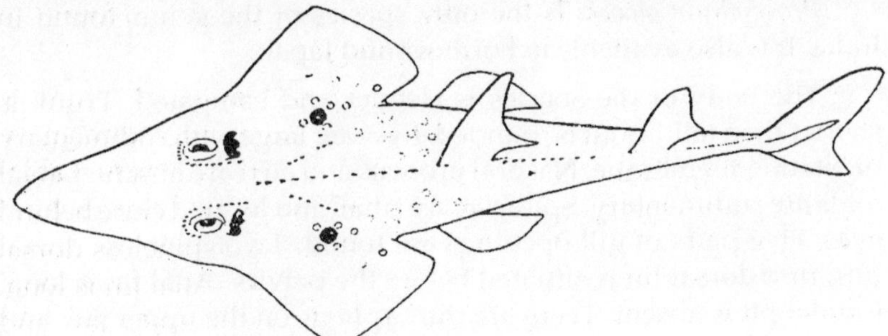

Rhynchobatus djiddensis

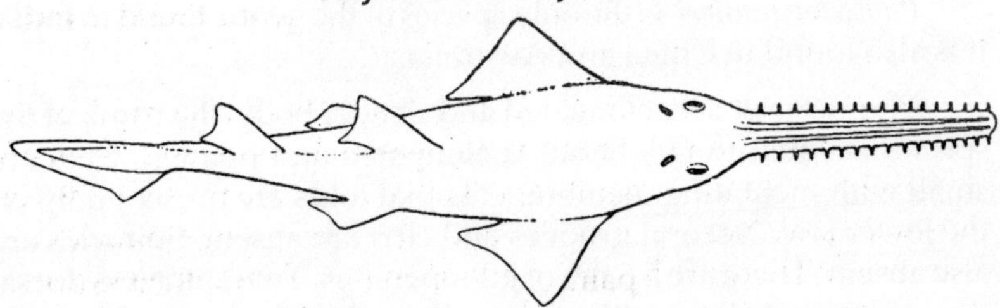

Pristis microdon

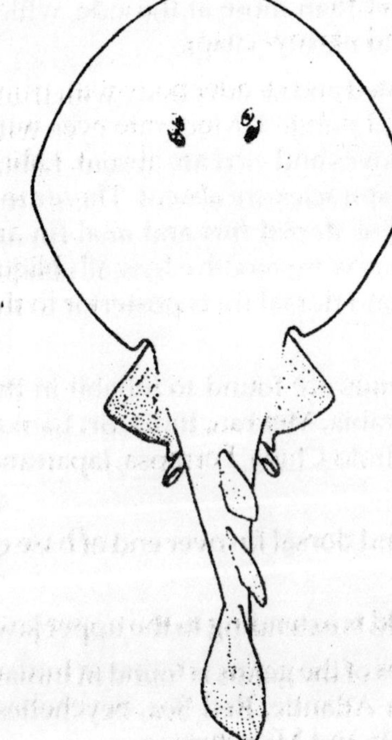

Narcine brunnea

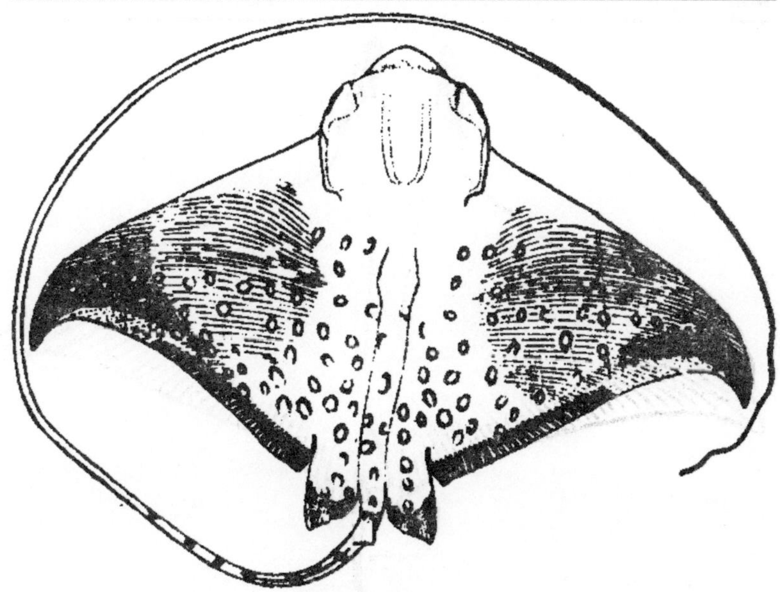

Actomylaeus milvus

Scoliodon laticaudus

The species is characterized by fusiform body, pointed snout and trunk slightly longer than tail. Eyes are moderate with nictitating membrane. Nasoral grooves and cirri are absent. Short labial fold is present at the corner of the mouth in the lower jaw. Spiracles are

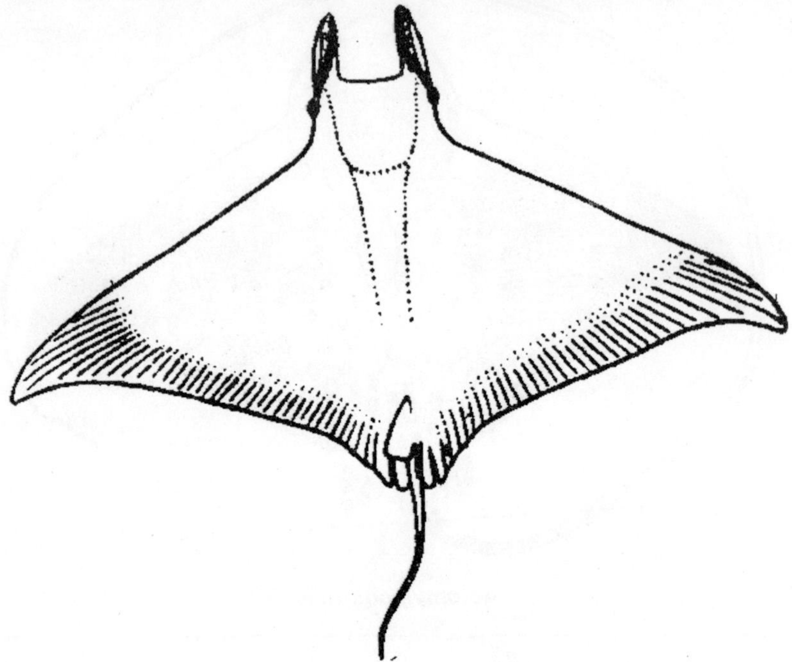

Mobule diabolus

absent, Five pairs of gill-openings are present. Two spineless dorsal fins and anal fin are present. Caudal pit is found. Teeth are small, narrow with broad bases, the lower one is erect and the upper erect or slightly oblique.

Both the species of the genus *Hypoprion* inhabit in Atlantic and Indo-Pacific Oceans. *Hypoprion macloti* has elongated and fusiform body with acutely pointed snout. Its trunk is slightly longer than tail. Eyes are moderate with nictitating membrane. Nasoral grooves and cirri are absent. Short labial fold at the corner of the mouth may be present. Spiracles are absent. Five pairs of gill-openings are present. Two spineless dorsal fin, anal fin and caudal pit are present. Both sides bases of upper teeth are serrated.

In *Hypoprion hemiodon* snout is rounded, Labial folds are absent and only outer sides of bases of upper teeth are serrated.

Twelve species of the genus *Carcharhinus* have fusiform and elongated body with trunk shorter or longer than tail. Their snout may be pointed or rounded. Eyes are moderate with well developed nictitating membrane. Nasoral grooves and cirri are absent. Their labial folds may be short or rudimentary. Spiracles are absent. Five pairs of gill-openings are present. Two spineless dorsal fins, anal fin and caudal pit are present. Teeth are serrated at both bases and cusps, but teeth in lower jaw are non-serrated in some.

The species are distributed in tropical Atlantic, Red Sea, Natal, Seychelles, Madagascar, India, Sri Lanka, Singapore, East Indies, Philippines, Indo-China, Australia, Micronesia, China, Melanesia and Hawaii.

In *Carcharhinus temminckii*, second dorsal fin is larger than anal fin.

In *Carcharhinus ellioti*, teeth in lower jaw are serrated. Teeth are distinctly owl-shaped and only the outer edge of cusp are serrated.

In *Carcharhinus gangeticus*, the snout is very short, eye 8 times in snout and depth of the body to sub caudal origin is 6 times.

In *Carcharhinus lamia*, snout is moderate, eyes are $4^{2}/_{3}$ to $4^{1}/_{2}$ times in snout. Depth of the body to sub-caudal origin is $4^{4}/_{5}$ to 5 times.

In *Carcharhinus menisorrah*, second dorsal fin is not larger than the anal fin. Teeth in lower jaw not serrated.

In *Carcharhinus limbatus*, second dorsal fin is distinctly smaller than the anal fin. Second dorsal and anal origins opposite.

In *Carcharhinus sorrah*, second dorsal fin origin behind anal origin.

In *Carcharhinus melanopterus*, preoral length is distinctly less than width of mouth. All the fins tipped with black. Tail is shorter than trunk.

In *Carcharhinus dussumieri* preoral length more than width of mouth.

In *Carcharhinus bleekeri*, preoral length is equal to width of the mouth.

Hemigaleus balfouri, the only species of the genus is distributed in tropical Indian and Western Pacific Oceans.

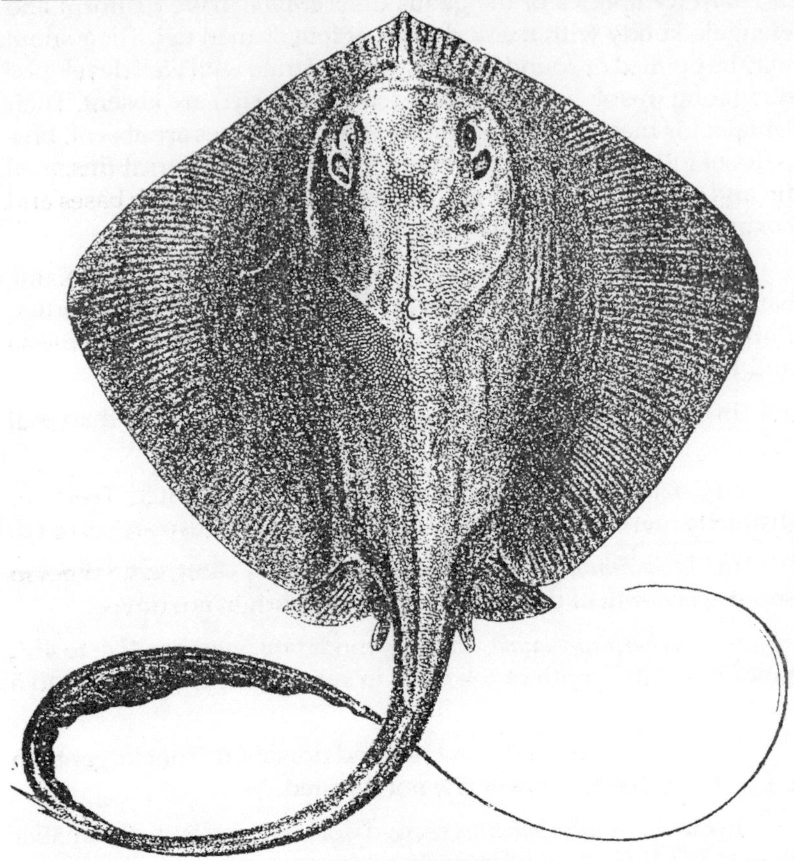

The Sting Ray
Trygon sephen

The species has got slender and elongated body with trunk shorter than tail and pointed snout. Eyes are moderate with nictitating membrane. Nasoral grooves and cirri are absent. Labial folds are present. Spiracles are minute and located behind eyes. There are 5 pairs of gill-openings. Two spineless dorsal fins and anal fin are present. Caudal pits are present. Teeth are dimorphous, upper inclined with denticles on basal part of outer edge and lower erect and smooth.

Hemipristis elongatus is the only species of the genus found in Red Sea, Sri Lanka and India.

The Hammer-Headed Shark
Zygaena blochii

Galeocerdo rayneri

The body of the species is slender and elongated with trunk shorter than tail and rounded snout. Eyes are moderate with nictitating membrane. Nasoral grooves and cirri are absent. Labial folds are present. Spiracles are small and located close behind eyes. Five pairs of gill-openings are present. Two spineless dorsal fin, anal fin and caudal pit are present. Teeth are dimorphous, upper teeth are large, broad, flat and serrated and the lower ones are smooth, slender and curved inwards.

Galeorhinus omanensis, the only species of the genus is found in India. It is also found in Atlantic Ocean, South Africa, Gulf of Oman, Formasa, Japan, Philippines, Australia, New Zealand, Melanesia. Polynesia, Hawaii and California.

With long and slender body the species has trunk equal to tail and obtuse and depressed snout. Moderate eyes with nictitating membrane. Nasoral grooves and cirri are absent. Labial folds are present. Minute spiracles located behind eyes. There are 5 pairs of gill openings. Two spineless dorsal and anal fins are present. Caudal pit is absent. Teeth are monomorphous, oblique, notched and smooth.

Galeocerdo arcticus is the only species of the genus found in India and Sri Lanka. They are also found in Arctic, Temperate and Tropical seas.

The fish with depressed head and elongated body has trunk more or less equal to tail. Snout is wide and short. Eyes moderate with nictitating membrane. Nasoral grooves and cirri are absent. Labial folds present on both jaws. Spiracles are small and located behind eyes. Five pairs of gill openings are present. Two spineless dorsal and anal fin are present. Teeth are large, flat, triangular, notched, oblique and serrated on both edges.

Myrmillo manazo, the only species of the genus is found in India and Sri Lanka. It is also distributed in Atlantic, South Africa, Red Sea, Indo-China, Korea, Japan, Australia, Tasmania, New Zealand.

With elongate and fusiform body the trunk of the species is more or less equal to tail. Snout is produced and pointed. Eyes are moderate with nictitating membrane. Nasoral grooves and cirri are absent. Well developed labial fold is found at each angle of mouth. Spiracles are small and located behind eyes. Possess 5 pairs of gill openings. Two spineless dorsal and anal fins are present. Caudal pit is present. Teeth are monomorphous, polyserial, pavement-like, smooth, obtuse and devoid of distinct cusps.

Triaenodon obseus, the only species of the genus is found in India and Sri Lanka. It is also available in Red Sea, Arabia, East Indies, Philippines, Melanesia, Polynesia and Hawaii.

The species has got an elongated body and trunk longer than tail. Snout is short and rounded. Eyes are moderate with nictitating membrane. Nasoral grooves and cirri are absent. Labial folds are short, not extending along the jaws. Spiracles are absent, but 5 pairs of gill-openings are present. Two spineless dorsal fins, anal fin and caudal pit are present. Sub-caudal are well developed. Teeth are minute, numerous in both jaws with central and lateral cusps.

Sphyrna group consisting of four species have elongated body, head 'T' shaped forming oculonarial expansions on either side. Trunk behind the head is compressed. Eyes are moderate with nictitating membrane, situated at the oculonarial extremities. Nasoral grooves and cirri are absent. Labial folds are rudimentary. Spiracles absent but 5 pairs of gill-openings are present. Two spine less dorsal fins, anal fin and caudal pit are present. Subcaudal lobe is produced. Teeth are monomorphic, oblique and notched, serrated or non-serrated.

The species is distributed in Atlantic Ocean, Mediterranean Sea, Indian Ocean and Malay Archipelago.

In *Sphyrna mokarran*, no groove along the front edge of head.

In *Sphyrna blochii*, eyes and nostrils are widely seperated, oculonarial expansion is long.

In *Sphyrna tudes*, a groove is present along the front edge of head. Eyes and nostrils are not widely seperated, oculonarial expansion is short. Anterior edge of oculonarial expansions are curved.

In *Sphyrna zygaena*, anterior edge of oculonarial expansions are straight.

Centrophorus rossi is the only species of the genus found in Eastern Atlantic, Mediterranean, South Africa, Arabian Sea, Japan, Philippines and New Zealand.

Its elongated and partly cylindrical body has trunk longer than tail. Snout is spatulate and much produced. Eyes are large without nictitating membranes and elongated orbit. Nasal grooves and cirri are absent. Labial folds are well developed in both jaws. Spiracles are large, nearly equal to eye and slightly larger than gill-slits. Five pairs of gill-openings are present. Two dorsal fins each with a spine present anteriorly. Anal fin and caudal pit are absent. Teeth are unicuspid, those of upper jaw are acute, triangular in two series and those of lower jaw are oblique, in single series.

Centroscyllium ornatum has elongated fusiform body, trunk longer than tail. Snout is depressed. Eyes large with nictitating membrane and elongated orbit. Nasoral grooves and cirri are absent. Labial folds are well developed in both jaws. Spiracles are large behind eyes, higher and superior. Five pairs of gill-openings are present. Two dorsal fins each with a well developed spine anteriorly are present. Anal fin is absent. Caudal pit absent and caudal fin is trunctated or pointed. Teeth are small, raptorial and multicuspid.

Distributed in Arabian Sea, Bay of Bengal, Japan and Hawaii islands.

Rhinobatos comprising of six species have elongated and depressed body. Their disk is triangular, slightly rounded and wider behind. Tail is depressed nearly equal to trunk. Snout is triangularly pointed. Nostrils are oblique and wide. Spiracles are wide, just behind the eyes, with folds on hind edge. Five pairs of gill-openings are on the ventral side. Two spineless dornal fins located behind the pelvics and closer to caudal than to snout end. Pelvics are closer to the pectorals than to the dorsals. The rayed portion of the pectoral fins are not continued on to the snout. Anal fin is absent. Teeth obtuse with indistinct transverse ridge.

Fishes of this group are distributed in West and South Africa, Red Sea, Arabia, India, Sri Lanka, Andamans, Mayanmar, Malay Peninsula, Malay archipelago Siam, Formosa, Philippines, Japan and Australia.

In *Rhinobatos thouniniana*, snout is long and pointed. Preorbital length $3\text{-}3^3/_5$ times the distance between spiracles.

In *Rhinobatos granulatus*, snout is not expanded at the tip. Length of nostril equal to internarial space and twice the width of mouth. Preorbital length is $3^1/_2\text{-}3^2/_3$ times the distance between the spiracles.

In *Rhinobatos armatus*, length of nostril is greater than internarial space and less than twice the width of mouth. Preorbital length is 3-$3^1/_5$ times the distance between the spiracles.

In *Rhinobatos obtusus*, snout is short, obtusely pointed. Preorbital length $2^1/_5$ times the distance between the spiracles. Length of nostril equal to internarial space and twice the width of mouth.

In *Rhinobatos annandalei*, snout is not obtusely pointed. Preorbital length $2^1/_3$-$2^1/_2$ times the distance between the spiracles. Length of nostril is greater than internarial space and less than twice the width of mouth. Base of first dorsal fin is $2^1/_4$-$2^2/_5$ times the distance between dorsals. The space between rostral ridges is rather narrow and a series of spines in the middle line of back.

In *Rhinobatos lionotus*, the base of first dorsal is $2^4/_5$ times the distance between the dorsals. Space between rostral ridges is broader and minute tubercles are present in the middle line of back.

Rhina uncylostoma the only species of the genus is found in India and Sri Lanka. It is also found in Red Sea, Arabia, East Africa, Seychelles, Penang, East Indies, Philippines, China, Japan and Australia.

The body is depressed and elongated. Disk is subtriangular, obtusely rounded in front. The tail is depressed nearly equal to trunk. Snout is broad and obtusely rounded. Nostrils are slighty oblique and wide. Spiracles are large, without posterior folds and about an eye diameter and a half behind the eyes. Five pairs of gill openings are located on the ventral side. Two spineless dorsal fins are present, first dorsal fin is situated opposite to pelvics and nearer to snout end than to caudal end. The rayed portion of pectorals extends only upto spiracles. Anal fin is absent. Teeth are obtusely rounded, each with several longitudinals ridges.

Rhynchobatus djiddensis is the only species of the genus recorded from India and Sri Lanka. It is also present in East Africa, Madagascar, Seychelles Zanzibar, Red Sea, Arabia, Andamans, Malay Peninsula, Malay Archipelago, Japan, China, Melanesia.

With depressed and elongated body the disk is triangular, longer than wide. Tail is depressed, nearly equal to trunk. Snout is triangularly pointed and nostrils are oblique. Spiracles are large, close behind eyes and with two small folds on hind edge. Five pairs of gill-openings is situated on the ventral side. Two spineless dorsal fins, first dorsal opposite to pelvics and nearer to snout end than to caudal end. The rayed portion of pectorals extend only upto spiracles. Anal fin is absent. Teeth are obtuse, pavement like, dental surfaces undulated.

Pristis, comprising of four species have elongated and moderately compressed body. The rostrum of the species is very much produced and saw-like and nostrils are oblique. Eyes without nictitating membrane. Spiracles are large and located behind the eyes. Five pairs of gill-openings are situated on the ventral side. Two spineless dorsal fins are present. Pectoral fins are moderate, front edge is quite free neither joining the head nor reaching snout. Anal fin is absent. Rostral teeth are strong and large, set in sockets on either edges of the blade-like snout. Oral teeth are small. Pavement-like.

The species of the genus are found in East Africa, Madagascar, Seychelles, Zanzibar, Red Sea, Arabia, India, Mayanmar, Sri Lanka, Andamans, Malay Peninsual, Malay Archipelago, Siam, Philippines, Japan, Melanesia, Queensland, and Tropical Atlantic.

In *Pristis microdon,* first dorsal fin originate clearly in front of pelvics. Rostral teeth are less in number, 17-20 on either side.

In *Pristis pectinatus,* first dorsal fin originate opposite to pelvic fins.

In *Pristis cuspidatus,* first dorsal fin origin distinctly behind pelvics. Rostral teeth are more in numbers 23-35 on either side. Upper margins of the dorsals are deeply concave with the posterior lobes produced and sub-caudal lobe is well developed

In *Pristis zijsron* upper margins of the dorsals are slightly concave with their posterior lobes not produced and sub-caudal lobe not well develops.

Zanobatus schoenleinii is distributed in India and Africa.

The disk of the species is wider than long, partly rounded and depressed, slender tail nearly half the total length. Rostral cartilage is small. Snout short, obtuse and nasal grooves are rudimentary. Nostrils are situated transversely and internarial space is about 2/3 of mouth width. Spiracles are close behind eyes and 5 pairs of gill-openings are located on ventral side. There are two spineless dorsal fins. The rayed portion of the pectorals continued to the snout to form subcircular disk. Anal fin is absent and teeth are very small.

Raja group consisting of five species are located in Atlantic Ocean, Gulf of Aden, Mediterranean Sea, South Africa, Arabian Sea, Indian, Andamans, East Indies, Formosa, Philippines, Korea, Japan, China, Manchuria, Kamchatka, Alaska, Australia, New Zealand, California and Washington.

Fishes of this group is characterized by subcircular to quadrangular disk. Tail not whip-like without spines and without fold along either side. Their snout are produced and pointed and eyes are prominent. Nasoral grooves are present and five pairs of gill-openings located on the ventral side. Two spineless dorsal fins are present and the rayed portion of the pectorals reaches beyond the eyes but not upto the snout. Anal fin is absent. Teeth are small, tessellate, flat to sharply pointed.

In *Raja mamillidens,* snout is about $3\text{-}3^1/_2$ times the interorbital distance. Dorsals are very close together. A single row of prominent spines on tail is present.

In *Raja resersa,* more than one row *i.e.* 3 rows of prominent spines on tail is present.

In *Raja johannis-davisi,* snout is about 4-5 times the interorbital distance. Dorsals are widely seperated by a distance of about the length of the base of the first dorsal fin. A single row of spines on tail. Second dorsal fin is situated away from the tip of the caudal fin by a distance equal to bases of both the dorsals.

In *Raja powelli,* more than one row of spines (three rows) are present. Second dorsal fin is situated nearer to tip of the caudal fin by a distance equal to or less than the base of the first dorsal fin. Interdorsal space is equal to base of first dorsal fin. There is no prominent rostral spine.

In *Raja andamanica,* interdorsal space is half of the base of first dorsal fin. Prominent rostral spine is present.

Genus *Taeniura* consisting of two species are distributed in Zanzipar, Mozambique, Mauritius, Red Sea, Arabia, India, Sri Lanka, Malay Peninsula East Indies, Siam, Philippines, Australia, Melanesia and Polynesia.

The fishes are characterized by rounded disk with head not distinct from it. Tail is compressed, longer than body, with midcaudal serrated spines and without any lateral folds. Mouth with buccal processes *i.e.*, small flaps of skin accross the floor of the mouth. No rostral cartilage is present. Nostrils are slightly oblique. Spiracles wide located behind the eyes. Five pairs of gill-openings located on the ventral side. Dorsal fin are absent. Rayed portion of the pectoral fins united anteriorly. Anal fin is absent. Sub-caudal rayless, below terminal end of tail. Teeth are small, tessellate, and grooved transversely.

In *Taeniura meyeni*, mouth is straight with five buccal processes.

In *Taeniura lymma*, mouth is curved with two buccal processes.

Dasyatis comprising of 14 species are distributed in Atlantic Ocean, Cape of Good Hope, Natal, Zanzibar, Red Sea, Arabia, India, Sri Lanka, Mayanmar, Malay Peninsula, Malay Archipelago, Siam, China, Formosa, Japan, Philippines, Australia, Melanesia, Micronesia, Polynesia and Hawaii.

Fishes of this group have oval to rhomboidal disk, elongate whip-like tail with serrated caudal spines, with or without dermal finfolds, not terminal in position but behind spines without lateral folds on caudal base. The species do not have rostral cartilage. Nasoral grooves are present and nostrils are slightly oblique. Spiracles are large and located behind the eyes. Five pairs of gill-openings are located on the ventral side. Rayed dorsals fins are absent. Rayed portion of the pectoral fins united anteriorly. Anal fin is absent. Teeth are flattened or with a central point or transverse ridge.

All the species of *Dasyatis* are carnivorous feeders and that *Dasyatis uarnak* appeared to be a voracious one, taking in a wide varieties of pelagic and benthic organisms. To cite an instance, one female *D. uarnak* of 1450 mm across the disk had in its stomack 88 numbers of medium sized prawns, 8 *Nemipterus* spp, 3 mackerels, 8 Leiognathids, 8 Apogonids, 3 lesser sardines, 3 *Anchoviella* spp, 2 flat fishes, 1 *Pentaprion longimanus*, 4 *Platycephalus* spp, 3 puffer fishes, 5 squids, 2 crabs and 2 molluscan shells. About 67 per cent of the volume of stomach of this ray contained fishes shows that this species is highly piscivorous. On the other hand the intensity of feeding on fish is less in *D. alcockii* as evidenced by the presence of fishes with

50–60 per cent of the total volume of the diet. This is further ruduced to 30.10 per cent in the case of *D. sephen*. *D. alcockii* and *D. sephen* appeared to prefer mostly benthic organisms like bottom dwelling polychaetes, gastropods, bivalves, crabs, sea squirts and flat fishes. In the youngest rays, where the yolk sac had just been absorbed, the food consisted almost without exception, of small crustaceans like Amphipods, Carangonids and Isopods. As the young ray grow up in size the food consisted of larger crustaceans and of small fishes.

In *Dasyatis pastinaca*, there is a dorsal cutaneous fold on tail.

In *Dasyatis sephen*, there is a ventral cutaneous fold on tail, which is well developed, broad and 4 times the length of caudal spine.

Fishes, crustaceans, molluscs and polychaetes form the major food items of *D. sephen*. Among fishes, leiognathids, *Nemipterus* spp and soles form the main components of the food of this species. Crustaceans constitute the second important item of food followed by crabs. Polychaetes, a majority of which are *Nereis* spp. constituted the most dominent item of the total diet of this ray. Among molluscs, gastropods forms a part of their food along with ascidians.

In *Dasyatis bennetti*, cutaneous fold on tail not well developed, narrow and about as long as caudal spine.

In *Dasyatis zugei*, there is no buccal processes.

In *Dasyati jenkinsii*, with dorsal and ventral cutaneous folds on tail, 4 buccal processes.

In *Dasyatis imbricatus*, tail short, scarcely as long as length of disk, with 2 buccal processes.

In *Dasyatis imbricatus* (Schneider) only the ovary and the uterus on the left side are functional. It has a prolonged breeding season from December to July and produces one or two embryos. The females reach maturity when they attain the size of 170-179 mm. The males in the 160-169 mm size group are mature as judged by the relationship between body length and clasper length. Sexes in the commercial catches vary from 51.25 per cent males to 48.75 per cent females.

In *Dasyatis marginatus*, tail is long exceeding the length of disk. Without ocelli on dorsal surface of the disk.

In *Dasyatis kuhlii*, blue ocelli is present on dorsal surface of the disk.

In *Dasyatis bleekeri,* tail is more than 3 times the length of disk. The species is with 2 buccal processes.

In *Dasyatis favus,* the tail is less than twice the length of disk.

In *Dasyatis microps,* the tail is short, nearly as long as length of disk. The species is with 4 buccal processes.

In *Dasyatis alcocki,* tail is not banded.

Fishes, crustaceans and molluscs constitute the major food items of fishes of this species. Polychaetes are also recorded in their stomach. In this species, fishes form the major constituent of food forming. 50.56 per cent of the total volume of the stomach content. Among the fishes in the diet, mackerels dominate. Leiognathids and young eels rank next in importance. The other fishes found in the stomach are Apogonids and *Platycephalus* spp.

Crustacean rank second in importance of which crabs are the major constituents among crustaceans. Prawns and *Squilla* are next to crabs. Among the molluscs, squids, gastropods and bivalves form importantant food items of the species. Polychaetes including free-dwelling and tube-dwelling forms form a minor constituent of their food.

In *Dasyatis uarnak,* banded tail. Teeth 25-38 rows in both jaws.

The species is a voracious carnivorous feeder, feeding on a variety of pelagic and benthic organisms. Teleostean fishes of wide range of families constituted the major item of food forming about 67 per cent of the total volume of stomach contents. They feed on fishes of Leiognathid family followed by *Anchoviella* spp, mackerels, *Nemipterus* spp, *Thrissocles,* Sciaenids, soles, *Platycephalus* spp and fishes of Apogonids and Polynemid families.

In *Dasyatis gerrardi,* teeth 13 rows in upper and 23 in lower jaw.

Urogymnus africana is found in India and Sri Lanka. It is also distributed in East Africa, Seychelles, Red Sea, Arabia, Malay Peninsula, Siam, Borneo, Java, Philippines and Melanesia.

The species has subcircular and profusely tuberculated disk, feeble tail, about as long as disk length without caudal spine. Nasoral grooves are rudimentary and cirri is present. Rostral cartilage is absent. Spiracles are large and found close behind eyes. Five pairs of gill-openings are situated on the ventral side. Rayed dorsal fins are absent. Rayed portion of pectoral fins united anteriorly. Anal fin is absent. Teeth are tessellated, flattened and rohmboid.

Gymnura comprise of 4 species have disk much wider than long. Tail is short, slender and serrated with spine. Rostral cartilage is absent. Nasoral grooves are present. Spiracles are large, located close behind eyes. Five pairs of gill-openings located on the ventral side. Rayed dorsal fins are absent. Rayed portion of the pectoral fins united anteriorly. Anal fin is absent. Teeth are minute, numerous in broad bands, each tooth with one to three cusps.

The species are distributed in South Africa, Natal, Red Sea, India, Sri Lanka, Mayanmar, Singapore, Malay Archipelago, Siam, China, Japan, Korea, Philippines, Australia, Polynesia.

In *Gymnura tentaculata*, a small dorsal cutaneous fold is present on the tail. Tentacles are situated behind spiracles.

In *Gymnura zonurus,* no tentacles are present behind spiracles.

In *Gymnura peocilura*, there is no dorsal cutaneous fold on tail. Tail about as long as length of disk.

In *Gymnura micrura*, tail is less than half the length of disk.

Aetomylaeus comprise of 4 species, are characterised by lozenge-shaped disk about twice as broad as long. Tail is whip-like, much longer than disk and without caudal spine. Head moderately conspicuous, rostral fins forming a unilobed snout. Eyes are lateral and nasoral grooves are present. Spiracles large located behind eyes. Five pairs of gill-openings on the ventral side. First dorsal fin is small, situated at basal part of tail. The second dorsal and anal fins are absent. Rayed portion of the pectorals is falciform and extending only upto the posterior region of the orbit. Teeth in 3 rows of which the lateral narrower than only upto the posterior region of the orbit. Teeth in 3 rows of which the lateral narrower than the central ones. The species are distributed in Red Sea and Indian Oceans bordering India, Sri Lanka Mayanmar, Malay Peninsula, Malay Archipelago, Siam, China, Japan, Philippines and Australia.

In *Aetomylaeus maculatus,* the origin of dorsal fin is behind ends of pelvic bases.

In *Aetomylaeus nichofii cornifera,* the origin of dorsal fin is opposite ends of pelvic bases. Orbital horns present.

In *Aetomylaetus nichofii*, about 5 blue cross bands on disk are present. Spiracles twice size of the eye and orbital horns are absent.

In *Aetomylaetus milvus*, green brown edged ocelli are present on hind part of the disk. Spiracles about the size of eye.

Aetobatus narinari, the only species of the genus is found in India, Mayanmar and Sri Lanka. It is also distributed in tropical Atlantic Ocean, Natal, Red Sea, Arabia, Seychelles, Malay Peninsula, Malay Archipelago, Siam, China, Philippines, Melanesia, Polynesia, Micronesia and Hawaiian groups.

The species is characterized by lozenge-shaped disk, about twice as broad as long with whip like tail, larger than the length of the disk with a serrated caudal spine. Its head is conspicuous, rostral fins forming an unilobed pointed snout. Eyes are lateral, nasoral grooves are present. Spiracles large about twice eye diameter and laterally situated about an eye-diameter and a half behind eyes. Five pairs of gill-openings are situated on the ventral side. First dorsal fin is small, situated at basal part of tail and second dorsal and anal fins are absent. Rayed portion of the pectorals is falciform extending upto the anterior margin of the spiracles. Teeth is situated in a single row.

The genus *Rhinoptera* comprising of four species are distributed in Mediterranean, Africa, Arabian Sea, Muscat, India, Sri Lanka, Malay Peninsula, Java, Siam, China, Philippines, Australia, Brazil and Lower California.

Their lozenge-shaped disk about twice as broad as long has whip-like tail, longer than disk with basal serrated spine. Head is somewhat conspicuous, rostral fins forming a bilobed snout. Eyes are prominent, lateral and nasoral grooves are present. Spiracles are large, located behind eyes and open laterally. Five pairs of gill-openings are situated on ventral side. First dorsal fin is above basal part of the tail. Second dorsal and anal fins are absent. Rayed portion of the pectorals are falciform and not joined with the rostral fins in the front. Teeth are wide, angular, flat in pavement, median row is widest.

In *Rhinoptera javanica*, teeth in 7 rows are present in upper jaw.

In *Rhinoptera sewelli*, teeth in 9 rows are present in lower and upper jaws.

In *Rhinoptera jayakari,* teeth are present in median row 8 times as wide as long.

In *Rhinoptera adspersa,* teeth are present in median row 3 times as wide as long.

Mobula diabolus, the only species of the genus is found in Red Sea and Indian Oceans bordering Arabia, India, Sri Lanka, Penang, Malay Archipelagos, Philippines and Queensland.

It is characterized by lozenge-shaped disk, about twice as broad as long with short whip-like tail in the young $1^1/_2$ times the length of disk and in the adults a little more than half the disk length, with or without serrated spine. Head is conspicuous, broad and flat with two curled, cephalic horns. Mouth is inferior, well behind the head, eyes large and lateral. Nasoral grooves are present. Spiracles moderate located behind the eyes. Five pairs gill-openings are located on the ventral side. First dorsal fin is small, triangular above and between pelvics. Second dorsal and anal fins are absent. Rayed portion of the pectorals are falciform extending upto the post orbital region. Teeth are small, numerous in both jaws or at least in the upper jaw.

Manta birostris, the only species of the genus is found in Atlantic Ocean. South Africa, Red Sea, India, Malay Peninsula, Malay Archipelago, Melanesia, North and South America, Galapagoes Islands and west Indies.

The fish is characterized by lozenge-shaped disk about twice as broad as long, with whip-like tail about as long as disk length without serrated caudal spine. Head is greatly depressed, broad and flat with two cephalic horns, rarely curled. Mouth is large, terminal in front of head. Eyes are prominent, lateral. Nasoral grooves are present. Spiracles are moderate located behind eyes. Five pairs of gill-openings situated on ventral side. First dorsal fin is small, above and between pelvics. Second dorsal and anal fins are absent. Rayed portion of pectorals are falciform extending upto the postorbital region. Teeth are small, numerous in pavement usually on lower jaw and sometimes on both jaws.

The occurrence of the species *Manta birostris* off Veraval, north-west coast of India, described as the largest of all rays growing to a width of 24 feet across the disk, weighing over two tonnes were caught in 25 m high-opening trawl net at a depth of 40 metres. This giant ray has a terminal mouth wide at front of the head and in between the cephalic fins (horns). Their skin is rough with small tubercles. The fish is dark grey in colour on the dorsal surface and white on the ventral surface. This unusually large "Devil Ray" had large liver weighing 160 kg. The fish is found to have habit of leaping repeatedly from the water causing booming splashes audible and visible many miles away on being frightened by approaching net in the fishing ground.

The genus *Narcine* comprising of three species is characterized by sub circular disk with head not distinct from it. Tail with lateral folds, slightly shorter than the length of disk and without serrated caudal spine. Snout is broadly rounded twice the interorbital distance. Rostral cartilage and nasoral grooves are present. Spiracles are large, situated close behind the small eyes. Five pairs of gill-opening are present on the ventral side between the electric organs. Two spineless dorsal fins are situated on tail. Anal absent, but pelvics are well developed. The rayed portion of the pectorals is continued to the orbital region. Teeth are in narrow bands.

Fishes of this genus are distributed in India, Sri Lanka, Malay Archipelago, China, Japan, Philippines, Australia and Tasmania.

In *Narcine brunnea*, the above part of the body is uniformly brown.

In *Narcine indica*, brown spots are present on the dorsal part of the body. Teeth are in 27 rows in upper jaw and 26 in the lower jaw.

In *Narcine timlei*, there are teeth in 23 rows in upper jaw and 21 in the lower jaw.

Benthobatis moresbyi, the only species of the genus is available in Indian Ocean and Arabian Sea.

The species has an oval disk with head not distinct from it. Tail is without lateral folds, slightly longer than disk length and without serrated caudal spine. Snout is broadly rounded, about twice the interorbital distance. Rostral cartilage and nasoral grooves are present. Eyes are obsolete, as two unpigmented spots. Spiracles are moderate, close behind the eye spots. Five pairs of large gill-openings are situated on the ventral side between the electric organs. Two spineless dorsal fins are present on tail and pelvics are well-developed. The rayed portion of the pectorals extends only to the orbital region. Anal fin is absent. Teeth are small as rhomboidal plate with the crown strongly and acutely produced.

Two species of the genus *Torpedo* are distributed in East Africa, Madagascar, Mauritius, Seychelles, Red Sea, India and Philippines.

They are characterized widely circular disk with head not distinct from it. Tail is very short with lateral folds and without serrated caudal spine. Snout is broadly rounded and equal to interorbital distance. Rostral cartilage is present but reduced. Nasoral grooves are present. Eyes are well-developed. Spiracles are moderate, located close behind eyes. Five pairs of gill-openings located on the ventral side between the electric organs. Two spineless dorsal fins are present on tail. Pelvics are well developed. The rayed portion of the pectoral extends to orbital region. Anal fin is absent. Teeth are small, in pavement, irregularly rhomboidal with crown obliquely pointed.

In *Torpedo mormoratus*, body is speckled with small light blotches and teeth are present in 18 rows.

In *Torpedo sinus-persici*, body is variably speckled with irregulars large and small dark spots and teeth are present in 20 rows.

Narke dipterygia, the only species of the genus is available in Arabian Sea, India, Sri Lanka, Malay Peninsula, Indo-China, Japan and Philippines.

The species is distinguished with circular disk with head not distinct from it. Tail is slightly shorter than the disk length, with lateral folds and without serrated caudal spine. Snout is broadly rounded and about one and a quarter times the interorbital distance. Rostral cartilage is present but in reduced form. Nasoral grooves are present. Spiracles are large and situated close behind eyes. Five pairs of gill-openings are present on the ventral side. There is one spineless dorsal fin. Pelvic and pectoral fins are well developed. The rayed portion of the pectorals extends to the orbital region. Anal fin is absent. Teeth are in narrow band, small, quadrangular with the crown not strongly produced.

Bengalichthys impennis, the only species of the genus is distributed in Indian Ocean. The species is characterized with oblong disk. Tail is slightly shorter than the disk length, with lateral folds and without serrated caudal spine. Snout is broadly rounded and one and a quarter times the interorbital distance. Rostral cartilage is present in reduced form. Nasoral grooves are present. Eyes are minute. Spiracles moderate located close behind eyes. Five pairs of small gill-openings present between electric organs. One spineless dorsal fin is present. Pectoral and pelvic fins are not well developed. The rayed portion of the pectorals extends only upto the orbital region. Anal fin is absent. Teeth roughly quadrangular, small, with the crown not much produced.

Chimaera monstrosa, the only species of the genus is distributed is South Africa, India, Sumatra, China, Korea, Japan, Philippines, Australia, Hawaiian groups, North and middle Atlantic Ocean.

The species has elongated body and shark-like in form tapering posteriorly to a point at tail. Head is large compressed and without proboscis or beak. Eyes are large or moderate and lateral. Mouth is inferior, nasoral grooves are present. Spiracles are absent. One gill-opening on either side of pharynx containing four gill-slits and four gills covered over by a skinny operculum. Two dorsal fins, the first dorsal with a strong spine anteriorly and the second dorsal is long and low. Pectorals are large, free and low. Pelvics abdominal and many rayed. Anal fin is small, distinct or not distinct from subcaudal. Mature males with trifid or rarely bifid claspers. Skin is naked, devoid of placoid scales. Teeth united to form bony plates, laminae or tritors, 4 tritors in the upper jaw and 2 tritors in lower jaw.

Harriotta indica is the only species of the genus found in India. It is also found in Gulf of Aden, Bay of Bengal, Japan and Atlantic Ocean.

The fish has got an elongated body tapering to a long tail with a filamentous tip. Head is provided with long rostral proboscis. Snout is depressed. Eyes are moderate, lateral. Mouth inferior. Spiracles are absent. One gill-opening on either side of pharynx, containing four gill-slits covered by a skinny operculum. Two dorsal fins with a spine anteriorly are present of which second dorsal is low. Pectorals are large and free. Anal fin is small and distinct from sub-caudal. Supra-caudal moderately high, upper edge without spines. Mature male is with a simple clasper. Teeth are with tritors.

Rhinochimaera sp has been recorded from Arabian Sea, off Travancore coast. It is also found in Japan.

The species has an elongated body tapering posteriorly to the long tail with a filamentous tip. Head is provided with long rostral proboscis. Snout is compressed, eyes moderate, mouth inferior and spiracles are absent. One gill-opening is present on either side of pharynx containing four gill-slits and four gills are covered over by a skinny operculum. Two dorsal fins, first dorsal with spine anteriorly and second dorsal is low. Pectorals are large and free. Anal is small not distinct from subcaudal. Supracaudal low, upper edge armed with spines. Mature male is with a simple clasper. Teeth are united to form grinding plate.

The Mottled Ray, *Aetomylus maculatus* (Gray)

The species appears to be viviparous with no placenta. Ovary and the uterus on the left side only are functional and their counterpart on the right side are rudimentary. The embryos may be liberated with the internal yalksac functioning as a nutritive organ until they are capable of searching for food. The parturition time of the species appears to be April-May. During June and July free swimming young ones of this species attain a size of 255-270 mm across the disk.

Eight embryos of *Aetomylus maculatus* in advanced stages of development (220-240 mm across disk) were recorded from adult females measuring 715-1040 mm across disk at Cuddalore during April-May.

The embryos resembled the parent in all respect except for the cephalic characters. Disk about as wide as long and spiracular openings were twice the diameter of eye. Fleshy snout was short with no horns. Brain case was not sufficiently hardened and the brain is visible from underneath. Teeth are somewhat soft, in several rows, of which median ones are broader. Fin rays are fairly visible and dorsal origin behind the ventral. Spineless tail is about four times the length of the disk and indistinctly banded. Yolk sac is completely absorbed and the stomach is filled with yolk.

Back side is smooth with no tubercles. Colour on the dorsal side is olive-green with rounded bluish brown spots on the posterior half of the body and whole of ventral fins, ventral side are creamy white.

Batoid Fishes

Cuddalore coast in Tamil Nadu extend 11 km with three estuaries, namely, Pennaiyar, Gadilam and Parvanar have been observed an ideal place of concentrations of Batoid fishes, skates and rays. These river mouths along Bay of Bengal coast attract 12 species of Batoid fishes of which *Rhynchobatus dijiddensis, Rhinobatus granulatus, Pristis microdon, Himantura bleekeri, Himantura uarnak, Dasyatis sephen, Actomylus nichofii Aetobatus flagellum* and *Rhinaptera javanica* contribute to the regular fishery, while *Rhinobatus obtusus* occur very rarely. Although *Dasyatis imbricata* and *Dasyatis zugei* occur in this coasts frequently, they do not form a fishery due to their smaller size.

These carnivores and predatory species concentrate near the mouth of Pennaiyar river on the north and Paravanar river on the south along with Gadilam river in between, which drain into the Bay of Bengal at Cuddalore.

Rhynchobatus djiddensis (Forsk) occurs only in certain months and grows 9 ft in total length. The maximum size recorded at cuddalore is 2.02 m in total length. The fish feeds on fishes, molluscs and prawns. The flesh is considered very nourishing whether taken fresh or salted and the fins are highly valued in the export market.

Rhinobatus granulatus (C) is more frequently fished in Cuddalore coast than *R. djiddensis*. Maximum size measures upto 1.48 m in total length and upto 12 embryos have been recovered from the uterus from a single fish. Pregnant female fishes occur from September to February.

Pristis microdon (Latham), the saw fish, occurs seasonally during the south-west monsoon period from July to September. Fishes upto 3.13 m in total length were recorded from the commercial catches. They are abundant in the southern region of the east coast of India and good potential grounds for this fishery is indicated along the north-east coast also. They feed on large numbers of fishes, like, the seer fish, mackerels, pomfrets, small sharks, sciaenids, ribbon fishes, crustaceans and molluscs. They give birth to young ones from May to July, during which period they move close to the shore in sheltered waters.

Himantura uarnak (Forskal) is caught roughly throughout the year. In the Indian oceans fishes ranging to 5 ft across the disk have been recorded. Much havoc is caused to the fishery by this species by devouring large number of fishes, molluscs and crustaceans.

Himantura bleekeri (Blyth) occurs throughout the year, and grows upto 85 cm across the disk along this coast. It feeds on the molluscs, crustaceans, polychaetes and ascidians. Upto two embroys have been found in the uterus during the winter months, that is, December to February.

Dasyatis sephen (Forsk) is observed to occur in the commercial catches all the year round and grows upto 2 m across the disk. The main items of its food are the fishes, prawns and crabs. The breeding season seem to be after the monsoon, as upto two embryos of nearly full term have been recovered from its uterus during February to April.

Aetobatus flagellum (Schn) contributes seasonally to the fishery. The maximum size across the disk varies from 1.15 to 1.58 m. Nearly full-term three embryos have been obtained from the uterus during August/September.

Aetomylus nichofii (Schn) occasionally make their appearances in the catches. They are prized very much as table fish by the fish merchants and grow to a size of 0.69 m across the disk. Fishes, molluscs and prawns forms the food items of these rays.

Rhinoptera javanica (Muller and Henle) is found to grow upto 1.5 m across the disk and contributes occasionally to the fishery. This species move in large shoals and bring havoc to the pearl oyster beds. They mainly feed on the molluscs, and occasionally on the crabs, prawns and fishes.

The elasmobranch catches are most abundant during the third quarter. The catches are fairly high in the first and second quarters, but the fourth quarter catches being the lowest probably due to north east monsoon. The fishing season for the saw fish (*P. microdon*) extend from July to March on the southern part of the east coast and from May to September on the northern part of east coast.

The ray fishery is of importance since they are good marketable food fishes. A very good ray fishery exists in the region of east coast. They are eaten fresh as well as in salt cured form.

Devil rays described by various lchthyologists as *Mobula diabolus, Raja diabolus* and *Dicerobatis eregoodoo,* was reported to attain big size as 5.5 m or more across the disk. 49 numbers of devil rays *(Dicerobatis eregoodoo)* were caught in a single gill net operation off Veraval in Gujrat coast at a depth of 25 m on 23.2.1978. Body measurements of the reported catch showed 1.2 m for the male specimen and 1.3 m for the female specimen across the disk. All the fishes were fully matured and majority of them were females. A fully developed young one was found inside a female fish.

Comparative body measurements of a male and female *Dicerobatis eregoodoo* is given below:

	Male	*Female*
Breadth of disk	1.2 m	1.3m
Length of disk	0.72 m	0.8 m
Total length	1.25 m	1.45 m
Length of tail	0.55 m	0.7 m
Weight	17 kgs	25 kgs.

The rays and skates are caught entangled in gill net. The net is made from nylon material and the mesh size varied from 6"-8" depending on the pieces place in position in the net. The net is operated from the catamaran along the Cuddalore coast at a depth range of 4 to 60 metres. While the net is lowered from the catamaran on the previous day, the catches are hauled up only on the succeeding day. After hauling the catch, the fishermen return to the shore leaving the net set up on the bottom of the sea. On the following day to bring the catches they again sail to the place where they had left the net on the previous day. The rays and skates are occasionally fished by hooks and line. Besides these methods, they are also caught using small trawlers.

Deep Sea Snake Fish [*Acanthocepola limbata* (Cuvier)]

Deep sea snake fishes are common in the coastal waters of the eastern Atlantic and widespread but rare in the central Indo-Pacific. Few of them have been recorded from Indian waters.

A male deep sea snake fish, *Acanthocepola limbata* (Cuvier) measuring 55.6 cm total length and weighing 140 gram was collected from boat seine landings at Karwar on 11th September, 1975. The species have been found to occur in seas around India, Japan, Formosa, Philippines, Indo-Australian Archipelago and Australia.

Fishes with elongated and laterally compressed body gradually taper to caudal region. Scales are cycloid, minute, present on head and opercle. When fresh its body is covered with slime. Eyes are red, large and lateral in anterior half of head. Middle part of eye is bright dark. Mouth is strongly oblique, large and slightly protractile. Median palatine teeth is present. Tongue is triangular in shape. Preopercle is bluntly serrated. Gill openings are wide and semi-circular. Lateral line ascending from the upper corner of the gill opening to dorsal and running very close to its base upto caudal continuous with the caudal fin. Anal fin is long, commences opposite to 8th dorsal ray. First dorsal and anal rays are short. Pectoral fins are short, round, branched and originate beneath 4th or 5th dorsal ray. The dorsal, anal, pectoral, ventral and caudal fins have 83, 93, 19, 6 and 10 fin rays respectively.

In fresh condition, colour of the entire body is fine red with golden yellow bands on sides. The first 15 bands are prominent, while the last 4-5 bands are faded. Membranous portion of dorsal and anal fins are pinkish red. Dark red oblong blotch on the dorsal fin is visible between 9th and 14th rays, covering little above half the length of rays. Black patches are present on sides of mouth under premaxillary and maxillary processes. A white band, 3 mm wide, runs at a distance of 18 mm from the bases of 7th to 16th anal rays.

Chapter 13

Muraenidae-Eel

Fishes of this family have elongated, cylindrical or band-shaped body. Their humeral arch is not attached to the skull. The branchial openings in the pharynx may be wide or narrow slits. Margin of upper jaw constituted anteriorly by the premaxillaries, which are more or less coalescent with the vomer and ethmoid; whilst laterally the sides of the upper jaw are formed by the maxillaries which are furnished with teeth. Vertical fins when present is confluent or seperated by a projecting tail. Pectoral fins may be present or absent. Scales when present are rudimentary. Vent may be situated close to the root of the pectoral fins or a long distance posterior to the head. The heart may be situated just or a long distance behind the gills. Stomach is with a blind sac. Ovaries destitute of oviducts.

In the first group of Muraenidae, heart is close behind the gills. Their tail is longer or scarcely shorter than the trunk. Nostrils are lateral or superior. Tongue is free. Caudal fin continued round the end of tail. Pectoral fins may be present or absent. Scales may be present or absent.

Genus-*Anguilla*

Fishes of this group have moderate gill openings situated near the base of the pectoral fins. Upper jaw is not projecting beyond the lower. Teeth are small and in bands. Dorsal fin commences at some distance behind the nape. Pectoral fins are present. Scales are small and embedded in the skin.

Anguilla bengalensis, a resident of Indian Ocean have head length of 3 to $3^1/_6$ in the distance between the snout and the vent. Length of tail is 3/7 more than that of the trunk. The distance between the gill-opening and the origin of the dorsal fin is 1/3 or 1/4 more than the length of the head. Head is broader than the body. Lower jaw is prominent. Length of the cleft of the mouth equals nearly or quite 1/3 in that of the head, while it extends behind the posterior edge of the orbit, the latter being $2^1/_2$ in the length of the snout. Lips are well-developed. The vomerine band of teeth does not extend posteriorly so far as the maxillary one. The mandibular teeth is divided by a longitudinal groove. Pectoral fins equals about 2/7 of the length of head. The species is brownish is colour superiorly, becoming yellowish on the sides and beneath. The whole upper surface of the body in some fishes are covered with black spots and blotches, some of which are continued on to the dorsal fin which has a light edging. Anal fin is with a dark marginal band and a light outer edging.

Residing in the coasts of India and Andaman seas *Anguilla bicolor*, is of dark olive colour superiorly, becoming yellowish beneath. Their dorsal fin commences above the vent or slightly before or behind it. Bands of teeth of nearly equal width are present, the vomerine reaching nearly as far backwards as those on the maxilla. Eye rather varies in size, usually about $2^1/_2$ diameters in the length of the snout. Head is slightly broader than the body, snout rather broad. Lower jaw is scarcely longer than the upper. Extent of cleft of mouth equal to above one third of the length of the head, and extending to at least one diameter of the orbit behind the eye in the adult to below it in the immature. Lips are thick.

Congromuraena anago living in Indian seas attain 30 cm in length. Length of their head is $2^1/_2$ to $2^2/_3$ in the distance between the end of the snout and the vent. Length of trunk is about $^1/_3$ less than that of the tail. Extent of cleft of mouth equals 3 to $3^1/_4$ in the length of the head. It ceases below the middle of the eye. Eyes are large, diameter, 1 to $1^1/_4$ in the length of snout, and 2/3 of a diameter apart. Lips are rather thick and upper jaw is longer. Teeth are of about the same size. The vomerine band extend backwards to about $^1/_2$ the length of the maxillary band. Dorsal fin commences slightly behind the gill-opening. Pectoral fin is nearly $^1/_2$ as long as head. Fishes are brownish along the back, becoming dull white beneath Fins are yellow. The vertical ones are with a narrow black edging. Upper half of pectoral fins are occasionally stained with black.

Genus-*Moringua*

Fishes of this group have sub-cylindrical body with the trunk considerably longer than tail. Gill-openings are rather narrow and inferior. Heart is located far posterior to branchiae. Cleft of mouth is narrow. The posterior nostril is situated in front of the eye. Teeth are in a single row. Vertical fins are limited to the tail. Pectoral fins if present are small. Scales are absent.

Living in the estuary of Ganges in Bay of Bengal *Moringua raitaborau*, the species attain 55 cm in length. Their length of head from snout to gill opening is 6 to $6^1/_2$ in the distance from snout to vent. Length of tail is from $^2/_5$ to 3/8 of the entire length. Height of body equals $^1/_3$ of length of head. Eyes are rather high up and 2 to 3 diameters from end of snout. Jaws of equal length in front or the lower is slightly longer. Cleft of mouth extends above 1 diameter of the orbit behind its posterior margin. Gill-opening is a slit at the side of the pectoral fin. Teeth in a single row is directed backwards and pointed. Dorsal fin is slightly developed. It commences about the length of the head posterior to the anus. The anal fin arises a short distance behind the orifice of anus. Both fins are interrupted in their middle, but posteriorly developed and join the caudal fin. Pectoral fins equals about one-sixth of its distance from the snout. Lateral-line is distinct. The fish is coppery, olive or even purplish above, becoming silvery underneath. Some black dots are present.

Genus-*Muraena*

Fishes of this genus have moderately or exceedingly elongated body with narrow gill openings. A tubular nostril on either side of the upper surface of the snout is present. The posterior nostril is a round foramen between the eyes or opposite the antero-superior edge of the eye. It may or may not be furnished with a tube. Teeth are well developed and acute or molariform, the maxillary teeth may be in one or two rows. Dorsal fin is either elevated or not so, the end of the tail surrounded by fin, which is occasionally rudimentary. Pectoral fins are absent.

Inhabitant of the seas of Bay of Bengal and estuaries, *Muraena tile* ascends the tidal rivers and attain 60 cm in length. Its length of head is $3^3/_4$ to $4^1/_4$ in the distance between the end of the snout and the vent. Their tail is rather shorter or about the same length as the trunk. Their eyes are rather small and situated slightly nearer the angle of the mouth than to the end of the snout. Length of the cleft of mouth equals about $3^1/_3$ in that of the head. Anterior nasal tubes are short. Gill-opening is about as wide as eye. Teeth are in two rows, except the lateral ones of the lower jaw. Canines are small. They are brownish yellow or greenish in colour, becoming lighter beneath. The body is covered with small white spots, specks or marks, which sometimes have a dark border or they may be entirely absent in adults.

In the Coromandel coast of India, *Muraena punctata* attains to a large size. Their head length is $3^1/_2$ in the distance between the end of the snout and the vent. Tail is longer than the trunk. Eyes are of moderate size, situated midway between the angle of the mouth and end of the snout. Anterior nasal tubes are about half as long as the orbit. Length of cleft of mouth equal to half the length of the head. The mouth can be completely closed. Gill-opening is about twice as large as the orbit. Teeth are in a single row, from 18 to 22 on either side of the mandible, the anterior of which are the longest. Canines are of moderate size. The fish is purplish black with black streaks radiating from the eye. The whole of the fish is covered with pure white spots having a dark edge and which are largest posteriorly, but nowhere exceed the size of the pupil of the eye.

Muraena punctatofasciata, an inhabitant of seas of India attain 33 cm in length. Their length of head is $3^3/_4$ to $4^1/_4$ in the distance between the end of the snout and the vent. Tail is slightly longer than the trunk. Their eyes are situated about midway between the end of the narrow snout and the angle of the mouth, 2/3 the length of snout. Mouth can be completely closed, the extent of its cleft equals $^1/_3$ of the length of the head. Gill-opening is about the same size as the eye. Body is rather slender. Teeth are pointed and in a single row, without any basal lobe. Occasionally there are 2 or 3 teeth forming an inner maxillary row. Canines are of moderate size, 16 to 17 teeths in either mandible. The fish are of reddish brown in colour, darkest along the back. From 28 to 35 dark, more or less complete rings of an irregular characters are present on the body and fins. Head and ground colour between the rings with blotches, spots and fine lines.

Muraena tessellata inhabits in Indian Ocean. Their length of head to gill-opening 3 to $3^2/_3$ in the distance between the end of the snout and the anus. Their tail is slightly longer or shorter than the trunk. Snout is compressed. Eyes are situated slightly before the middle of the distance between the angle of the mouth and the end of the snout. Length of cleft of mouth is $2^1/_3$ to $2^1/_2$ in the distance between the end of the snout and the gill-opening. Anterior nasal tubes are $^1/_2$ as long as eye. Teeth are large compressed and pointed, in a single row in the adult. In the young, occasionally is a short internal row in the maxilla, Dorsal fin commences above the gill-opening. Fishes have dark polygonal or round spots on the head, body and fins, which are seperated by narrow light lines or interspaces. Most or all of the spots are wider than the ground colour.

Muraena fimbriata an inhabitant of Indian Seas and coast of Andaman Islands have length of head 3 to $3^1/_4$ in the distance between the end of the snout and the vent. Tail is one third longer than the trunk. Eyes are rather nearer the angle of the mouth than the end of the narrow snout. Extent of cleft of mouth equal to 1/3 of the length of the head. Gill-opening is smaller than the eye. Teeth are pointed situated in a single row, without any basal lobe. Occasionally there are 2 or 3 teeth in a second inner maxillary row. The fishes are olive brown in colour with a few black spots on the head, and numerous irregularly formed ones on the body, dorsal and anal fins and many take a vertical direction and few are larger than the eye. Fins are with a white edge. In some specimens the spots are in 2 or 3 regular longitudinal rows.

Muraena pseudothyrsoidea have length of head $3^1/_2$ to $3^3/_5$ in the distance between the end of the snout and the vent. Tail is little shorter than the trunk. Eyes are nearer the end of snout than angle of mouth. The extent of cleft of mouth $2^1/_4$ the length of the head. The gill-opening about 1/3 wider than the eye. Teeth are present in a single row, about 18 or 20 in either mandible, the 2 anterior being canines. Canines are of moderate size. The mouth can be completely shut, 1 or 2 of anterior vomerine series are larger than those in the premaxillaries. Fishes are brownish in colour and covered with fine dark spots on the head and body, amongst which are reticulated yellow lines, most distinct in the caudal region. Sometimes a white edge is present to fins. Gill-opening usually surrounded by a black spot. The species inhabit in the coasts of India, Sind to the Malay Archipelago.

Resident of Indian Ocean, *Muraena undulata* is of light brownish is colour covered with irregular sized blotches and usually light reticulated lines over the body more distinct posteriorly. No black spot at gill-opening and no white edge to fin is noticed. Their head length are $3^1/_4$ to $3^1/_3$ in the distance between the end of the snout and the vent. Tail is little longer than the trunk. Eyes are $1^3/_4$ to 2 diameters from the end of the snout and about midway between the angle of the mouth and the end of the snout. Length of mouth cleft is $2^1/_3$ to $2^1/_2$ in the length of the head. Gill-opening is about as wide as the eye. Anterior nasal tube is short. Mouth cannot be shut completely. Snout is pointed. Teeth are in a single row, occasionally one or two additional ones forming an inner row in the maxilla. Normally there are four pairs of canines in the mandibles and 18 to 20 teeth in either ramus of mandibles and 2 canines in the maxilla.

Muraena macrura residing in Seas of India attains more than 3 metres in length. Their length of head from snout to gill-opening is 1/4 of the distance between the end of the snout and the anus. Tail form $1^3/_4$ to twice as long as the trunk. Their eyes are situated in the front 1/2 of the distance between the angle of the mouth and the end of the snout and about twice the diameters from the end of snout. Teeth are pointed, those in the maxilla and mandibles are in two rows. Canines are badly developed. The dorsal fin is anteriorly low and densely enveloped in skin. It commences midway between the gape of the mouth and the gill-opening. The fish is uniformly brown and the fins are tinged with black.

Residing in Indian Ocean, *Muraena thyrsoidea* attains 36 cm in length. They are of a light brown in colour covered all over with closely set purplish spots amongst which are light lines forming a net-work. Gill-opening sometimes with a black mark around it. No white edges is present in fins. Dorsal fin is more than 2/3 as high as the body. Teeth are conical and biserial laterally on the maxilla, about 23 on either ramus of the mandibles. Vomerine teeth are globular and in two rows. There is no large canines. The mouth can not be shut completely. Head length of the fish is $3^1/_2$ in the distance between the end of the snout and the vent. Eyes are $1^1/_3$ diameters from the end of snout, which are nearer to than the angle of the mouth. Anterior nasal tubes are equal in length 2/3 of the vertical diameter of the eye. Length of mouth cleft is $3^1/_2$ in the length of the head. Gill-opening are rather larger than the eye.

Muraena picta inhabit in Indian Ocean. Their head length is $3^2/_3$ in the distance between the end of the snout and the vent. Tail is about as long as the trunk. Their eyes are small, about twice the diameters from the end of the snout and situated over about the centre of the cleft of mouth, which latter is about $^1/_3$ of the length of the head. The mouth cannot be completely closed. Anterior nasal tube is not quite so long as the vertical diameter of the orbit. Gill-opening is about as large as the orbit. Maxillary and premaxillary teeth are in a single row. The vomerine ones are posteriorly rounded and generally anteriorly in a bifurcated row. The frontal 2 or 3 vomerine teeth are rather curved, sharp and about the same size as those in the pre-maxillaries. Mandibular teeth are in one row, except anteriorly where it is doubled. There are many different forms of colour found in the fish, usually of a grey or greyish-yellow ground work covered with black spots, which are connected together by a net work of dark lines causing the fish to appear marbled.

Muraena nebulosa residing in Indian Ocean attains nearly 2 metres in length. Their head length is $4^1/_2$ to $4^2/_3$ in the distance between the end of the snout and the vent. Tail is a little shorter than the trunk. Eyes are nearly 2 diameters from end of snout and situated midway between it and cleft of mouth, which last equals about $^1/_3$ of the length of the head. Gill-opening is smaller than the eye. Most of the teeth are obtuse or molariform. Vertical fins are well developed and commencing a little in front of gill-opening. The fish is brownish or olive, darkest along the back. A row of 20 to 25 black blotches are present along the upper surface of the head and back extending on to the dorsal fin; and nearly as wide as the ground colour. There are some white spots in the centre of each blotch. A similar row of blotches are present along the abdominal surface. Intermediate ground colour of fish is covered with small black stars, spots or vermiculated lines.

Living in the seas of India *Muraena polyzona* is characterized by deep brown colour, encircled with fine narrow 25 to 30 yellow whole or half bands which usually increase in width as they descend. Their head length is $^1/_4$ of the distance between the end of the snout and the vent. Tail is about $^1/_3$ longer than the body. Their eyes are of moderate size, placed about midway between angle of mouth and end of the snout. Length of mouth cleft is $3^1/_2$ in the length of the head. Gill-opening is small, scarcely so large as the eye. Their teeth are rounded crowns and the form change considerably with the age. Dorsal fin of the fish is rudimentary, commencing a little behind the vertical from the branchial opening.

Muraena nigra, an inhabitant of Andaman Seas attain 40 cm in length. Their length of head is $4^1/_2$ in the distance between the end of the snout and the vent. Tail is nearly half of the total length. Eyes of the fish are situated nearer to the snout than to the angle of the mouth. Eyes are small, diameter half that of the snout. Anterior tubular nostril is of moderate length. Gill-opening is as wide as the eye. Cleft of mouth equals $^1/_3$ of length of the head. The mouth can not be completely closed. Teeth are biserial, except in the mandible where there are three rows in some places. All the teeth are obtuse except the inner maxillary row, which are pointed and finer than the outer row. Premaxillary and vomerine teeth are of equal size and with globular heads. Mandibles are provided with 20 teeth in either side. Dorsal and anal fins of the fish are moderately developed. The dorsal fin commence just behind the vertical line from the gill-opening. The fish is uniformly black in colour without light edge to the fins.

Genus-*Gymnomuraena*

The fishes of this group have gill-openings of moderate or narrow width. Two pairs of nostrils are present on the upper surface of the snout, the posterior being a round foramen or with a short tube. Teeth are small, pointed and numerous. Fins are absent except a rudimentary one round the end of the tail. Their scales are also absent.

Gymnomuraena marmorata inhabit in Andaman Seas and attains a length of 26 cm. Head length of the fish is 4 to $4^1/_2$ in the distance between the end of the snout and the vent. Their tail is rather longer than the trunk. Eyes of the fish are small, about $1^1/_2$ to $1^2/_3$ diameters from end of the snout to which it is nearer to the angle of the mouth, Extent of cleft of mouth equal from $2^1/_2$ to $2^3/_4$ in the length of the head. The gill-opening is wider than the eye. Anterior nasal tubes of the fish are short, the posterior nostrils are with a raised edge in adults. Teeth of the fish are pointed, placed in a band in either jaw. The inner row is larger with no large canines. Vomer is in a single row, the anterior of which are enlarged. The fins are rudimentary. The vertical fins only exist round the end of the tail. The fish is of a brownish grey in colour, marbled all over with arborescent dark lines.

There are in total ninteen species of eel in the world. Two species are distributed in the Atlantic ocean, one type inhabiting the side of the European continent and the other the side of the American continent. As for the other species 12 species are distributed in the Pacific Ocean, 3 in the Indian Ocean and 2 in both Pacific and the Indian Oceans.

Eels around Indian oceans consists of *Anguilla bengaleusis, A. bicolor, A. marmorata, A. mossambica* and *A. nebulosa.*

Eels of Indian Oceans are elongate and serpentine with a pointed head and numerous needle like teeth on its jaws. There is a single small gill opening on each side just in front of the paired pectoral fins. Pelvic fins are absent and the caudal fin is continuous with the dorsal and anal fins extending from a dorsal point about one third of the way down the body from the head to just posterior to anus. The female when approaching maturity may be 75 to 105 mm long and weigh 1 to 1.5 kg. Female specimens of 120 mm long weighing 8 kg have been recorded. The male rarely exceeds 60 mm in length and is frequently less than 55 mm long at maturity.

Eels may be yellowish to olive-brown (yellow eels) or black above and silvery below (silver eels). The yellow eels are the residents in fresh water, estuaries or the sea. They represent non-migrating condition, and undergo a transition to the silver eel at the time of their reproductive migration from rivers to the ocean. Colouration is variable between these extremes since, in response to changing illumination, eels can alter their skin colouration by pigment redistribution within hours. A copious amount of mucus (slime) is secreted. Scales develop after eels have spent a few years in fresh water, but they are microscopic and easily overlooked.

Eels are born in the sea and grow up in the fresh waters of rivers lakes and ponds.

Eels spawn in the middle layer zone of the ocean at 400 m in depth. Each female is said to produce upto 15-20 million eggs. The eggs are about 1 mm in diameter. After fertilization by the male they develop into surface dwelling leptocephali. These are like transparent ribbons, shaped like willow leaves, with small pointed heads and long teeth. Because they are very weak swimmers the ocean currents carry them passively towards the coasts.

Leptocephalis, while moving up and down during day and night will spread out across the ocean riding on the surface of the current and will grow up as they migrate. The larvae approach the coast about 1-3 years after they are hatched, transform to elver (fry) near the sea bottom, and soon after transformation they enter the rivers. The water temperature strongly influence the migration to the rivers; and the elvers start to ascend rivers choosing the period when the difference of the water temperature between river and coast is smallest.

Approximately one year after hatching when they measure 8-10 mm in length, they arrive at the edge of the continental shelf. In the winter months they undergo a dramatic transformation (metamorphosis) to glass eel, which resemble a small eel but is completely devoid of pigment. As they approach the shores pigmentation develops and they are called elvers. The elvers, 8-10 mm in length, invade the shores for upstream run during summer months.

The elvers penetrate upstream, climbing falls and other apparently insurmountable obstacles. They may gain entrance to lakes with no visible water connection to the sea by use of underground channels. Elvers may penetrate many miles inland. Large populations of eels are found in warm shallow lakes; young eels do not enter cold streams in larger numbers.

Only a certain proportion of the elvers enter fresh water. The remainder populate estuarine areas like, tidal marshes, harbours and areas behind the barrier beaches or live in the coastal regions of the sea itself, usually in the vicinity of eelgrass (*Zostera*). It is stated frequently that migrants into fresh water are females, while those residing in brackish or sea water are males.

The eels that have moved into fresh water, rivers, lakes or ponds, that are their destination and stay there from 5-10 years. They will burrow between rocks or into holes or into the mud during the day and will go out into the waters to feed at night. In fresh water they grow slowly. They are capable of eating a wide variety of organisms, particularly, insects, fish, crustaceans, snails and worms. Their remarkably acute sense of smell presumably enables them to locate food easily. Eels do not usually eat rotten or decaying matter, their food must be fresh. They are active on warm dark nights and their activity is largely determined by light intensity. In the day they are usually hidden in holes, buried in mud or occasionally, draped over vegetation. In winter, eels becomes inactive and lie in morbid state in the mud, where there presence are recognized by a ventilation hole.

The male eel becomes sexually mature in 3-4 years, where as it takes 4-5 years for the female eel to reach sexual maturity.

In the fall, some of the larger and older yellow eels become silver eels. Towards the autumn the eels which have reached maturity become bluish black on their backs, light golden on their sides and light pinkish on their abdomen, signifying that they have reached the breeding age. In addition to the marked change in colouration, there is a thickening of the skin and accumulation of fat.

Other changes include reduction in production of mucus, enlargement of the eye and changes in its internal structure, alternation of body flexibility and of locomotory behaviour, pointing and narrowing of the pectoral fins, increased endocrine activity, development of gonads, cessation of food consumption, reduction of the digestive tract, increased number of chloride cells in the gills and changes in osmoregulatory capacity.

Sometime between late August and mid-November the silver eels move downstream swimming with the current. It is reported that they may even travel at night through damp grass or shallow ditches to reach the sea. Their down stream movement is determined by changes in the environment. A dark night combined with an increase in water level is most favourable. Large runs of silver eels usually occur on warm, dark, stormy nights in late summer or fall. Light intensity is particularly critical.

Immediately after entering the sea, the parent eels proceed to their particular spawning grounds. They take 2 or 3 months swim to the spawning area. After spawning and fertilization they are believed to die. They have never been seen returning to the streams.

Eels are captured in commercial quantities as elvers, yellow eels or silver eels.

Adult eels inhabiting either rivers or lakes are caught in autumn as they start to descent the rivers.

The fishing method is based upon the trapping using a knowledge of the eel's ecology, and different kinds of traditional gears have been used in various regions since old times.

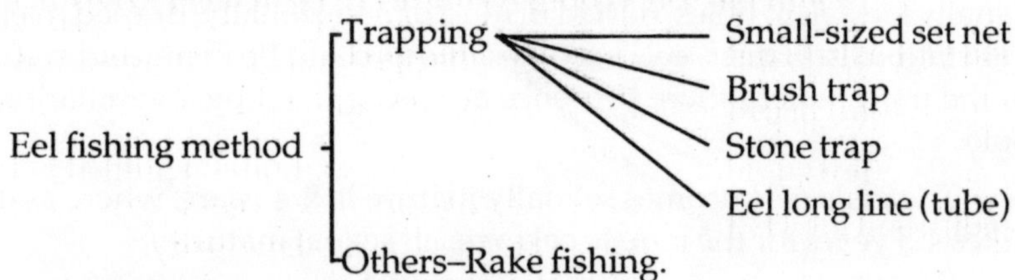

Eels dislike light and prefer to live in dark places. The fishing methods making use of the eel's habit of burrowing in narrow places such as holes or between piled up rocks are eel long lines, small-sized set nets brush traps, stone traps.

The fishing methods to attract eels with a paste bait, a mixture of powdered larvae of silk worm and rice bran making use of eel's highly developed sense of smell are; eel long lines, small-sized set nets.

The fishing methods to catch eels by changing the kind of bait according to seasonal changes in the diet of eels are, eel long lines.

The fishing method making use of eel's habit of hibernating in mud during the cold season is rake fishing.

In Japan most eels are artificially raised. A very tiny elver (fry) weighing 0.2 g grows into an adult of about 200 gram with the present eel culture techniques. The survival rate of the eel during the culture period from elver to adult eel is 50-80 per cent. The marketing size of the eel in Japan is 180-250 g in weight.

The fishermen normally capture elver when they have approached the coast and are just entering the rivers in the bay or estuary areas. Fry approach the coast after sunset and start ascending the river. They become active with flowing tide and gradually decreases with the ebbing tide. Towards the dawn, the fry stop ascending the river and bury themselves in the sandy mud. Generally scooping nets are used to catch eel fry. The present eel culture industry relies 100 per cent on natural fry. The appropriate water temperature for rearing eels is 15°-25° C, and when it drops below 10°C eels start hybernating by burrowing into the mud. On rising water temperature to 13°C they start looking for feed. Exceeding 28°C mass death occurs due to asphyxia. The oxygen content of culture media must be maintained and the decomposition of organic substances has to be avoided by controlling the amount of feed given into the water and removing the excretions of eels. A total of 25 tons of fresh water is required to produce 1 kg of adult eel from elver. Selective feeding with fresh and frozen low priced fish, like mackerel, sardine, anchovy, horse mackerel, saury pike can be used as their feed. In hot house eel culture system by maintaining 10°C water temperature for 8-12 months the production increases to 30–40 kg/3.3m^2 as against in poor production in out door pond culture (2-13 kg/3.3m^2)

Females produce spherical eggs of 3.14-3.22 mm in diameter. Yolk is colourless, transparent and coarsely vacuolated without oil globule. Perivitelline space is narrow. No pigment is noticed on yolk or embryo. In eggs collected in advanced stage of development, eyes are demarcated, heart pulsating and 58-62 myotomes could be counted.

After about 28 hours, the embryo occupied all the space, the tail end reaching the head.

Newly hatched larva measures 6.2 mm in length. Yolk almost absorbed. Anus opens below 59th myotome. 45 post-anal myotomes are present. Eyes are not pigmented. Mouth has opened.

First Day Larva

The larva is transparent with rounded cephalic and caudal ends. Mouth and gill openings have not yet formed. Eyes are devoid of pigmentation. The heart is distinct with a large pericardial space. At the broad anterior portion of the yolkmass is a group of large oil globules. Portions of the incompletely formed alimentary canal are seen above the yolk. Pectoral fins are not formed, but a callous area, one on either side above the yolk, indicates their position. The vent, which is situated in the 64th myotome, is still closed. Post-anal myotomes are very indistinct. The dorsal fin arises from the nape and is continuous with the caudal and anal fins. Total length 7.29 mm. Maximum depth 0.896 mm.

Second Day Larva

The transparent larva has a conspicuous head with a wide open mouth and a rounded caudal end. There are four teeth on either side of each jaw, the anterior of which are larger than the posterior. Eyes are well pigmented and black. Gill openings have appeared with the pectoral fins immediately behind them. The yolk has been absorbed except in the region of the liver below the gut. The anal opening which has not yet formed is indicated in the 65th myotome. Myotomes in the caudal region are still indistinct. The median fins become broader than in the previous stage. Total length 8.64 mm, Maximum depth 0.992 mm.

Third Day Larva

The transparent larva has a pointed snout and a rounded caudal end. The head is contained about 6.6 times in the body and the eye 4.7 and 1.7 times in the head and snout respectively. There are fine teeth on either side of the upper and lower jaws and those at the anterior end, especially in the upper jaw, are large and conspicuous. Eyes are black. There is a large prominent chromatophore at the tip of the lower jaw. The well formed alimentary canal is seen as a wide tube. There is still a trace of unabsorbed yolk close to the liver. The vent is opposite the 72nd myotome.

The chromatophores are very characteristic. In addition to one at the tip of the lower jaw there are 13 on the body, of which 10 are pre-anal in position. Of these, the first one is immediately below the pectoral fin where the gut is constricted. Among the trunk chromatophores, 7 are large and 3 are comparatively small. The posterioı of the three caudal chromatophores is conspicuously branched. In between the anterior pairs of chromatophores in the trunk faint streaks of black pigment are seen. Total length 9.99 mm. Maximum depth 1.17mm.

Leptocephalus

The leptocephalus is 105 mm long, narrow and pointed at both extremities was collected from Bay of Bengal off Chennai coast. Head is about 26 and height 16.6 in total length. Eye is about 7.8 in head and 2.4 in snout. Cleft of the mouth extends upto the posterior border of the orbit. The anterior nares are situated on the lateral side, a little behind the tip of the snout and immediately above the margin of the upper jaw. The laterally situated posterior nares are at a distance of about half the orbit diameter in front of the eyes, almost in a line with their ventral margin. In the upper jaw there are 12 larval teeth on each side, of which the 4 posterior ones are relatively very small. Similarly, in the lower jaw there are 10 teeth on each side of which the 6 anterior ones are more prominent than the rest. Pectoral fins are present. The gut shows at intervals 12 hump-like thickenings, over which are discernible dense groups of chromatophores. The vent is anterior in position to the middle of the body. The dorsal fin originates far behind at a distance of about 8 mm from the tip of the tail. The caudal end is bluntly pointed and, though there is no well developed caudal fin, a few very short, almost vestigial rays could be seen covering the blunt tip. There are 223 myotomes, of which 80 are pre-anal. Elongate streak like chromatophores are present near the tip of the lower jaw and on the lateral margins of the upper and lower jaws. From the sixth myocomma onwards prominent dendritic chromatophores arranged in a line are present immediately below the midlateral line along most of the myocommas. At the base of the anal fin regularly arranged chromatophores are present. The general shape of the body, the nature of the caudal end and the distribution of chromatophores indicate that this leptocephalus belongs to the family Ophichthyidae.

82 mm long leptocephalus, collected in shore seine net off Kovalam, 7 miles south of Trivandrum is thin and moderately broad. The head is about 28 and height 13 in total length. The eye is about 4.6 in head and 1.7 in snout. Cleft of the mouth extends nearly upto the posterior margin of the orbit.

An olfactory pit with one opening is discernible immediately before the eye. In addition to the prominent tooth situated a little above the tip of the upper jaw, there are 14 teeth on each side, of which the 5 anterior ones are relatively larger. Similarly in the lower jaw there are ten teeth on either side in addition to two big ones arising a little below the tip of the jaw. Pectoral fins are formed. The dorsal originates far behind at a distance of nearly 5 mm from the caudal end. The caudal fin is vestigial and its rays indistinct. In the gut are seen alternate constrictions and distensions. The body has 174 myotomes of which 107 are pre-anal.

The head is devoid of any chromatophores, while the region of the heart has five conspicuous and highly dendritic chromatophores. Along the dorsal and ventral margins of the first four hump-shaped distensions in the gut, 5 to 8 dendritic chromatophores are discernible. Along the dorsal margin of the 5th and 6th humps and on the portion of the gut immediately preceding the vent, minute, closely set, dendritic chromatophores forming a dark streak are present. A line of closely arranged chromatophores is present immediately below the mid-lateral line along the myocomma after the 32nd myotome. Regularly arranged chromatophores are present along the base of the anal fin.

Relatively broader and longer (103.5 mm) leptocephalus obtained from Kovalam has only 141 myotomes. The vent is situated in the 67th myotome. Head is about 27 and height 14 in total length. Snout is twice as long as the orbit and the head nearly 5 times. Cleft of the mouth extends upto or a little beyond the middle of the orbit. An olfactory pit with a small anterior opening is present immediately in front of the orbit. In addition to the prominent median tooth originating a little above the tip of the upper jaw there are 15 pointed teeth on each side directed forwards of which the five anterior are bigger than the rest. In the lower jaw there are in all 11 teeth of which the two anteriormost originate a little below the tip of the lower jaw. The vent is situated in the middle of the body. Pectoral fins are present. The dorsal fin originates far behind in the caudal region at a distance of 6 mm from the tip of the tail, while the caudal is absent. While only two rather indistinct chromatophores are present on the outer side of the upper jaw at the base of the 2nd and 3rd teeth, a few are present on the gut, at the dorsal and ventral margins of the humps as well as in the intermediate region. Most of the chromatophores present on the top margins of the humps are small, closely set, and linear in character, whereas those on the humps at their lower margins are conspicuous, isolated and dendritic in nature. Congregation of pigment spots forming thin streaks in between the adjacent rays, are present along the entire length of the base of the anal fin. No chromatophores are discernible at the base of the dorsal. Closely set pigment spots forming conspicuous lines are present along most of the myocommas, immediately below the mid-lateral line from the fifteenth myotome onwards.

Leptocephali of *Pisoodonophis hijala* (Ham)

In coastal waters of Bay of Bengal, off Mahanadi estuary in Orissa two leptocephali of *P. hijala* measuring 68.0 and 73.1 mm long were collected in tow-net haul. They are transparent, rather narrow and short with the head 20 and depth of body 13 in total length. The eye is about 7 in head and 2 in snout. The anterior nares are situated on the lateral side immediately above the margin of the upper jaw at a little distance behind the tip of the snout and the posterior ones, also lateral, some distance above the margin of the upper jaw, almost in a line with the lower margin of the orbit, at a distance of one third the diameter of the orbit, anterior to the front border of the eye. Cleft of the mouth extends upto the posterior border of the orbit and is endentulous. Pectoral fins are present. The gut has six humps.

The body has 155 myotomes of which 60 are pre-anal. The dorsal fin originates rather far behind in the body and extends only upto a short distance before the tip of the tail. The caudal end is blunt without a fin.

Head is devoid of chromatophores. Three small chromatophores are present on the gut immediately below the gill openings. Three small rows of chromatophores are present in the region of the distensions on the alimentary canal, along the dorsal and ventral margins and in between them. The central row is absent in the region of the two posterior humps. Chromatophores, present along the dorsal surface of the gut immediately before the vent, are linear. Conspicuous, dendritic chromatophores, varying in 1 to 11 numbers are present on most of the myocommas immediately below the mid-lateral line. Regularly arranged pigment spots are found along the base of the anal fin.

Indian Short-Finned Eel, *Anguilla bicolor bicolor*

The species was caught in the west coast of India at 6-7m depth in the inshore areas off Karwar during its migration to spawning ground. The migrant eel was 866 mm long, 1150g in weight having a head length of 13.97 per cent in total length. Head is flat with a small muscular dome on the dorsal side. Snout is broad and round. Eyes are large and slightly oval in shape. Lower jaw slightly projecting beyond the upper jaw with thin lips. Pectoral fins are long, broad with round edges. Dorsal fin commences above vent. The dorsal surface is grey in colour. Ventral surface and sides below the lateral line is silvery white with tinges of golden yellow. Pectoral fins are dark black except at the origin where it is golden yellow in shade. Lateral line is prominent.

The ovaries weighed 34 gram. The gonadosomatic index was 2.96. The size of the ovarian eggs ranged from 0.13-0.27 mm. The fish was in a maturing stage and just commenced its migration to the spawning ground.

A. bicolor bicolor undertakes migration in the end of July when the oocytes measure 0.13-0.27 mm in diameter. The eel might have descended from the Kali river, which is about 4 km south from the point of collection, into the sea and started south ward migration. The probable spawning grounds for *A bicolor bicolor* are in the eastern Indian Ocean near Sumatra in Mentavene deep between 4°S and 2°N latitudes and in the western Indian Ocean near the northeast of Malagasy between 10°S and 20°S latitudes and 60°E and 65°E longitudes. Considering the gonadal condition and the site of collection, the eel apparently was on its way to the oceanic spawning ground 3000-4000 km away from the place of collection.

Fishes of Lophobranchii family have a dermal, segmental skeleton, with the opercular pieces reduced to a single plate. Their gill-openings are small and gills consists of small rounded tufts attached to the branchial arches. Their muscular system is poorly developed. Their snout is produced with terminal small mouth. Teeth are absent.

Genus-*Syngnathus*

Body of fishes of this group have more or less distinct ridges. The dorsal edge of the trunk not being continuous with that on its caudal portion. Humeral bones is firmly united into the breast ring. The opercle may be entirely crossed by a distinct ridge or it may be only at its base or the ridge is absent. Dorsal fin is situated either opposite or nearly so to the vent. Its base may be raised. Pectoral fins are well developed and caudal fin is present. An egg-pouch is present on the tail of the males. The eggs being covered by cutaneous folds.

Living in the seas of India *Syngnathus serratus* are light brown in colour with light spots along the side and the body is banded in rings. Tail fin is black with a white lower edge. A dark mark is present at the base of anterior dorsal fin rays and a dark line along the middle of the fin. The lateral line of the fish bent downwards and passes into the lower edge of the tail. Their length of head is from 5 to 6 in the distance between the end of snout and vent. Their eyes are large, situated in the middle of the length of the head. A spinate or serrated crest is present along the upper edge of the posterior 2/3 of the snout. Length of snout is less than 1/2 of length of head. Interorbital space is broad with the prominent orbital edges which is smooth, occiput and nape with a median ridge. Opercle is finely striated. Body is scarcely deeper than broad, shields are without spines. Vent is situated nearly below the middle of the dorsal fin, the base of which is elevated. It stands on 5 or 6 rings, 2 of which are anterior to the anal ring. Caudal fin is extremely small. Egg pouch is more than half as long as tail.

Syngnathus longirostris occur in Tamil Nadu coast of Indian Oceans. Their length of head is $3^3/_4$ in the distance between the end of the snout and the vent. The length of the trunk from the end of the snout to the vent is scarcely more than $^1/_3$ of the total length. Egg-pouch is not $^1/_2$ as long as the tail. Eyes are $^1/_4$ of length of snout and situated in the posterior $^1/_2$ of the distance between the end of the snout and base of the pectoral fin. Body is deeper than broad. No elevated ridge is present along the upper surface of the snout. Length of snout is more than half the length of the head. Interorbital space is nearly flat. A low and blunt ridge is present at the occiput and nape. Opercle is covered with fine radiating lines. Vent is situated below middle of the dorsal fin. Body is prominent inferiorly from the sixth to the eleventh body rings. Dorsal fin is low, it commences on the third body ring before the one carrying the vent. It is situated on seven rings and has its base elevated. It is greyish brown in colour with dark rings. The under surface of snout is light coloured with dark spots.

Captured from Tamil Nadu coasts, *Syngnathus intermedius* is greyish in colour and banded. Its dorsal fin is spotted. Its length of head is $3^3/_4$ to $4^1/_4$ in the distance between the end of the snout and the vent. The length of the trunk is between the end of the snout and the vent and equals $1^2/_3$ in the total length. Egg pouch is 2/5 of length of tail. Its eye diameter is $3^1/_2$ in the length of the snout and situated in the posterior 1/2 of the head. Body is deeper than broad. Length of snout is 1/2 of that of the head and without any elevated ridge along its upper surface except a projection above the nostrils. Interorbital space is slightly concave, owing to the supraorbital ridges being well developed. A low and blunt projection on occiput continued as a sharper ridge along the nape. Opercle is with fine radiating lines. Dorsal fin is of moderate height. It commences on the third body ring before the one carrying the vent. It is situated on 6 or 7 rings, having its base elevated. Length of caudal fin is equal to half that of snout.

Genus-*Tetrodon*

Fishes of the genus *Tetrodon* have broad back or compressed into a ridge. Their either jaw is provided with a medium suture. When a conspicuous nasal organ exist there may be two on either side in a papilla, or a single tubular one, or an imperforate one having a fringed edge and the spiny body or a simple round cavity and smooth body or two imperforate, tentacles on either side or the nasal organs may be inconspicuous and the back compressed into a keel. Dorsal and anal fins are with few rays. Body wholly or partially covered with fine dermal spines or such may be absent. There may be a more or less distinct fold along the lower part of the tail and very apparent nasal organs or when the fold is absent the body will be spinate and the nasal organs very distinct, or the fold may be absent with smooth skin. A portion of the oesophagus is dialatable and able to be distended with air.

Occuring in the seas of India *Tetrodon inermis* attains 44 cm in length. Their head length equals its distance from the base of the dorsal fin. Eyes are rather large and situated nearer the gill-opening than the end of the snout. Interorbital space is broad equalling $1^1/_3$ diameter of the eye. Teeth are present in the two jaws of about the same size. Dorsal fin with its longest anterior rays is situated in the last third of the distance between the nostrils and base of the caudal fin. Its height is rather above twice the length of its base. Anal fin is of similar size and shape. Caudal fin is emarginate. Spines are soft and widely seperated. They only exist along the abdominal surface, not extending so far as the vent. A moderately well marked fold of skin is present along the side of the tail. Skin on the summit of the head is with short longitudinal rugae. Upper surface of head and back is plumbaceous in colour. A broad dull brown band passes from the eye nearly to the tail, beneath which it is silvery white. Gill-opening is black internally. Upper two-third of dorsal fin is dark. Caudal fin is dark in its last half, margined with white.

Tetrodon lunaris inhabits in seas of India and attain at least 30 cm in length. The fish is greenish-olive in above, sides and abdomen of a white satin having a yellowish line from the eye to the centre of the caudal fin. End of caudal fin is dark with a light outer edge. Its head length equal to or rather less than the distance between it and the base of the dorsal fin. Eyes are large and situated much nearer to gill-opening than to the end of the snout. Interorbital space is flat and equal to $1^1/_2$ diameter of the eye. Teeth are of same size present in jaws. Dorsal fin with its anterior longest rays are situated in the last third of the distance between the front edge or middle of the eye and the base of the caudal fin. Anal fin is similar to dorsal fin. Caudal fin is emarginate. Small spines either entirely covering the back or only anteriorly from opposite the front margin of the eyes. Abdomen is similarly protected. Snout, sides and tail are spineless.

Available in Tamil Nadu coast of seas of India, *Tetrodon hypselogenion* attains 15 cm in length. Its head length equals its distance from the base of the dorsal fin. Eyes are situated behind the middle of the length of the head. The osseous interorbital space is very much narrower than the shortest diameter of the eye. Same size of teeth are present in both jaws. Dorsal fin is situated in the hind third of the distance between the front edge of the eye and the base of the caudal fin. The length of its base is only two-fifth of its height. Its anterior rays are the highest. Anal fin is of similar form. Caudal fin is slightly emarginate, the outer rays being the longest. Spines are of moderate size and are widely seperated. They commence on the dorsal profile from a slight distance posterior to the eyes and are continued half the distance or even more to the base of the dorsal fin. They are also continued along the abdominal surface almost to the vent. Some are likewise present on the sides behind the pectoral fin. The species is of a yellowish-brown in colour, becoming white stained with yellow beneath. The colour of the back is often sharply defined from the white of the sides. The whole of the upper surface is covered with light dots, which have usually dark edges. Three to five vertical bars are present under the eye. Caudal fin with about seven vertical bands most distinct in its upper lobe.

Occuring in the seas of Tamil Nadu coast *Tetrodon oblongus* is of light brown colour along the back, becoming white tinged with yellow on the sides and beneath. Bands of dark brown pass from the back down the sides to the middle of the body enclosing light spaces of irregular sizes and shapes or round spots. From the snout to the dorsal fin round or oval light spots are predominant. Occasionally a few bars descend from the eye over the cheek. Fins are yellow stained with orange. Sometimes a black shoulder spot is present. Head length of fish equals its distance from the base of the dorsal fin. Eyes are in or a little before the middle of the length of the head and are small in adults. The interorbital space is nearly flat and very broad in adults, in which it exceeds the length of the snout. Lower teeth are a little larger than the upper pair. Dorsal fin with its central ray is somewhat the highest. It is situated in the last third of the distance between the front edge of the eye and the base of the caudal fin. Its height equals about twice the length of its base. Anal fin is situated beneath the dorsal fin and is of similar shape and size. Caudal fin is truncated Spines are with two roots and present on the back from the nostrils to the dorsal fin. A few are present on the lower part of the cheek and numerous along the abdominal surface almost as far as the vent.

Occurring in the seas of India, *Tetrodon patoca* attains 32 cm in length. Its head length is rather less than its distance from the base of the dorsal fin. Eyes are situated in about the middle of the length of the head. Interorbital space is broad. Teeth are of about the same size in either jaw. Dorsal and anal fins are rounded. The caudal fin is truncate. Height of dorsal fin equals the length of the head excluding the snout. Spines are small with three or four roots, extending from a short distance behind the nostrils to the base of the dorsal fin. Interiorly they extend over the chest nearly as high as the base of the pectoral fin and posteriorly to the vent. Upper half of the fish is brown or black in colour with numerous round or oval white spots. Sides and abdomen are silvery. A yellowish tinge divide the dark back from the white side. Caudal fin in the young is stained dark in its outer half. The caudal fin is rather longer, the back is darker, the light spots are more distinct in some specimens. In some especially in marine forms and immature specimens two to three distinct dark bands descend from the back to the middle of the sides, and a fourth passes across the head.

Inhabiting in the seas of India, *Tetrodon nigropunctatus* attains 22 cm in length. Its length of head is nearly equals its distance from the base of the dorsal fin. Eyes are rather small, about two diameters from the end of the snout and about three apart. Interorbital space is nearly flat and the osseous portion is less than two diameters of the eye wide. Teeth are about the same size in both jaws. Dorsal fin is situated in the last fourth of the body excluding the caudal fin. Dorsal and anal fins are rounded. Caudal fin is also rounded and equal to about 1/5 of the total length Spines are small and usually cover the entire body. They are either much projecting or almost concealed in the skin. The extent over which they extend also varies. In fact they appear to be absent in some specimens. The fish is brown colour above, becoming lighter on the sides and beneath. Some scattered black dots are present over the body and the vent in a black spot. The snout and round the mouth are black or very dark brown. The spots are absent in some specimens. In one variety, namely *Citrinella* the entire body is of a beautiful lemon colour. Besides the scattered spots on the sides, there are some irregular small and large black spots on the back. One large blotch is present round the base of the dorsal fin, which is also black. A large round black spot is situated round the eye and gill opening.

Tetrodon reticularis is found in seas of India and attain 42 cm in length. Its length of head equals its distance from the base of the dorsal fin. Eyes are situated in about the middle of the length of the head. The interorbital space is broad and flat. Teeth are of about the same size present in either jaw. Fins are rounded. The dorsal is situated in the last fourth of its total length excluding the caudal fin. Its height somewhat exceeds half the length of the head. Length of caudal fin equal to $4^1/_2$ in the total length. Small spines cover the entire body except the forepart of the snout. The upper surface of the body is deep grey or brown in colour, becoming white below with eight to ten longitudinal black stripe, which under the eye, as well as round the mouth and pectoral fin are concave. On the back are darker blotches anteriorly, where blackish bands surround spots of white or grey. It is chequered with black posteriorly. Caudal fin is reticulated with black on a white ground. In the young the bands on the cheek are wanting.

Tetrodon hispidus living in seas of India attain 50 cm in length. Its length of head equals its distance from the base of the dorsal fin. Eyes are of moderate or rather small size, from $2^1/_2$ to 3 diameters from the end of snout. Orbit is prominent and interorbital space is concave. Upper profile of the snout is concave. Fins are rounded. Caudal fin is about one-fifth of the total length. Spines are small. Those are on the abdomen with two or three roots. They cover the body from the snout to a short distance behind the base of the dorsal fin. In some specimens they appear not to extend so far. Colour of the fish is brown superiorly with a moderate number of bluish-white spots. One or two bluish-white bands sometimes exists round the orbit. Another well marked ones are round the gill opening and pectoral fin. Sometimes there are lines or black spots on the abdomen, but they do not ascend to the cheeks. Some have several black blotches along either side of the abdomen or bands descending on to it.

Tetrodon leopardus inhabit in Indian Ocean. Its length of head equals its distance to the base of the dorsal fin. Eyes are of moderate size, $1^1/_2$ diameters apart and the same distance from the end of the snout. Interorbital space is flat. Teeth are about the same size in both the jaws. Dorsal fin is highest in front situated in the last third of the distance between the front edge of the eye and the base of the caudal fin. Anal fin is similar to dorsal. Caudal fin is rounded and is $4^2/_3$ in the total length. Spines are two rooted, widely seperated along the back, extending from the front edge of the eyes of the dorsal fin. Inferiorly they are closer together and reach to the vent. The fish is olive coloured superiorly, extending two-thirds of the distance down the sides, with an interrupted black network surrounding white spots. There are three black cross bands, one over the head with a 'V' shaped light interorbital base posterior to it. The second above the pectoral fin, the posterior one from the base of the dorsal fin. Reticulated narrow black lines enclosing large white spots on the caudal and dorsal fins are present. The latter having like-wise a narrow black basal band.

Occurring in Malabar coast of Indian Ocean, *Tetrodon viridipunctatus* has head length equal its distance from the base of the dorsal fin. Caudal fin is 1/5 of the total length. Eye diameter is 1/5 of length of head rather nearer to gill-opening than to the end of the snout. Nostrils are with two solid tentacles. All fins are rounded. Spines are two rooted, short and some distance as under. They commence from the occiput and pass along the back two-thirds of the way to the commencement of the dorsal fin and none on the sides. Inferiorly they begin below the orbit, surround the lower and posterior margin of the pectoral fin and are continued backwards as far as the anus. The fish is light green colour in back, abdomen silvery-white, back and sides covered with emerald green spots. A bar of the same colour passes across the vortex from one eye to the other and also goes backwards in the median line towards a second irregular band of the same colour, which passes across the back more posteriorly. Eyes are brown, with a golden rim surrounding the iris. Caudal and anal fins are tipped with black. Dorsal fin is yellowish. Four black spots are present under the throat.

Occuring in seas of India, *Tetrodon fluviatilis* attain 16 cm in length. The fish is greenish-olive superiorly, becoming white along the sides and below. Back and sides are with large black blotches, leaving very little of the ground colour apparent. One or two irregular light bands, one crossing between the eyes and another between the pectoral fins are observed. Often one or two more over the back and abdomen is covered with round or angular black spots and blotches much wider than the ground colour. Sometimes it is quite black beneath. Fins are yellowish. End of caudal fin is stained dark and sometimes with black spot. Length of head equals its distance from the base of the dorsal fin. Eyes are rather large. The interorbital space is broad and slightly convex. Teeth are of about the some size in both jaws. Dorsal fin is situated in the last third of the distance between the eye and base of caudal fin. Caudal fin is truncated, its length being one-fourth of the total. Spines are with two roots, widely seperated, concealed in the skin and extending from the hind edge of the eyes to the base of the dorsal fin. Spines are also present over cheeks and abdomen as far as the vent.

Genus-*Diodon*

The body of the fishes of genus *Diodon* are nearly globular. Jaws are without median suture. Body covered with stiff and erectile dermal spines each having a pair of lateral roots. A portion of the oesophagus is dialatable and can be distented with air. No pelvic bones and air-vessel is present.

Diodon hystrix occurs in seas of India. Its body is covered with coarse spines which are longest on the sides of the body behind the pectoral fin, where they become about 3/4 the length of the fin. They are dialated at their bases and usually with a pair of basal grooves. Two or three pairs of immovable spines are present at the upper and lower sides of the tail. The whole fish except ventral surface is light brown in colour covered with round blue or brown spots, rarely above one to the base of a single spines. The fins are similarly spotted.

Genus-*Belone*

Fishes under the genus *Belone* have wide gill-openings, elongated, sub-cylindrical or compressed body having lateral eyes and numerous branchiostegals. Their jaws are prolonged into a beak, the upper of which is formed by the premaxillaries. Fine teeth are present in both jaws with a single row of long, widely-set conical ones. Their anterior dorsal rays may or may not be elevated, forming a lobe to the fin, whilst the middle and posterior ones may be short or elongated. Their caudal is usually forked. Theys have small scales, lateral line on free portion of tail with or without a keel.

Belone annulata inhaviting in Indian Ocean grows to at least 60 cm in length have strong teeth. They look green on the upper surface of head and back with steel-blue reflections minutely dotted with black fading into silvery white on the abdomen. The side of the upper jaw is dark olive and lower silvery. A black line passes from the upper third of the angle of the mouth along the base of the teeth. Cheeks and opercles are silvery. Membranes of the dorsal, caudal and pectoral fins are light greenish-grey, minutely dotted with black, especially at their margins. Anal and ventral fins are anteriorly white, dotted with black. Iris is silvery, the margin of the orbit is black. The colour vary in young fishes, the upper surface is more of a yellowish-green. There is a large black spot on the opercle and the lower jaw is also black. A narrow silvery greenish band passes longitudinally above the lateral line. The upper portion of the dorsal is black and the centre of the caudal is stained blackish.

Its dorsal fin is situated in the last third of the distance between the head and the base of the caudal fin, elevated in front, and its middle and last rays are elongated. Pectoral fin is longer than the body height. Ventral fins are situated midway between the front edge of the eye and the base of the caudal. Anal fin commences slightly in advance of the dorsal fin and is of somewhat similar shape. Caudal fin is lobed, the lower lobe in longer.

The bluish-green *Belone choram* living in Indian Ocean has a wide shallow groove along the summit of the head. Its superciliary region is striated and maxilla concealed by the preorbital. Its dorsal fin commences in the posterior 1/3 of the length of the body and its hind rays reach nearly the base of the caudal fin. Pectoral fin is as long as the head behind the middle of the eyes. Ventral fins are inserted midway between the front edge of the eye and the base of the caudal fin. Anal fin is same as the dorsal. Caudal fin is forked and lower lobe is longer. Free portion of the tail is as wide as height.

The fish is bluish-green dorsally and lighter below. A silvery stripe is present along the side and the upper edge of dorsal and pectoral fins are usually black.

Belone cancila inhabiting in Indian Ocean grows to 30 cm in length. It has got a row of large sharp teeth, widely seperated in both the jaws with an external row of numerous fine ones.

Its dorsal fins commences opposite the anal fin and is rather more than twice as far from the anterior extremity of the orbit as it is from the posterior extremity of the tail. Pectoral fins, equals half the distance of the head behind the front edge of the eye. Ventral fin is inserted rather nearer the base of the caudal fin than the hind edge of the eye. Caudal fin is slightly emarginate.

Small scales cover the body in irregular rows, some over front end of groove on head and sides.

Dorsally greenish-grey, becoming whitish along the abdomen, a silvery streak having a dark margin extends along the body from opposite the orbit to the centre of base of caudal fin. The whole upper two-thirds of the body is closely marked with fine black spots, though there are 4 to 5 larger blotches along the side between bases of pectoral and anal fins, which are absent in the young. Dorsal and caudal fins tipped darker posteriorly and anal fin whitish with a greyish margin. Its eyes are golden in colour.

Belone strongylurus inhabiting in Indian Ocean attain more than 60 cm in length has double lateral line. Its free portion of the tail is compressed much deeper than width and without any distinct lateral keel. It has got small scales. The head top and back of the species is yellowish-green in colour with minute brown dots, fading into silvery on the sides and white on the abdomen. Its cheeks and opercles are silvery. A deep blue longitudinal band, bordered beneath by another broad silver band passes along the posterior half of the sides. Dorsal fin tinted with a little orange colour along its upper edge, the anal fin is bright yellow and the rays dotted with brown. Pectoral and ventral fins are diaphanous, the latter occasionally with a black spot at the base. Caudal fin is yellowish or greenish, minutely dotted with black and having a round bluish-black spot in the centre. Iris silvery and the upper surface of the eye is bluish-black.

Ventral fin arises midway between the orbit and the base of the caudal fin. Anal fin is situated in the last third of the body. Pectoral fin is nearly as long as the postorbital portion of the head. Caudal fin is rounded.

Genus-*Exocoetus*

Fishes of the genus *Exocoetus* have moderately oblong body with very wide gill-openings. Their jaws are short with seperate premaxillaries and maxillaries. Mandible in some species exist with a tubercular prolongation. Teeth when present are rudimentary and minute. Pectoral fins are elongated and used for flying. Air vessel is large and pyloric appendges are absent.

Exocoetus paecilopterus grows to about 33 cm length are found abundantly at Madras coast. They are bluish along the back, becoming lighter on the sides and beneath. Their pectoral fins are with many rounded and oval spots, sometimes transversely arranged in bands. Seven rows of scale are situated between the base of the dorsal fin and the lateral line. The dorsal fin of the fish commences some way in front of the anal fin. The distance between the first dorsal ray and the base of the caudal fin rather exceeds the length of the head. Ventral fin arises some what nearer the base of the caudal than to the hind edge of the head and reaches to the end of the base of the root of the anal fin. The pectoral fin reaches as far as the end of the dorsal fin.

Exocoetus splendens are found abundantly in Andaman Sea and also in Bombay coast. They grow to about 23 cm. The dorsals fin of the fish arises above or slightly in advance of the origin of the anal fin. Pectoral fin reaches to the base of the caudal fin. Its ventral fins are short, about half as long as the head, reaching half the way to the base of the anal fin. Caudal fin is lobed, the lower lobe is much longer.

The fish is bluish along the back, becoming lighter on the sides and beneath. Pectoral fins of the fish are grey or black in colour with lighter edge.

Exocoetus bahiensis inhabiting in Indian Ocean has darker pectoral fins than the other fins and sometimes nearly black in its last third. Dorsal fin sometimes has a dark mark on its summit. Rudementary teeth are present and barbels are absent. The dorsal fin of the fish commences between the front edge of the eye and the posterior extremity of the lower lobe of the caudal fin. Its anterior rays are half as long as the head. Ventral fin arises midway between front end of snout and the end of upper caudal lobe and reaches to the centre of the base of the anal fin. Pectoral fins reaches to the end of the base of the dorsal fin. Anal fin commences below the last third of the dorsal fin.

Exocoetus mento, inhabitants of Indian Ocean, abundantly appear in Tamil Nadu coast during winter season. They are bluish in colour, becoming silvery along the abdomen. Dorsal fin of the fish is black in colour. Upper half of the pectoral fin is also black having a white edge. Ventral fins are white. Anal fin is also white with a dark mark along the base. The upper lobe of the caudal fin is white with a black bar along its base, lower lobe being greyish.

The species have villiform teeth in both the jaws, in several rows in the upper jaw and only one row in the lower jaw. Lower jaw of the fish is little longer, slightly produced in the form of a tubercle at the symphysis. There is no barbels.

The dorsal fin of the fish is elevated, as high as the body, and its membrane is deeply cleft. Seventeen to eighteen rows of scales are present between the occiput and origin of dorsal fin, which commences opposite to or slightly before the first anal ray. Ventral fins in some specimens reaching to the anus, in others to the anal fin. The fin arises midway between the anterior edge of the orbit and the base of the caudal fin. Pectoral fin reaching to opposite the middle or end of the dorsal fin. Ventral fin does not quite reach the anal fin. Lower caudal lobe is longer.

Flying Fish

Flying fish (Exocetidae) are ocean-living pelagic fish that leap out of the water and glide long distances using elongated pectoral fins and winged gliders. While swimming on the surface or close to it, the pectoral fins are held close to the body. They prefer warm temperatures and are therefore found mainly in tropical and sub-tropical areas.

Several species of flying fish are caught in various parts of world's ocean. The main ones are small-sized four-winged flying fish, *Hirundichthys affinis*. At least ten species of flying fish have been recorded from Indian ocean. Of these, a small variety, a coastal species, known as Coromandel flying fish (*Hirundichthys coromandelensis*) has been landed on the east coasts of India and Sri Lanka for half a century or more, Fishermen aggregate the fish with lures and catch it with scoopnets.

Almost all other species are larger in size and oceanic in habitat. These include the two varieties that predominate in the gillnet catches, Sutton's flying fish (*Cypselurus suttoni*) and Indian spotted flying fish (*Cypselurus poecilopterus*). They exhibit a size range of 23-35 cm. Coromandel flying fish usually belongs to the length group of 18-24 cm.

The larger varieties of flying fish are generally scattered along the pelagic zone close to and beyond the continental shelf, along the southwestern part of the Bay of Bengal. They are more visible along the east coasts of India and Sri Lanka, than along the west coasts.

The common flying fish spawn in the water surface around floating objects. Since flying fish eggs do not float, unlike the eggs of most surface-living fish, the eggs get deposited by the females on floating objects through the sticky filaments that surround the eggs. Male flying fish deposit their sperms over the eggs to induce fertilization. About 100 hours after fertilization the eggs hatch out as larvae. The larvae float near the surface, rapidly becoming juvenile flying fish. *H. coromandelensis*, lay 5000 to 12000 eggs, depending on the size of the fish.

Flying fish feed mostly at night, and mainly on juvenile fish. They live for one to three years.

The national production of flying fish in India is around 4000 tonnes. Tamil Nadu and Pondicherry contribute 76 per cent and 22 per cent respectively of this total. As in other parts of the world, capture of flying fish is a traditional small-scale activity. Fishing techniques used to catch flying fish vary on the size of fish and spawning season. The most common small-sized species are caught mainly during the spawning season; by small-mesh short gillnets and scoop nets used in conjunction with various man made floating objects that serve as fish aggregating devices. To enhance the aggregating effect of the device, chopped fish and oil is poured into the water. The larger flying fish are caught mainly during the non-spawning season, using specially designed gillnets and hook and line.

Fishing craft used in flying fish fishery vary in type and size and range in length from 4 to 15 m. As flying fish are caught mainly in the near offshore areas the craft engaged in this fishery function as day–boats in most countries, leaving for the fishing grounds early morning and returning to landing centres late afternoon.

In Tamil Nadu, artisanal fishermen go upto 40 nautical miles off shore in locally built 7 log catamarans. According to one estimate, about 5 per cent of the total number of catamarans in the area venture out to catch flying fish. During May–July; small varieties of flying fish, mainly the Coromandel flying fish (*H. coromandelensis*) come close to the continental shelf for spawning, mainly off Nagapattinam in Thanjavur district and Cuddalore in South Arcot district. Fishermen use traditional small-mesh surface gill nets and large scoop nets in conjunction with shurb lures, to capture the flying fish. Flying fish are landed and marketed fresh, or salted and dried.

The main varieties caught are Sutton's flying fish (*Cypselurus suttoni*), Indian spotted flying fish (*C. poecilopterus*), large-scale flying fish (*C. oligolepis*) and tropical two-wing flying fish (*Exocoetus volitaus*). Average size of the catch was 28 cm and all the species were non-spawners. Where and when the larger species are spawning is not clear.

Many eggs of flying fish are lost when fishermen discard egg-laden lures after removing fish from them. An international market does exist for flying fish eggs. Japan imports flying fish eggs from Indonesia. There is a possibility of commercial utilization of flying fish eggs after having resource knowledge.

The resource of flying fish in both India and Sri Lanka, bordering Indian Ocean is spread widely. The catch is mainly of adults, rather than juveniles. If the bigger varieties of flying fish spawn at the same size, like the smaller variety does, there is no danger of the flying fish stocks getting overfished, provided a sufficient quantify of eggs is allowed to develop in the sea.

Monacanthus monoceros occurs in seas of India and attain 45 cm in length. They are brownish or blackish in colour with yellow coloured fins. Vertical fins are low, dorsal spine is weak. There is no ventral spine. Caudal fins are truncated or very slightly convex. Their length of head is about 4 and height of body from $2^3/_4$ to $3^1/_3$ in the total length. Eyes are rather small situated between the upper end of gill-opening and first dorsal fin. The body of the fish is oblong, snout moderately produced with its upper convex profile.

Monacanthus scriptus living in Indian seas attain 30 cm in length. Body of the fish is oblong, snout produced with its upper concave profile. Their head length is 4 to $4^1/_2$ and height of body is $4^1/_4$ in the total length. Their eyes are of moderate size, situated between the upper end of the gill-opening and the first dorsal fin. They have got eight compressed and pointed teeth. Vertical fins of the species are low, dorsal spine is weak and of moderate length. There is no ventral spine. Caudal fin is wedge-shaped. The fish are of buff colour, covered with small brown spots and blue lines

Genus-*Anacanthus*

Fishes of this genus have compressed and elongated body. A fleshy barbel is present below the symphysis of the lower jaw. Incisoriform teeth are present in both the jaws in two rows in the upper and one row in the lower jaw. The first dorsal is in the form of a single flexible spine, the second dorsal and anal fins are with many rays. Ventral fin is absent. Body of the fish is covered with fine asperities.

Inhabiting in seas of India, *Anacanthus barbatus* attains 25 cm in length. The young ones of the species are very common in Tamil Nadu coast. They are dull brown or grey in colour. Their fins are yellow in colour. The caudal fin is with about six vertical or angular dark bands not so wide as the ground colour. The second dorsal and anal fin of the species are low. Caudal fin is wedge-shaped and its central ray are the longest. The head length of the species is about one third of the body excluding the caudal fin. The eyes are high up, behind and above the branchial opening. Body is strongly compressed, with a fleshy barbel below the symphysis of the lower jaw. In the male there is a skinny prolongation from the throat and continued nearly as far as the anal fin. It is supported by a prolongation of the pelvic bone.

Genus-*Ostracion*

Fishes of this group have shortened and angular body with the integuments modified into a solid carapace composed of angular osseous plates juxtaposition with one another, but leaving the snout, bases of the fins and the hind portion of the tail, covered by soft skin. They have six branchiostegals and may be destitude of spines or have them variously situated. The carapace form three to five ridged and closed behind the anal fin. They have small mouth, premaxillaries and maxillaries coalescent. Teeth are slender and in one row. A single spineless dorsal fin is placed opposite the anal fin. Ventral fins are absent.

Occuring in Indian seas, *Ostracion turritus* has three–ridged carapace. The superior or dorsal ridge being elevated and superiorly compressed into a sharp triangular spine. Each lateral or ventral ridge is well developed, and armed with four triangular, flattened and strong spines directed backwards. The body is about as wide as high, being of a triangular shape the apex is above. The carapace forms a moderately broad bridge across the back of the tail. A compressed supraorbital spine is directed upwards or a little backwards.

Twelve teeth in the upper and eight teeth in the lower jaw is present. They are conical, weak and of brownish in colour. Dorsal and anal fins are highest anteriorly. Caudal fin is rounded or truncated. Nine to eleven scutes are present from the gill-opening to the tail. Nine or ten scutes transversely and about eleven scutes are present across the ventral surface. The fish is olive brown in colour with three badly marked dark bands in the lower one third of the body and a fourth just behind the base of the dorsal fin. A light blue spot occur in the centre of each scute. Fins are straw-coloured. The caudal fin has two dark vertical bands, one at its base and the other at its outer extremity. The young have several dark blotches and bands over various parts of the body. A ridge extends from the orbit to the upper part of the bridge over the tail.

Inhabiting in seas of India, *Ostracion cubicus* attains 45 cm in length. Its carapace in four-ridged and spineless. Ridges are blunt, but the ventral is more prominent than the dorsal ones. Its body is about as wide as high, and the back is convex, but without any distinctly raised ridge along the median line. The carapace forms a broad bridge across the back of the tail. Inter-orbital space is concave. The fish has ten teeth in either jaw, conical, weak and of reddish brown in colour. Dorsal and anal fins are highest anteriorly. Caudal fin is truncated and equal to about one-fifth of the total length. There are ten scutes between the gill-opening and the tail, 5 or 6 transversely and 5 or 6 across the ventral surface. The fish is brownish in colour with a single blue black-edged ocellus in the centre of most of the scutes. Sometimes these spots are absent from some parts of the body. On the head they are frequently black, and have more than one spot on each scute. On the abdomen the spots may be white, with or without black marks or the spots may be black or even absent. The fins are yellowish in colour and covered with black spots or immaculate.

An inhabitant of Indian Ocean *Ostracion punctatus* is reddish-brown in colour covered with numerous white dots which are sometimes confluent into lines. The carapace of the fish is four-ridged and spineless. Ridges are blunt, but the ventral ones are more prominent than the dorsal ones. Its body is not quite so high as wide. Back is rather convex, but without any ridge along the median line. The carapace forms a moderately broad bridge across the back of the tail. Superior profile of snout is concave. Interorbital space is concave, sometimes flat. There are about 10 teeth in either jaw, conical, weak and of a reddish brown in colour. There are about 10 scutes across the ventral surface. Dorsal and anal fins are highest anteriorly, caudal truncated and equal to nearly one-fifth of the total length.

Ostracion cornutus occurs in seas of India. Its carapace is four ridged and form a bridge across the back of the tail. A long anteriorly directed and conical supraorbital spine is present. Each ventral ridge terminates posteriorly with another somewhat similar one. Dorsal ridge is rather elevated about the middle of its length with a low ridge, but no spine. Interorbital space is very concave. There are about 8 or 10 teeth in either jaw, conical and weak. About 10 scutes are present between the gill-opening and the tail, 5 or 6 transversely and 7 across the ventral surface. Dorsal and anal fins are highest anteriorly. Caudal fin is truncated and equal to about half the total length. The colour of the carapace is of light brown with round blue spots all over it in moderate numbers, about one on each scute, some however, being destitute of any. Caudal fin is blue spotted.

Family–Gymnodontes

Fishes of this group have more or less short body. Some possess the means of dialating an elastic portion of oesophagus or an abdominal sac with air or this power may be entirely wanting. Bones of the upper and lower jaw is in the form of a beak, having a cutting edge and being covered with a layer of ivory-like substance in which a median suture may be present or absent. A spineless dorsal, anal and caudal fin exists. Pectoral fins are present but ventral fins are absent. Dermal coverings modified into small or large spines or laminae. Pelvic bones and air vessel may be present or absent.

Genus-*Triodon*

In the genus *Triodon*, there is six branchiostegals. The upper jaw with and the lower jaw without a median suture. Two seperate nasal orifices are present on either side. Tail is elongated, terminating in a bilobed fin. Dorsal and anal fins are with few rays. Abdomen possessing a large sac, the upper portion of which can be dialated with air, whilst it is kept distended by elongated pelvic bones. The dermal covering consist of spine, osseous laminae, which are not imbricate. Air-vessel is present.

Inhabiting in seas of India, *Triodon bursarius* attains 53 cm in length. It is dull-brown in colour. A large irregularly shaped, black yellow or blue-edged spot is present on the upper portion of the sac. Fins are yellow-coloured. Their eyes are situated in the posterior half of the head. Interorbital space is rather concave. Dorsal fin of the fish is situated in the last half of the distance between the eye and the base of the caudal fin and slightly in advance of the anal fin. Both dorsal and anal fins are highest anteriorly. Caudal fin is emarginate.

Genus-*Xenopterus*

Fishes of the genus *Xenopterus* have five branchiostegals. Their back is rounded and jaws divided by a median suture. Nostrils are funnel shaped with fringed margins. Dorsal and anal fins are with many rays. Parts of the body are covered with fine dermal spines, having a double or treble roots.

Inhabitant of Indian Oceans, *Xenopterus naritus* have head length more than its distance from the base of the dorsal fin. Eyes are rather small, interorbital space is broad and flat. Lips are thick and fringed. Lower teeth is little larger than upper pair. The length of the base of the dorsal fin exceeds the length of head. It is highest in the middle, but its highest point is not equal to that of the body. The anal similar to but smaller than the dorsal. Caudal fin slightly emarginate. Spines are with 2 or 3 roots, large, rather widely seperated, situated between the eye and base of the pectoral fin, above which they are continued a short distance behind that fin and along the abdominal surface nearly as far as the vent. The fish is pale yellow in colour, darkest along the back and in the lower two-third of the dorsal fin.

Syngnathus spiciefer

Inhabitant of Indian oceans the species ascends rivers high above the influence of the tides and even into fresh water. It grows to about 13 cm in length. Its length of head is $2^2/_3$ in the distance between the end of the snout and the vent. The length of the trunk from the end of the snout to vent is 2/5 of the total length. Eyes are nearly 1/6 of the length of the head. Gill-cover crossed by a raised longitudinal keel, an elevated ridge along the upper edge of the snout and extending to the nape. Interorbital space is concave, when ends posteriorly in a ridge which is continued to the nape. Its body is compressed, higher than wide and ventral edge is very prominent. Length of egg pouch is more than half of the tail. The dorsal fin is situated on the first five rings of the caudal fin and the anal rays are minute and sometimes imperceptible. Rings are with smooth edges, the division between one and the next is little apparent. The species is generally light brownish in colour with a dark brown streak extending from the orbit to the angle of the mouth and a second from the posterior angle of the eye over the opercle. A few black spots are present on the under surface of the lower jaw. Body is inferior with fine brown bars. Dorsal fin is barred with brown spots. Caudal fin is blackish with a light brown base.

Genus-*Ichthyocampus*

Fishes of this group have the concave back and tail are continuous to the caudal fin. Dorsal fin is nearly or quite opposite to the vent. Pectoral and caudal fins are present. Males are with an egg-pouch having a cutaneous covering situated on the tail.

Living in sea, estuaries and fresh waters of India, *Icthyocampus carce* attain about 13 cm in length. Its length of head is $3^1/_2$ in the distance between the end of the snout and the vent. The distance between end of snout and vent is $2^2/_3$ in the total length. The diameter of the eye is 1/6 of length of head. Body is higher than it is wide. Snout is slightly elevated in its anterior third. Interorbital space is slightly concave, with a sharp low median ridge passing along the snout. Another ridge exists in the occipital region. A very sharp straight ridge crosses the opercle. Egg-pouch is about 1/2 as long as the tail. The dorsal fin is situated on the second to the eighth caudal rings. Rings are with smooth edges. They are deep brown in colour with one occasionally two milk-white spots in the centre of each body ring along the infero-lateral ridge. The lower surface of the snout is dotted with black spots. Dorsal fin is yellow in colour while pectoral and caudal fins are leaden-coloured. Eyes are greenish blue.

Inhabitants of Andaman Seas, *Hippocampus trimaculatus* are pale yellow in colour. Two rows of blackish spots are present along the dorsal fin and occasionally three large brown blotches along the edge of the back on the first, fourth and seventh body rings. In some specimens there are numerous fine black dots over the body. Sometimes light markes exist on the side of the back and the body is banded. Spines are very diversified, exceeding low in some specimens and not so in others. The dorsal fin of the species is situated upon the last two rings of the trunk and the first of the tail. Eye diameter of the species is 1/8 of the length of the head. Length of snout equals the distance between the front edge of the eye and the gill-opening. Tubercles are variously developed. In some long and acute and in others are low. Coronet is similarly high or low with a tubercle anteriorly and surrounded by five rudimentary or well-developed spines. A low supraorbital spine is present, one or either side of the throat, which are about the same shape, pointed and curved backwards.

Occurring in Indian Ocean, *Hippocampus guttulatus* attain 30 cm in length. Their body colour vary, generally greyish, marbled with darker and covered with light or dark spots, or brown with black spots or cross bands. Dorsal fin may have a dark intra-marginal band with a white outer edge. The dorsal fin of the species stands on the last two body rings and the first two tail rings. The diameter of eye is 1/6 to 1/7 of the length of the head. The anterior margin of the orbit is situated midway between the end of the snout and the posterior extremity of the head. There are two supraorbital spines directed backwards and outwards. The coronet is rather low with from four to six blunt tubercles. The tubercles on the body and tail are generally obtuse.

Inhabitant of Andaman Seas, *Hippocampus hystrix* are grey or yellowish white in colour with numerous brown and also smaller white dots. Six or seven patches are present on the tail, covered with dark spots and one or more similar spots on the body. Snout is dark with a light ring. Each spine is black at the tip. Coronet of the species is high with five spines, and a tubercle anteriorly in front of which there is another spine. Snout is slender, as long as the distance between the front margin of the orbit and the first nauchal spine, Two strong spines are above the posterior and one above the anterior margin of the orbit, 2 spines below the centre of the orbit and one behind the middle of its posterior edge. All the tubercles are in the form of slender and pointed spines except those on the end of the tail.

Fishes of Sclerodermi have compressed or angular body. Their snout are somewhat produced. Small numbers of distinct teeth are present in the jaws. Fishes of one group possess a barbel. The elements of spinous dorsal and ventral fins are generally present but variously modified. Their skin is rough or spinate or the scales in the form of a firm carapace.

Fishes of *Triacanthus* group have oblong and compressed body ending in a somewhat elongated caudal portion. Eyes are lateral, high up, situated near the hind edge of head. Teeth are in two rows in both jaws, the outer ten in number, being incisor-like. The inner is from 2 to 4, being more molariform. Their first dorsal fin consist of a long and strong spine, followed by from 3 to 5 smaller and weak ones. Ventral fins are formed by a pair of strong spines articulated by a joint or ossified to the pelvic bones. Caudal fin is lobed. The fishes have minute and rough scales. Lateral line is present. Air vessel is strong.

Triacanthus brevirostris inhabit in Indian seas and attain a length of 24 cm. They are silvery in colour with a black spot on the first dorsal fin, which extends a short distance on to the back, usually a dark supraorbital blotch. Its length of head is $4^1/_2$ to $4^2/_3$ and height of body is $3^1/_4$ to $3^1/_2$ of the total length. Eyes are high up, diameter one fourth of length of head. A considerable rise from the snout to the first dorsal fin and an elevation opposite the orbits exists. Outer row in upper jaw consisting of 8 compressed cutting teeth. The inner row has six rounded teeth. The two teeth present at the centre are the longest. In the mandibles there are ten in the outer and two in the inner row, smaller than but otherwise similar to those in the upper jaw. The first dorsal spine is very strong, longer than the head in the immature and of about equal length or shorter in the adult. Second and third spines are weak and of about equal length. Dorsal fins are close together in the very young stage, but no so in the adult. Anal fin is highest anteriorly, its lower edge is concave. The length of its base is 2/3 of that of the head. Ventral spine is slightly shorter than that of the first dorsal. Pectoral fin is short and rounded. Caudal fin deeply lobed.

Triacanthus strigilifer occurs in Tamil Nadu coast. Their length of head is 4 to $4^1/_2$ and height of the body is $3^3/_4$ of the total length. Their eyes are situated high up and diameter is $3^1/_2$ to $3^3/_4$ of the length of head. The upper profile between the eye and snout is rather concave. The fish have eight compressed and cutting teeth in the outer row of upper jaw, whilst the inner row consists of four rounded teeth. Ten similar ones in the outer and two in the inner row are present on the lower jaw. First dorsal spine is very strong and longer than the head. The remaining spines are weak, the second spine being more than half as long as the head and three times as long as the third spine. Fins are otherwise similar to those of *T. brevirostris*, except that the length of the base of the anal fin is only about half of that of the head. The fish is silvery in colour with some irregular yellow lines and blotches on the head and the body.

Fishes of *Monacanthus* group have compressed body. Barbels are absent. Incisoriform teeth are present in both jaws, in two rows, in the upper with six teeth in the outer row and a single row of six in the mandibles. The first dorsal fin composed of a spine which may be feeble or strong and merely rough or provided posteriorly or laterally with barbs, occasionally a second rudimentary one. Ventral fin, when present is reduced to a single osseous process, sometimes rudimentary and either movable or fixed. Scales are minute and rough. Cutaneous filaments are present in some species. The side of the tail may be peculiarly armed in adult males but less apparently so or not at all in the females.

Monacanthus choirocephalus is found in Tamil Nadu coast of Indian Ocean. Their length of head is $3^1/_2$, caudal fin 5 and height of body is $2^2/_3$ in the total length of the fish. Eyes are situated in the posterior half of the head over the gill-opening and below the dorsal spine. Body of the fish is rather elevated. Profile from snout to first dorsal fin is almost straight. The origin of the second dorsal fin is at nearly the highest point of the dorsal profile. Dorsal spine is situated over the last half of the eye, rather weak and rough anteriorly and with a row of small barbs on either side of its posterior surface. Second dorsal and anal fins are with the first few rays rather shorter than the succeeding ones. Ventral spine is movable without any enlarged spines at its termination. Caudal fin is rounded, sometimes with a setiform prolongation of one of its upper rays. Scales are indistinctly visible. Skin is not very rough, numerous small fleshy tentacles are present on the sides of the body. The fish is of a grey stone colour, with irregular and rather large black blotches. Two dark bands across the caudal fin are present.

Chirocentridae

Fishes of this family have much elongated and compressed body. Margin of upper jaw formed by the premaxillaries mesially and the maxillaries laterally. Opercular apparatus is complete. Barbels and pseudobranchiae are absent. A single rayed dorsal fin belonging to the caudal portion of the vertebral coloumn. Body of the fish is with thin deciduous scales. Stomach with a blind sac and intestinal canal is short. Air vessels are present. Branchiostegal with 8 rays.

Genus–Chirocentrus

Fishes of this genus have abdomen with sharp, but not serrated margin. Gill-membranes united for a short distance and gill-opening is wide. Eyes are subcutaneous. There are eight branchiostegals. Cleft of mouth is oblique and deep and the lower jaw is longer. A row of canines in the mandible and a horizontal pair in the premaxillaries are present. Minute teeth are on the palatines, pterygoids and tongue. A single short dorsal fin is placed far backwards opposite to a long anal fin. An elongated osseous appendage is present in the axilla. Ventral fins are very small. Scales are thin, small and deciduous.

Chirocentrus dorab, an inhabitant of Indian Oceans is bluish-green in colour along the back and silvery on the sides and abdomen. Its upper lip is terminating anteriorly in a short mesial flap. Lower jaw is longer, The maxilla reaches to below the hind edge of the eye. The fish has one pair of long, sharp and straight teeth near the centre of the premaxillaries. The remainder of which and the whole extent of maxilla is armed with sharp straight teeth of irregular lengths and becoming smaller at the posterior extremity of the jaw. Each mandible has a row of about 12 sharp laterally compressed teeth of which the first two are the shortest, and those most anterior have an oblique anterior direction, while the posterior ones gradually become directed more and more backwards until the last form an acute angle with the jaw. Five or six card-like teeth are present on the palatine bones, and a small oval group of velvety ones are on the pterygoid. The ventral fin commences midway between the end of the snout and the base of the caudal fin. The dorsal fin is in about the posterior third of the body, above the anal fin. Along the whole extent of the lower margin of the abdomen are short hair like rays. Scales are small and deciduous and the lateral line is indistinctly marked. Air vessel is small, elongated and cellular. It attains a length of at least 3.5 metres and bites at everything near it when captured.

Chapter 14

Threadfin Bream (*Nemipterus*)

The species of *Nemipterus* reported to occur in the seas around India are, *Nemipterus marginatus, N. hexodon, N. japonicus, N. luteus, N. mesoprion, N. tolu* and *N. delagoae*.

Nemipterus metopias, which has not been recorded so far from Indian waters, were caught in hook and lines off Vizhinjam during February 1980. The known distribution of this species are east coast of Africa, Sri Lanka, west coast of Malay Peninsula, Indonesia, Philippines and south-west coast of India.

The body of *Nemipterus metopias* is slender. Head convex and anterior part is scaleless. Snout of the fish is somewhat blunt. Inter-orbital space is almost flat. Mouth of the fish is some what oblique with rows of small conical teeth. In the upper jaw, before the band of small teeth are eight strong curved canines, while in the lower jaw seven canines are present. Pectoral fins are short. First ray of the pelvic fin is elongated. Caudal fin is deeply forked with the first two rays of upper lobe are prolonged into a short filament. The dorsal and anal fins have ten spines, nine rays and three spines seven rays respectively. The pectoral and caudal fins have 13-17 and 16-17 rays respectively.

In fresh condition the upper half of the body is pink and the lower half is silvery. Margin of pre-maxilla is yellow in colour. The head is pinkish brown with two bright yellow bands, the first one starts from the nostrils and extends across the eye upto posterior extremity of the orbit, but not the pupil; the second band starts from the front part of the upper jaw and extends backward to a point in line with the hind end of the orbit. Six faint horizontal lines along the body is present, the first one running just below the lateral line. A broad yellow band, present on the ventral surface between isthmus, base of pelvic fin bifurcates into two narrow bands below the base of the pelvic, and passes along side the anal to meet at the base of the caudal fin. Dorsal fin, pale yellow, with a narrow yellow margin immediately followed by a narrow pinkish band, and a bluish band along the base. Pelvic fins are whitish with yellowish axillary area and axillary scales. Pectoral fins are pale pinkish with a pinkish base. Anal fin is whitish with a row of yellow blotches (one blotch each between two spines or rays). Caudal fin is pink with deep red at the middle of the fork, distal part of upper lobe of caudal and its filament is yellow. In fish preserved in 5 per cent formalin, body colours faded gradually.

N. metopias resembles *N. japonicus* and *N. luteus* in having the tip of upper lobe of caudal fin is drawn into a filament which is bright yellow. It differs from *N. japonicus* in not having the bright orange red blotch just below and near origin of lateral line and yellow lines on anal fin which forms reticulate pattern, but has single median row of yellow spots on the anal fin. Body of *N. metopias* is very slender being less than head depth, while in *N. japonicus* body is as deep or deeper than head.

N. metopias differs from *N. luteus* in not having the first and second spines of dorsal fin very close together forming a single long filament of bright yellow in colour. *N. metopias* has two bright yellow bands on head which are absent in *N. luteus* and *N. japonicus*.

The threadfin breams constitute an important group of commercial fishes in India. Of the twenty species of the genus *Nemipterus*, in the Tamil Nadu coast six species namely, *Nemipterus delagoae, N. japonicus, N. nematophorus, N. mesoprion, N. tolu* and *N. metopias* have been found to occur. *Nemipterus metopias* is distributed in the coast of Sri Lanka, west coast of Malaya Peninsula, Indonesia and Philippines.

Nemipterid fishes in the trawler catches off Kakinada along east coast of India collected specimens of *Nemipterus tolu*

(Valenciennes), *Nemipterus japonicus* (Bloch) and *Nemipterus mesoprion* (Bleeker)

Nemipterus mesoprion

The species is distributed from east and west coasts of Sumatra and Singapore and extend westwards towards the east coast of India.

Body of the fish is pink coloured dorsally and on upper half of flanks. Lower flanks and belly silvery coloured. Three to four longitudinal yellow bands are present on sides below lateral line. Thick yellow longitudinal bands are observed on the ventro-lateral borders. Dorsal fin with closely packed yellow pigment dots on the lower three fourth of the fin forming a longitudinal yellow band which divides into three small bands on the last few rays; upper margin is pink in colour. A reddish blotch below lateral line is present near its origin. Pectoral and pelvic fins are pink in colour, but with the first ray whitish all along its length. Anal fin is whitish with a distinct narrow longitudinal yellow band in the middle. Caudal fin is pink coloured including the elongated filament on the upper lobe.

The mouth of the fish is oblique, maxillary reaching to below anterior border of pupil. Teeth in jaws are present in several rows, pointed, upper jaw with 3-4 caniniform teeth anteriorly in each ramus. But such teeth are absent in lower jaw. Scales are ctenoid and three rows of scales are present on preopercle. Height of suborbital is nearly half of vertical diameter of the eye. Hind border of preoperculum is crenulate. Dorsal spines are strong, first spine is shortest, fourth to tenth more or less of same length. Soft dorsal is deeper than spinous one. First anal spine is shortest and third is longest. Soft anal is less deep than soft dorsal; the last two rays of anal are longer than the preceding ones. Pectoral fins are falcate, extend to above first or second anal ray. Pelvics with the first ray produced reaching to first or second anal rays. The second branched ray of upper caudal lobe produce into a long filament, the length of which is more than the caudal fin itself.

Colour is the most important character in distinguishing species of Nemipteridae. The colour of *N. mesoprion* is described among other things, as; head with yellow streaks from eye to middle of upper jaw. Dorsal fin is with a broad median yellow longitudinal band which sub divides towards tail into three yellow bands seperated by blue lines. Pelvic fins are pink with elongated first ray deep red. Caudal fin is reddish; median rays are yellow and outer rays and filament are red. The differences in the colour pattern is due to the differences in the habitat.

The head length in *N. mesoprion* is more than body depth. The snout length is less than horizontal eye diameter. Suborbital height is about half of vertical eye diameter. The first pelvic ray is produced and reaches beyond anal origin upto second anal ray. Colour on anal fin is only one longitudinal yellow band in the middle. The tip of upper caudal lobe and filament is pink in colour.

Nemipterus mesoprion forms a seasonal fishery in Andhra coast, India. The fish attains 140mm, 185mm and 205 mm at the completion of first, second and third years respectively.

The testis in *N. mesoprion* is very small even in matured fish, thus making it difficult to study the process of maturation in males.

Fishes above 90 mm length show mature ovaries in females. 100 mm is taken as the length at first maturity of female *N. mesoprion*. Ova are distributed around two modes in mature ovaries. One mode at 0.152 mm and the other at 0.418mm ova diameter. The former group constitutes maturing translucent ova and the later constitutes mature opaque ova. In running ripe females a group of ova are seperated and form modes at 0.722 mm. These constitute the ripe ova with translucent yolk and distinct oil globule.

The species spawns in the sea off Andhra coast during December-April period with peak during January. The fact that ripe oozing females occur in the catches in considerable numbers during certain months indicates that *N. mesoprion* spawns in the trawling ground off Kakinada.

Nemipterus japonicus

Even though five species of *Nemipterus* have been taken from the Indian seas, *Nemipterus japonicus* is dominant one along Kerala coast. The average catch of *Nemipterus* spp at Cochin is about 35 tonnes a year.

In the early months of the year upto May, the bulk of thread fin bream catches are obtained within 30 m depth. In the latter parts of the year better catches are obtained from 30-50 m depth. The height of fishing season is August to October with peak landing in September.

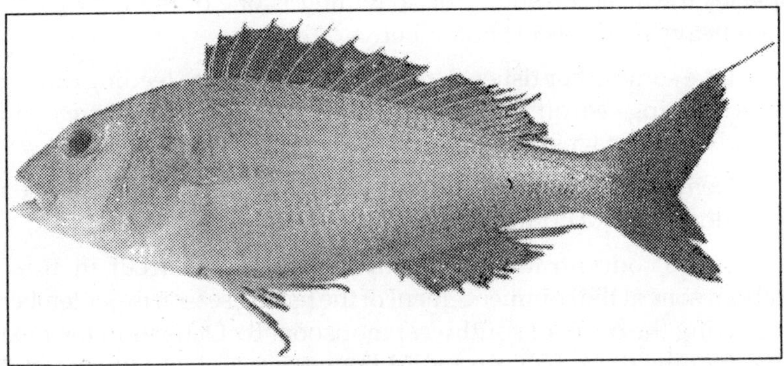

Nemipterus japonicus

N. japonicus is a carnivore, voraciously feeding on crustaceans and fishes. The major food items of the species are crustaceans with *Acetes indicus* as the main individual item followed by squilla and crabs. During monsoon, fishes, mainly anchovy and detritus are the dominating item of food. The intensity of feeding is poor in the months of January to April. In August the intensity of feeding is very poor. From September onwards there is increase on the feeding of *N. japonicus*. October to December witness a very good feeding activity of the species.

Though *Nemipterus* is a seasonal fishery, it is available throughout the year, almost all the time, to a certain extent and mixed size groups are seen.

Upwelling influences the *Nemipterus* fishery. The demersal fishes migrate from the oxygen depleted mid water towards the shore during upwelling. In the monsoon season, the catch is better than the earlier months. The rough seas of the monsoon months prevent most of the vessels from fishing, otherwise a better catch could be expected during the monsoon season. In September, as the monsoon declines, the landings are exceptionally good. As larger size groups with advance stages of maturity are observed at this time, the inshore migration may also be connected with breeding.

N. japonicus form schools. In September large schools of the species are observed south west of Cochin. They are comming from the south and moving north to Calicut. The fishes in advanced stage of maturity are caught in September-October. The movement of the species. towards the shore for spawning is one of the reasons for their heavy landings in September.

Like some other fishes, *Nemipterus* do not cease feeding during the spawning season. The feeding intensity is found to increase in the advanced stages of maturity.

Nemipterids are abundant on muddy bottoms, but least common on sandy or rocky bottoms.

Large concentration of *Nemipterus japonicus* occur in 16-27 fathom zone at the commencement of the fishing season in September following the onset of southwest monsoon. By October in the next five months these fish are found between 8 and 23 fathoms, but mainly around 10-23 fathom zone. During April-May they again move to the deeper 24-27 fathom zone, completely disappearing from the shallower 8-15 fathom zone. Along the west coast of India *N. japonicus* is the most dominant species in depth zones of 75-100 and 101-179 metres. *Nemipterus* was also recorded from depth range of 180-450 metres. The catches from off Goa region showed that *Nemipterus* was in good concentration in 41-50 m depth.

Both in east and west coast of India, *N. japonicus* have comparable growth of an average of 150mm (140-170) in its first year, 210 mm (200-230) at the end of second year and 240 mm (220-260mm) at the end of third year.

Threadfin bream forms an important fishery along the Andhra-Orissa coasts, sometimes forming as high as 13.8 per cent of the total catches. The fish breeds over a short and definite period from December to February. It appears that the fish breed for a second time in June-July period.

Fishes, ranging in size from 130 to 209 mm length have on an average 13900 to 47200 ova in their ovaries weighting 1.5 g to 4.9 g respectively.

In the stage I the ovary of *N. japonicus* is small and slender. Eggs are transparent and not visible to the naked eyes. The transparent ova possess a distinct nucleus and a protoplasmic layer devoid of any yolk deposition. The size of ova varied from 1 to 5 micrometer division (1 micrometer division = 0.016 mm).

In the stage II the ovary is slightly enlarged, eggs transparent and not visible to the naked eye. The maturing ova, though small and still transparent, have started yolk deposition. A central semi-transparent portion can be made out. The diameters of the ova vary from 6 to 12 micrometer division.

The ovary in the third stage is visible to the naked eye. Eggs are not completely opaque. The maturing ova is small with centre fully yolked and opaque, but the periphery is still transparent. The diameters vary from 12 to 15 micrometer division.

Most advanced eggs in the stage IV have become completely opaque. The diameters of the largest eggs are 30 micrometer divisions. The matured ova are opaque and fully yolked but still contained within the follicles. The size ranges from 16 to 25 micrometer divisions.

Ovary in the stage V is large. Most advanced eggs have become partly transparent. The maximum diameter of the most advanced eggs are 40 micrometer divisions. Fully matured ova at this stage are large, free, partly transparent eggs, which have burst from follicles. The diameter range being 26 to 38 micrometer divisions.

Ovary in the stage VI is greatly enlarged. Eggs are fully transparent. The maximum diameter of the most advanced eggs are 55 micrometer division and ready to be shed. Ovas at this stage are free and large, fully transparent with deposition of one or more drops of oil globules and ready for liberation. The size ranged from 40 to 55 micrometer divisions.

In the seventh stage the ovary is spent, flabby and contracted.

Pomfrets

The pomfrets constitute about 2.32 per cent of the total marine fish landings in India. The bulk of the catch comes from Maharashtra and Gujrat States, which jointly contribute to about 61 per cent of the total all India pomfrets landings.

Maharashtra landed about 36 per cent, closely followed by Gujrat (26 per cent) and Andhra (15 per cent). Kerala, Tamil Nadu, West Bengal and Orissa and Karnataka also land small quantities of pomfrets, their percentage to total all India catch being about 8, 7, 6 and 3 in the order of abundance. During 1976, however, West Bengal and Orissa landed 11,285 tonnes which is a bumper catch and the highest recorded ever in that state.

Maximum landings of pomfrets are recorded during October to December in Maharashtra and the minimum during July to September. Shore seine and gill nets are predominantly employed for the exploitation of pomfret fishery. In addition, trawl nets and 'Dol' nets are also used to catch pomfrets. *Pampus argenteus* and *Parastromateus niger* are the dominant species comprising the pomfret fishery.

In Gujrat the pomfret fishing season is in second and fourth quarters (April to June and October to December). In some years the maximum catch was during the fourth quarter, while in the other years, the highest landings were obtained in the second quarter. Similarly the lean fishing season for pomfret fishery is in the first and third quarters (January to March and July to September).

Gill nets, Dol nets, trawl nets and drift nets are operated for pomfret fishery in Gujrat.

The dominating species of pomfrets along Gujrat coast are *Pampus argenteus, Pampus chinensis* and *Parastromateus niger*. In Umbergaon to Bilimora of Gujrat coast, the size of *Pampus argenteus* is in the range of 201 to 335 mm, with the dominant species being in the range of 311-315 mm. In Cambay coast however, the size range of species is in the range of 77-165mm.

Silver Pomfret, *Pampus argenteus* (Eupher)

P. argenteus forms about 0.3-0.5 per cent of the dol-net landings at Nuabunder, Gujrat. Large quantities of small sized *P. argenteus* were caught in dol net in April-May and September-October, mostly in the size range of 20-99 mm length, while they were to the tune of 84-86.3 per cent. The fish in the size range of 100-199 mm and above 200 mm were 13.1-15.5 per cent and 0.5-0.6 per cent respectively. Though the young ones are present throughout the year in the narrow coastal belt of 45 km, at a depth of 15-35 meters, large number of these, however, enter the fishery during January-March and September-October. The juveniles grow 50 mm in 4 months and 60 mm in five months.

Though estimated 50.7 and 40.7 tonnes of *P. argenteus* were landed at Nawabunder during 1979-80 and 1980-81 respectively forming 0.3 to 0.5 per cent of the dol-net landings, the abundance of juveniles has been high, particularly in the size range of 20-29 mm to 90-99 mm (84-86.3 per cent). The abundance of young ones in the region of Saurashtra coast suggests that this area is a nursery ground for. *P. argenteus. P. argenteus*, best table fish, grows to a size of 210 mm weighting 300 g. The maximum size recorded in commercial catch was 311 mm weighting 1.1 kg. During February-May there is largest concentrations of juvenile pomfrets along Saurashtra coast forming 80-95.6 per cent of pomfret population.

Grey Mullet, *Mugil cunnesius* (Val)

Mugil cunnesius is common on the east and west coasts of India ascending tidal rivers though not to as great a distance as *Mugil persia*. The coastal waters of Bombay and Visakhapatnam form the typical localities of this species and in both these areas they occur in fairly large numbers. It is common on the coast of Kerala and appears to be present in the Gulf of Mannar also. The species is fairly abundant all along the Madras and Andhra coasts. In the Chilka lagoon it is found in shoals in the outer channel during the summer and winter months. In Bengal this species is very common, being second in abundance only to *Mugil parsia* and is found in large numbers in the coastal waters as well as in the lower estuaries. Besides the coastal waters and estuaries of India, *M. cunnesius* has a fairly wide distribution in the Red Sea, the Arabian Sea and the seas off Philippines, Queensland (Australia) and the South Pacific.

There appears to be only one spawning season in a year, which lasts from May–June to July–August. There is no conclusive evidence to show that the individual fish spawns more than once during a spawning season. The smallest maturing fish obtained is 130 mm long. The fecundity of the species ranges from 14913 to 56486.

The maximum length of the species was found to be 300 mm standard length. The fish attain a mean length of 84 mm in first year, 131 mm in the second year, 175 mm in third year, 201 mm in the fourth year and 233 mm in the fifth year of its life.

There does not appear to be any clearly defined or long range migration of *M. cunnesius*. The spawning of the species must be taking place some where in the sea, presumbly near the mouth of estuaries. The breeding season of the fish coincides with the monsoon season. Fry of the species 10-19 mm in length, appear along the coastal waters of Bengal by about June, and continue to occur until late August.

The fishery of *M. cunnesius* lasts only for a period of six months. During this period young of the species are caught in large numbers. Both in coastal waters and estuaries the fry of the species predominate in the mullet catches during July-August. The length-frequency distributions show that the bulk of the catches consists of the 0 group and the one-year old fish. *M. cunnesius* leave the estuaries by about October and do not return until the onset of south-west monsoon. It is likely that some fish do remain in the estuaries, but these might have grown too big to be carried into the embanked bheris along with the tide. Though *M. cunnesius* does not naturally enter bheris, it is well suited for culture in enclosed water as evidenced by the fact that it is successfully cultured in fresh water ponds on the Contai coast, West Bengal with other mullets and carps.

Mugil cephalus (Linnaeus)

Mugil cephalus has been found to breed between September and January in the sea. Young larvae lives through the winter and enters the lagoon by summer *i.e.,* April to June where they continues to grow. The span of life for the *Mugil cephalus* is six years and a few months. The first year fish ranges in length upto 120 mm, the second year fish from 121 to 431 mm and the third year fish from 439 to 493 mm. The largest mullet of 570 mm belong to fifth year class.

Mugil parsia (Hamilton)

M. parsia is the most abundant mullet in estuaries of Bengal. The species occurs in the fore shore areas of sea also.

In the marine environment, decayed organic matter and diatoms form the predominant items of food. Algae and miscellaneous matter occured only in small quantities. Miscellaneous matter, which is of

Mugil cephalus **(Linnaeus)**

animal origin form only a very minor portion of the total food consumed. In the estuarine environment decayed organic matter figured as the most important food item. Large quantities of such matter are brought into the estuaries by the flood waters and are available to the fish in appreciable quantities. Algae and diatoms are present only in comparatively small portions in the guts. The miscellaneous matter from estuarine environment consisted mainly of polychaete moults. The fish in the brackish water bheri consume mainly algae (Chlorophyceae and Myxophyceae). In the winter months lumps of filamentous alage often constitute the only food item in the stomachs, particularly of large-sized fish. Decayed organic matter come next in importance. But diatoms form only a minor portion of the diet, because of rare availability in the benthic zones of this environment. The young ones feed on appreciable quantities of diatoms from the surface plankton. Miscellaneous matter form a rare item in the diet of fish from this environment. Copepods constitute the main food of *M. parsia* in Bombay waters, which is supplemented by diatoms and organic matter taken along with mud.

The young of *M. parsia* from 5-10mm feed at or near the surface. The miscellaneous matter, consisting of various copepodid stages, nauplii and rotifers form the main food and occur in 100 per cent of guts of post larvae 5-10 mm in length. Sand grains and decayed organic matter appear in appreciable proportions in the guts of fry of about 20 mm length, which may be considered the beginning of adult feeding habits. There is a gradual change over from the surface feeding habit of the post-larva to bottom feeding in adult.

There is a remarkable enhancement in the intensity of feeding during winter, and a lowering of it in the rainy months. The increased feeding intensity in winter corresponds to the period of general abundance of algal blooms in these waters. During rainy season there is considerable agitation and disturbance of the benthic flora and this is probably the reason for the fall in feeding intensity at this time.

Mugil speigleri (Bleeker)

M. speigleri, though not so abundant as *M. parsia* is caught in considerable numbers in the estuaries and in the fore shore areas of sea.

Decayed organic matter form a total of 37.7 per cent of food when all the environments are considered together. Diatoms constitute 14.8 per cent and algae 7.7 per cent. Miscellaneous matter consists of polychaete remains, copepod remains, mysids and nauplius larvae of which the first is the most predominant. This item constitute 8.6 per cent of the total stomach contents. Sand grains form 31 per cent of the stomach content as in *M. parsia*.

Minute sand grains and decayed organic matter is present is small proportions in the guts of size group. 17-22 mm, besides diatoms and miscellaneous matter consisting of copepods, nauplii and mysis. Adult feeding habits of consuming food materials from the bottom appear to get established in *M. speigleri* by the time the fry are about 25 mm in length.

Silverbellies

The fishes of the family Leiognathidae, popularly called silverbellies contribute to an important fishery in the states of Tamil Nadu, Kerala, Andhra Pradesh and Karnataka bordering Indian Oceans. Of these states, Tamil Nadu accounts for the bulk of the catches of silverbellies. Within Tamil Nadu, the south-east coast comprising Palk Bay and Gulf of Mannar regions yield very high catches of silverbellies. In these regions the fishery is almost continuous throughout, generally from April to October along the Palk Bay and November to March along the Gulf of Mannar.

In this area, of the 19 species of silverbellies known to occur in the seas around India, except, *Leiognathus blochii, L. elongatus* and *L. indicus,* all the other 16 species occur in varying proportions throughout the year. Of the sixteen species occuring in this area, *L. jonesi* is the most dominant in Palk Bay followed by *L. brevirostris* and *Secutor ruconius*. In the Gulf of Mannar, the relative abundance of various species varies considerably. *L. dussumieri* being one of the dominant species. Other species commonly found in the Gulf of Mannar are *L. bindus, Secutor insidiator, S. ruconius* and *Gazza minuta*.

Leiognathus dussumieri (Valenciennes)

Length-weight relationship of male and female *L. dussumieri* was found to be different. The males weigh comparatively less than the females. Among the females, the immature, maturing and spent females weigh more than the mature females. This may be due to the accumulation of fat in fish other than the matured ones.

The species mainly spawns during April-May and November-December. Minimum size of maturity for males were found to be 78 mm and for females 83 mm. The following maturity stages were recognised.

Stage-I

Ovaries small, transparent, occupying a very small portion of the body cavity. Ova is not visible to the naked eye, measure a maximum 0.126 mm. Testis are similar in appearance as the ovaries.

Stage-II

Ovaries occupy about one third of the body cavity and are semitransparent. Granular ova is visible to naked eye, measure a maximum of 0.315 mm Testis are semitransparent and occupy about one third of the body cavity.

Stage-III

Ovaries occupy about half the body cavity, yellow in colour. Ova measure a maximum of 0.378 mm. Ova become opaque with full deposition of yolk. Testis occupy about half the body cavity and become creamy-white in colour.

Stage-IV

Ovaries occupy about three fourth of the body cavity, pale yellow in colour. Ova measure a maximum of 0.504 mm and become semi-transparent. Testis are creamy-white and occupy nearly three-fourth of the body cavity.

Stage-V

Ovaries are small, blood-shot, occupy less than one third of the body cavity. Maximum size of ova is 0.378 mm. Testis are small and occupy less than one third of the body cavity.

Generally, females predominate over males in the commercial catches. The counts of mature ova ranged from 805 to 41683 per female.

The fish spawn in April, May and November, December. The males mature at any size between 78 to 87 mm. Females mature at any size from 83 to 92 mm. Therefore the minimum size of maturity for male is 78 mm and for female 83 mm.

The most important items of food of the species include polychaetes, copepods, amphipods, bivalves, gastropods and foraminiferans. No significant variations in the food of fish from different places in different seasons have been found. The feeding habits of this fish also do not change with age. The fish was found to feed more actively in the month of January, February, April and November. The fish attains a length of 99 mm at the end of first year, 114 mm at the end of second year, 128 mm at the end of third year, 138 mm at the end of fourth year and 145 mm at the end of fifth year. The life span of the species was found to be five years, the maximum length recorded in Gulf of Mannar was 161 mm.

Good catches are obtained from deeper waters of 20 to 40 meters in Gulf of Mannar. The species is mainly landed by drift gillnets and trawl nets. Zero, one year and two year old fish contribute the bulk of commercial catches. The commercial catches are dominated by fish of 93 to 117 mm size. Bulk of the catches are landed by trawl nets during the period November to March. The species is also known to move in schools. Occasionally, but it is usually captured with other silverbellies. The species prefer a sandy, coral-sandy area rather than muddy areas. Since the species is short lived and breeds by the time it is one year old and the present method of exploitation leaves enough brood to replenish the future stocks, it is advisable to catch the fish of all sizes for the best utilization of the resource.

The species, *Leiognathus splendens* is widely distributed in the Indo-Pacific region. Its occurrence is reported in the seas of India, Malaya, Indonesia, China, Philippines, Queensland and the Fiji Islands.

The body of the fish is oblong and compressed and the mouth is protrusible. It is a small fish, attaining an average length of about ten centimeters. Their colour is silvery and hence the name, silver belly.

Although there are two species of silver bellies, namely, *Leiognathus splendens* and *Leiognathus ruconis*, in India, *L. splendens* forms an important fishery at Thangachimadam, on the Rameswaram Island, the annual landings being more than 500 tonnes.

The fish is captured by means of a bag net (*Mada valai*), which is of conical shape and of small mesh of 1.25 cm. The net varies in length from 12 to 13 metres with a mouth opening of about same diameter. The net is operated from two boat catamarans and at a depth of 3 to 6 fathoms.

Leiognathus splendens matures at an average standard length of 60 mm and spawns from March till September, the peaks of spawning period being in April and August. Three stages in the development of the eggs were recognized. Stage–I consisted of immature eggs only, ranging in diameter to about 0.135 mm; stage–II, an intermediate group of eggs ranging from 0.135 to 0.375 mm and finally stage-III, matured eggs ranging in diameter from 0.445 to 0.775 mm. The immature eggs are present in the ovary at all times of the year, but the intermediate and maturing groups are found only during the spawning season. It takes nearly two weeks for the intermediate eggs to differentiate into matured eggs. When the eggs are fully matured, they burst from their follicles and become segregated in the lumen of the ovary. The matured eggs are translucent and contain a single oil globule.

Immature ovaries contain no eggs visible to the naked eye. Maturing ovaries contain larger eggs which are translucent rather than opaque. Since the intermediate and maturing groups of eggs are never found in the gonads except during the spawning season, the presence of either one or both of these classes indicate that the fish would spawn within the next few days. A great majority of adult females mature during the months of March and April and in August and relatively fewer fish mature during the months of May, June and July. Obviously the fish spawns from March till August or September but the peaks of spawning occur during the months of April and August. The males during these months are also fully matured.

Leiognathus splendens spawns more than once in a season *L. splendens* contains an average of 7566 eggs. The larger fish (74 mm SL) may contain upto 11000 eggs.

It is probable that the fish do not live for more than three years and that they die sometime in the third year of their life. No fish was found to attain a standard length of more than 85 mm. Fish which hatched in March, attain an average modal length of 17 mm in April, 22 mm in May, 32 mm in June, 35 mm in July, 47 mm in August and 65 mm in November of the same year. The growth of fish in the first year is very rapid, the fish attaining a length of 65 mm at the end of the first year of its life. But the growth in the second year is very slow. The fish attain a length of only 72 mm at the end of the second year of its life.

It is seen that the weight of *L. splendens* increases at a rate slightly greater than the cube of the length. The fish with average length of 35.20mm, 40.19 mm, 44.78 mm, 49.35mm, 51.31mm, 60.50mm, 65.20 mm, 70.04 mm, 74.71 mm 78.73 mm and 83.00 mm weighs on an average, 1.29 gm, 1.98 gm, 2.82 gm, 4.20 gm, 6.21 gm, 8.12 gm 10.26 gm, 12.38 gm, 15.00 gm, 17.68 gm and 19.50 gm. respectively.

Among coastal fisheries of India, silver bellies occupy an important place forming about 4 per cent of the total marine fish catch. Nearly half of the total landings of silverbellies is obtained from Tamil Nadu, bulk of which is obtained from the Palk Bay. *Leiognathus splendens* form a dominant fishery in the Palk Bay. *Leiognathus jonesi*, which closely resembles *L. splendens*, form the dominant species accounting for as much as 90 per cent. The species has been subjected to heavy exploitation off Mandapam in the Palk Bay after introduction of mechanized trawlers.

L. jonesi attains an average length of 72 mm at the end of one year and 108 mm at the end of two years. It reaches an average length of 116 mm when it is 28 months old.

The rate of mortality of fish may vary according to size and habit of the fish. Sometimes, small fishes may not be subject to fishing mortality as they may escape through the meshes of the net if the mesh size is big. Further they will not be subject to fishing mortality if they are not available in the fishing ground. Thus two factors come into play, one being the behaviour of the fish and the other being the properties of gear selection.

The estimate of natural mortality in regard to important marine fishes of India poses a great problem for the fishery biologists, on account of various factors like determination of age, short life span and standardization of effort.

In case of oil sardine the estimate of natural mortality ranged from 0.67 to 1.45. For mackerel, the same ranged from 0.65 to 0.9. The natural mortality of ghol is estimated as 0.87. The estimate of natural mortality (2.28) derived for *L. jonesi* is higher when compared with the values obtained for oil sardine, mackerel and ghol. This can be attributed to the observation made that the life span of oil sardine, mackerel and ghol is longer than that of *L. gonesi* whose life span is less than three years only.

Silverbellies form an important fishery on the coasts of southern maritime states. It form a good demersal fishery along the coasts of Andhra Pradesh, Tamil Nadu and Kerala.

A review of all-India total marine fish landings and landings of silverbellies for the years 1957-75 reveal that the catch of silverbellies, which was of the order of 13208 to 18268 tonnes in 1957-63 showed substantial increase during the subsequent period of 1964-75, when the annual landings ranged from 27213 tonnes in 1965 to 51240 tonnes in 1974. Their percentage contribution to the total catch which was in the range of 1.75 to 2.84 during 1957-63 increased from 2.81 to 5.47 in 1964-75 period. This increase can be attributed to the introduction of increasing number of mechanized fishing boats since 1964.

Tamil Nadu records higher landings of silverbellies. The maximum landings were in 1970 (27145 tonnes) and the minimum in 1957 (4720 tonnes). The highest percentage of silverbellies to the total catch in Tamil Nadu was recorded in 1970 (17.45 per cent) and the lowest in 1958 (4.46 per cent)

The trends in the catch alone will not give any clear picture as the catch depends on fishing effort put in and other factors like salinity, temperature, availability of food and other environmental factors. The study on parameters of population characteristic may give some idea about the distribution pattern. For this, the fishing effort in man hours and the catch per unit effort (the index of abundance) is taken into consideration to find a relation between fishing effort and the index of abundance in respect of both total mechanized catch and the catch of silverbellies by mechanized boats in Tamil Nadu during the years 1971 to 1975. The man hours refer to the product of number of operations, trawling hours, and the number of crew. The relation between fishing effort in 1000 man hours and the index of abundance (catch per unit effort in kg) has been calculated by using the equation of the form.

$$Y-ax^{b},$$

where, x represents the fishing effort in 100 man hours and y the index of abundance in kg.

The relation between the index of abundance and the fishing effort reveals that the indices of abundance have fallen steadily, the fall in respect of silverbellies being sharper than total landings. The analysis of monthly catch trend of silvebellies in the six zones of Tamil Nadu coast (Royapuram Kasimedu zone, Cuddalore zone, Porto Novo zone, Adirampatnam zone, Mandapam zone and Rameswaram zone) reveal that excepting Royapuram and Adirampatnam, the catch declined sharply in all other zones.

Higher catches of silverbellies are observed on full moon nights than on new moon nights.

Thread-fin, *Polydactylus indicus* (Shaw)

The thread-fins of family *Polynemidae* constitute an important group of commercial fishes in Indian waters of which *Polydactylus indicus* forms the most important species in Bombay and Saurashtra coasts.

Six species of thread-fins recorded from Indian Seas, namely *P. indicus, P. plebeius, P. sextarius, P. heptadactylus, P. paradiseus* and *Eleutheronema tetradactylum,* have also been reported from Bombay waters. Of these, *P. paradiseus* and *P. sextarius* inhabit inshore or coastal waters, while *P. plebeius* is very rare. From the view point of commercial fisheries *P. indicus* is the most important, followed by *E. tetradactylum* and *P. heptadactylus*. Of the threadfins, is the most highly esteemed in Bombay, being often classed with pomfrets in palatability. *P. indicus* is very popular in Bombay markets.

Polydactylus indicus though having five pectoral filaments as in *P. plebeius*, is easily distinguished from the latter by the large number

Eleutheronema tetradactylum

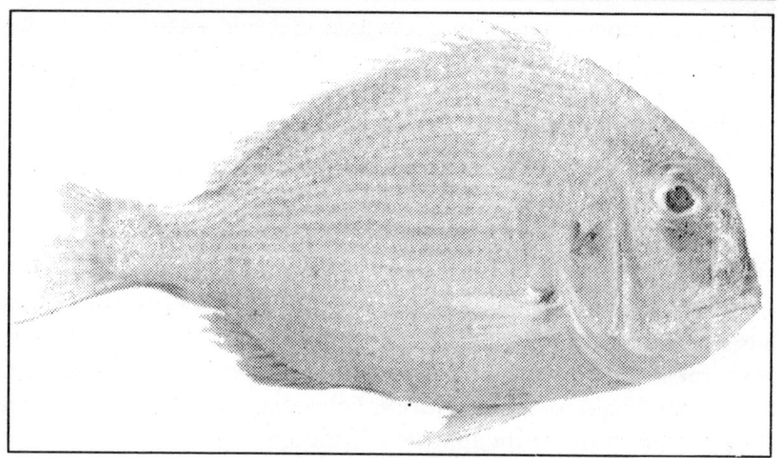

Argyrops spinifer

of (68 to 79) scales along the lateral line (56 to 61 in *P. plebeius*). The eyes have a thick transparent covering which in older fish becomes tough and comparatively opaque. The dorsal profile undergoes considerable change in its curvature as the fish grows old. In specimens upto about 50 cm in standard length, the tips of the dorsal and ventral lobes of the caudal fin are produced into long filaments, each filament being formed of two of the outer rays. This prolongation gets broken off in the later stages, with the result that it is difficult to get the correct length of the fish. The lateral line, formed of a row of scales with prominent sensory tubes, runs almost straight from the upper edge of the opercle to the base of the caudal fin, where it takes a downward bend and continues in the form of a row of minute scales on to the edge of the lower lobe of the caudal fin.

In the fresh condition, the body is dark in colour, particularly on the dorsal side and on the fins. Ventrally it is less dark. In large specimens the colour is often replaced by a golden yellow tinge, often tending to turn reddish.

Like most of the bottom living marine fishes *P. indicus* is predatory in habit and takes fairly large quantities of food in the adult stages. The juveniles or immature fishes are predominantly crustacean feeders. Besides they also take fishes like small sciaenids as food. In the case of larger specimens, the piscivorous habit is more pronounced and only a few feed on crustacean diet. But the crustaceans constitute the largest single item of food consume by both in the juvenile and adult fishes. Among the crustaceans food, prawns dominate, followed by stomatopods. Crabs are rather rare.

Eleven species of fishes are found in the stomach of *P. indicus*. Bombay duck, *Harpodon nehereus* and *Coilia dussumieri* are predominating. The sciaenids as a group constitute second largest food item of the fish.

Appreciable catches of *P. indicus* are obtained only from regions lying off the mouth of the Gulf of Cutch and the Gulf of Cambay. In both these places, particularly in the Gulf of Cambay, mature specimens begin to appear during February. In March most of the fishes are mature. By the middle of June spent fishes predominate. The peak period of breeding therefore appears to be from March to May, before the onset of the monsoon. The largest specimens of this species are caught in the mouth of large rivers. The possibility of this species breeding inside the Gulf of Cutch and in the upper reaches of the Gulf of Cambay near the mouths of the large rivers cannot be entirely ruled out.

The smallest mature female of this species observed is 78.4 cm in length. The smallest mature male is 82.0 cm.

P. indicus has been reported from both the coasts of India, eventhough it seems to occur as a commercial fishery only from Bombay northwards up to Karachi. Though it has been recorded from the east coast, it does not figure in the trawl catches made off the Bengal and Orissa coasts in the Bay of Bengal. The occurrence of the species on both the coasts of Tamil Nadu has been observed. On the west coast, near Calicut, *P. plebeius, P. heptadactylus* and *Eleutheronema tetradactylum* occur in the inshore waters. Small immature fish (31.4 to 37 cm) was obtained from the inshore waters off Travancore Cochin during July-August. The occurrence of *P. indicus* and *P. plebeius* in the trawl catches off Sind coast is also reported.

Good catches of *P. indicus* were obtained at depths varying 15 to 19 fathoms in the Cutch region and from 17 to 22 fathoms in the Cambay region. The species are scarce beyond 30 fathoms. The size range of the catch varied from 25.3 to 107.2 cm.

The breeding season of the fish seems to be from March to May. Sexual maturity is attained at a size of 78.4cm and above.

The species forms about a fifth of the total landings of the trawl catches in Bombay. Though small quantities are caught in bag nets, the chief method of capture of this species is by drift nets. Five dominant size-groups are met in these catches. Of these only the smallest and largest size groups are found in the Kutch region. This peculiar distribution of size groups indicates the possibility of migration in the life history of the species.

Maximum catches are obtained during the months of December, January and February. *P. indicus* occurs in large numbers in Gulf of Cutch region, where the maximum catch per hour actual trawling realised has been 104 kg.

Indian Salmon *Eleutheronema tetradactylum* (Shaw)

Eleutheronema tetradactylum is widely distributed in the seas of India, Malay Archipelago and China. Along with other polynemids, like, *Polynemus indicus, Polynemus sextarius,* Indian salmon has entered into Chilka lagoon, the largest brackish water with an average area of 1040 sq. km. The lagoon is connected with the Bay of Bengal through a sandy narrow channel of 29 km long and 365 metre wide, showing high tidal influence. The species is of considerable commercial importance, constituting about 7 per cent of the total fish production of the lagoon. Between 1957-60, the yield of the species varied between 193 to 364 tonnes.

The growth of *E. tetradactylum* is very rapid during the first year and declines gradually with the advancement of age. The fish of the early-brood grows to about 300 mm in its first year, to about 475 mm in its second year and to about 600 mm in the third year of its life. The fish of the late-brood grow to about 260 mm, 430mm and 570mm at the same ages. The maximum length of the fish encountered in the lagoon is 1000mm.

E. tetradactylum is a voracious carnivore. Its main food items, namely, mysids, prawns and forage fishes, are very abundant in the lagoon, which makes the lagoon an especially suitable habitat for *E. tetradactylum* to live in.

Sexes, in this species, cannot be distinguished externally. By internal examination sexes are recognizable in specimens measuring

140 mm and above in total length. The smallest mature female observed was 285 mm and the male 225 mm long. Hundred per cent maturity is attained at 350 mm among males and at 450 mm among females.

The fecundity of *E. tetradactylum* varies from 226541 to 3826683 of fish between size-range 347 and 840 mm in length.

Immegrent *Elentheronema tetradactylum* breeds in the Chilka lagoon during January to July.

Chapter 15

Reef Fishes

All the three major types of coral reefs, atoll, fringing and barrier occur in India Ocean. In India the reefs are distributed along the east and west coasts at restricted places. Fringing reefs are found in Gulf of Mannar and Palk Bay. Platform reefs are seen in Gulf of Kutch. Patchy reefs are present along Ratnagiri and Malvan coasts. Fringing and barrier reefs are located in Andaman and Nicobar Islands. Atoll reefs are found in Lakshadweep. The absence of reef in Bay of Bengal (North East coast) is due to huge quantity of freshwater and silt brought by Ganga, Krishna and Godavari rivers. The total area of coral reefs in India is estimated to be 2375 sq. km.

The exact number of fish species associated with coral reefs of Indian Ocean is not known. However the number of fish species dwelling in Indian Ocean is 1367. The Lakshadweep Islands have a total of 603 + species of fishes, over 1000 + species are found is Andaman and Nicobars Islands and in Gulf of Mannar Biosphere Reserve it is 538 +. The category of fishes occurring in coral reef ecosystem of India includes groups, such as the damselfishes (52 species), butterfly fishes (32 species), sweet lips (16 species), angel fishes (16 species), parrot fishes (14 species) snappers (42 species), wrasses (53 species) groupers (43 species) and surgeon fishes (18 species). Another 20 per cent are cryptic and nocturnal species that are confined primarily to caverns and reef crevices during day light. This assemblage includes families like, the cusk eels, some groupers and their relatives, most of the moray eels and some scorpion fishes, wrasses and nocturnal families including the squirrel fishes, cardinal fishes and sweet lips. Another 10 per cent of fishes including snake eels, worm eels, various rays, lizard fishes, grab fishes, flat fishes and some wrasses and gobies dwell primarily on reefs covered with sand and rubble. A relatively small percentage (about 5 per cent) of the fish fauna is transient mid water reef species that roam over large areas. This group includes most sharks, jack, fusiliers, barracudas and a scattering of representatives of other families.

Coral reefs are highly productive communities which are important to the fisheries of tropical Indo-Pacific countries. There is a positive correlation between fish diversity and living coral cover. It was found that the greater the percentage of living coral cover, the higher the number of fish species and abundance of Chaetodontids. The butterfly fishes of the family, Chaetodontidae are conspicuous members of ichthyofauna and the wide spread species are always found in close association with coral reefs. The species richness of coral reef fishes is similar to that in corals. Coral reefs harbour more species and diverse fish communities than any environment on earth. The high diversity of species on reefs is due to the great variety of habitat that exists on reefs. Reefs, areas of sand, various caves and crevices, areas of algae, shallow and deep water and different zones progressing across the reef offers habitat diversity.

Coral reef fishes differ between day and night. The majority of fish species are visible in day. At night these diurnal fishes seek shelter in the reefs and are replaced by a small number of nocturnal species not seen during the day. Since some of the nocturnal species are ecologically similar to certain diurnal species (Apogonidae replace Pomacentridae), this is another way of permitting a greater number of species to exist on the reef without competing directly. All nocturnal species of reef fishes are predaceous, but the diurnal fishes span nearly all trophic categories (Carnivores, planktivores, omnivores etc.) Carnivores constitute 50-70 per cent of coral reef fishes. They are opportunistic taking whatever is available to them. Herbivores and coral grazers make up the next largest group of fishes (about 15 per cent of the species) and the most important of these are the families Scaridae and Acanthuridae. The remaining fishes are omnivorous and include representatives from virtually all families of fishes on the reef (Pomacentridae, Chaetodontidae,

Pomacanthidae, Monocanthidae, Ostraciontidae, Tetraodonidae). Only a few fishes are zooplankton feeders and they are mainly small schooling fishes of the families Pomacentridae, Clupeidae and Atherinidae. Feeding of some fishes such as grunts (Haemulids) has side effect of enhancing the nutrition and growth of corals. A large number of reef fishes have toxic secretions on their body surface Parrot fish (Scaridae), wrasses (Labridae) and surgeon fish (Acanthuridae) to deter predation by abundant carnivores. Colour is an outstanding characteristics of coral reef fishes. The bright warning colouration seems to advertise that the species is toxic or otherwise distasteful and predators will avoid them.

Reef fishes sampled by various methods namely beach seining on sand flats and seagrass beds, gill netting on reef edges, fishing with lines and traps, visual observations in transects along reef edges and underwater photography and video recording reveal that of the total number of 204 species, 43 species were caught from both habitats, namely, reef flats and reef edges. Two species of the family Nemipteridae, *Scolopsis trilineatus* and *Pentapodus caninus* belong to the group of dominant species in both habitats as they prefer sandy habitats. *Scolopsis bilineatus* and *Scolopsis cancellatus* are very common along the reef edge but rare on reef flat. They prefer deeper water. The other four species that are dominant in both habitats are sand dwellers. The two mullids mentioned in the list and *Mulloides auriflamma*, which is also equally common in both habitats, feed in loose sediments, as do all members of their family. The labrid, *Halichoeres scapularis* is associated more with sand, rubble or sea grass bottoms nearer reefs than the reefs themselves.

Reef Flats	*Reef Edges*
Spratelloides gracillis	*Holocentrus operculare*
Plotosus lineatus	*Kutaflammeo sammara*
Saurida gracilis	*Myripristis murdjan*
Hemiramphus far	*Holocentrus pralinus*
Tylosaurus strongilurus	*Cephalopholis argus*
Atherina eendrachtensis	*Cephalopholis miniatus*
Atherina forskali	*Carangoides oblongus*
Fistularia petimba	*Caranx sexfasciatus*

Reef Flats	Reef Edges
Syngnathoides biculeatus	*Caesio caerulaureus*
Corythoichthys fasciatus	*Monotaxis grandoculis*
Apogon quadrifasciatus	*Scolopsis margaritifer*
Scolopsis trilineatus	*Scolopsis trilineatus*
Gerres macrosomas	*Scolopsis bilincatus*
Gerres oyena	*Scolopsis cancellatus*
Gerres abbreviatus	*Plectorhynchus chaetodontiodes*
Gerres oblongus	
Pentapodus caninus	*Penlapodus caninus*
Lethrinus harak	*Lethrinus kallopterus*
Lethrinus miniatus	
Lethrinus microdon	
Upeneus bifasciatus	*Upeneus bifasciatus*
Parupeneus barberinus	*Parupeneus barberinus*
Parupeneus trifasciatus	
Chaetodon Kleinii	
Chaetodon triangulum	
Chaetodon vagabundus	
Chaetodon auriga	
Chaetodon melanotus	
Chaetodon raflesii	
Chaetodon trifasciatus	
Chaetodon operculum	
Chaetodon ephippium	
Chaetodon mesoleucus	
Amblyglyphidodons curacao	
Amblyglyphidodon leucogaster	
Paraglyphidodon melas	
Heniochus varius	
Centrophyge vroliki	
Megaprotodon trifascialis	
Dascillus trimaculatus	
Dascillus aruanus	

Reef Flats	Reef Edges
Abudefduf coelestinus	*Abudefduf coelestinus*
Premnas biaculeatus	*Abudefduf saxatilis*
Liza oligolepis	*Chromis caerulea*
Mugil cephalus	*Chromix xanthura*
Acanthochromis polyacanthus	
Pomacentrus tripunctatus	
Sphyraena jello	
Halichoeres scapularis	*Halichoeres scapularis*
Halichoeres argus	*Thalasoma hardwiki*
Halichoeres maragaritaceus	*Thalasoma lunare*
Halichoeres hoeveni	*Thalasoma chlorurus*
Stethojulis strigiventer	*Hemigymnus melapterus*
Stethojulis interupta	*Cheilinus fasciatus*
Cheilinus undulatus	
Cheilo inermis	
Novaculichthys taeniopterus	*Epibulus insidiator*
Hemipteronotus pentadactylus	*Choerodon anchorago*
Leptoscarus vaigienis	*Scarus sordidus*
Parapersis cylindrica	*Scarus bicolor*
Siganus canaliculatus	*Scarus dimidiatus*
Siganus spinus	*Scarus blochi*
Bothus affinis	*Scarus lunula*
Monacanthus hajam	*Scarus niger*
Arothron immaculatus	*Ctenochaetus strigosus*
Canthigaster valentini	*Ctenochaetus straitus*
Naso lituratus	
Acanthurus nigricans	
Zebrasoma veliferus	
Zanclus cornutus	
Siganus chrysospilos	
Lo vulpinus	
Bolistapus undulatus	

The same holds for *Halichoeres argus*. The species of this genus are commonly found in close association with the living coral, and sheltering beneath slabs of dead coral clinker. Several other common labrids, namely, *Cheilinus, Choerodon, Cheilo stethojulis* are found in both habitats. Two important genera on the list of species occurring in both habitats are *Lutjanus* and *Lethrinus*, each with four species. These snappers and emperors are all non-territorial predators, often moving around in groups. They feed on crustaceans, sea urchins and fish and consequently are not strictly bound to one of the habitats. The rabbit fishes, *Siganidae* represented in the list by three species *Siganus* are mainly herbivorous and may therefore have slight preference for the reef flat, but they are known to be common in coral dominated habitats as well. Non-demersal fishes present in reef flat and reef edges are given below.

Group	*Species*	*Reef flat*	*Reef edge*
Epi-pelagic	*Cypselurus brevis*	Not present	Present
	Hemiramphus far	Present	"
	Hemiramphus quoyi	"	Absent
	Hemiramphus commersoni	"	"
	Tylosurus annularis	"	Present
	Tylosurus strongylurus	"	Absent
Pelagic	*Rastrelliger brachysoma*	Absent	"
	Rastrelliger kanagurta	"	Present
	Carangoides oblongus	"	"
	Caranx crumenophthalmus	"	"
	Caranx mate	"	"
	Caranx sexfasciatus	"	"
	Decapterus macrosoma	"	"
	Elagatis bipinnulatus	"	"
	Chorinemus toli	Present	Absent
Free swimming	*Chirocentrus dorab*	Absent	Present
	Sphyraena jillo	Present	"
	Sphyraena langsar	"	"
	Sphyraena forsteri	"	"

The ordinary pelagic fishes are restricted to the reef edge. Often they can be observed in great numbers, but usually they avoid the close proximity of the reef and do not normally swim over the reef flat. The epipelagic species do not exhibit such reluctance. Species of free-swimming category may be found in both habitats. Comparing fish fauna of reef flats and reef edges it is obvious that the species they have in common mainly belong to groups that are dependent on sandy bottoms for feeding and groups of more or less free-moving predators. The occurrence of a regular interaction by daily or tidal migrations is likely to occur at least to some extent.

Bright colour, interesting behaviour and ability to adapt to captive conditions are mainly responsible for the popularity of ornamental fishes. They appear in a variety of shapes, sizes and colours in coral reefs ecosystem. The ornamental fishes occuring in coral reef ecosystems includes the groups such as those of clowns, damsels, angels, lions, butterflies, triggers, puffers, snappers, hawk fishes, wrasses, groupers, gobys, and some others have both ornamental and economic values.

Commonly Available Ornamental Fishes in Indian Waters

Sl.No.	*Group of Fishes*	*Common Name*	*Species Name*
1.	Pomacentridae	Sebae clown	*Amphiprion sebae*
		Blue damsel	*Pomacentrus caeruleus*
		Yellow tail damsel	*Neopomacentrus nemurus*
		Green choromis	*Pomacentrus pavo*
		Three spot damsel	*Dascyllus trimaculatus*
		Sergeant major	*Abudefduf saxatilis*
2.	Chaetodontidae	Pakistani butterfly	*Cheetodon collare*
		Vagabond butterfly	*C. vagabundus*
		Eight-banded butterfly	*C. octofasciatus*
		Racoon butterfly	*C. lunula*
		Lined butterfly	*C. lineolatus*
		Threadfin butterfly	*C. auriga*
		Banner fish	*Heniochus acuminatus*
3.	Pomacanthidae	Blue-ringed angel	*Pomacanthus annularis*
		Koran angel	*P. semicirculatus*
		Emperior angel	*P. imperator*

Contd...

Sl.No.	*Group of Fishes*	*Common Name*	*Species Name*
4.	Labridae	Cleaner wrasse	*Labroides dimidiatus*
		Moon wrasse	*Thalassoma lunare*
		Checkerboard wrasse	*Halichoeres hortulanus*
5.	Acanthuridae	Surgeon fish	*Acanthurus mata*
		Powder blue surgeon fish	*A. leucosternon*
		Convict surgeonfish	*A. triostegus*
		Sailfin tang	*Zebrasoma veliferum*
6.	Scorpaenidae	Lion fish	*Pterois volitans*
7.	Pseudochromidae	Dotty back	*Pseudochromis fuscus*
8.	Balistidae	Orange-lined trigger	*Balistapus undulatus*
		Redtooth trigger	*Odonus niger*
9.	Scaridae	Parrot fish	*Scarus ghobban*
10.	Tetraodontidae	Puffer fish	*Canthigaster coronta*
		Guineafowl puffer	*Arothron meleagris*
11.	Haemulidae	Oriental sweet lips	*Plectorhinchus orientalis*
		Painted sweet lips	*Diagramma pictum*
12.	Zanclidae	Moorish idol	*Zanclus cornutus*
13.	Ephippididae	Bat fish	*Platax teira*
14.	Gobiidae	Gobi fish	*Istigobius ornatus*
15.	Holocentridae	Red coat	*Sargocentron rubrum*
16.	Muraenidae	Moray eel	*Gymnothorax meleagris*
17.	Lutjanidae	Mangrove jack	*Lutjanus argentimaculatus*
		Black spot snapper	*L. fulviflamma*
18.	Holocentridae	Crown squirrel	*Sargocentron diadema*
		Sabre squirrel	*S. spiniferum*
19.	Serranidae	Squaretail grouper	*Plectropomus areolatus*
		Camouflage grouper	*Epinephelus polyphekadion*
		Dwarf-spotted grouper	*E. merra*
		Coral grouper	*Cephalopholis miniata*
20.	Cirrhitidae	Dwarf hawkfish	*Cirrhitichthys falco*

Butterfly Fish

The body of the butterfly fish (*Chaetodon vagabundus*) is compressed and somewhat rounded and is very deep, the depth being 1.5 to 1.6 times of its standard length. Ctenoid scales cover their body. There are 29 to 35 scales along the midline. The majority of the trunk scales are relatively large and rhomboid. The fins include a single continuous dorsal fin with 12 to 13 spines and 23 to 26 rays; an anal fin with 3 spines and 19 to 22 rays and a rounded caudal fin. The spinous dorsal base is longer than the soft dorsal base. The mouth is terminal and protruding and bears 5 bristle-like teeth in both jaws. There are 4 to 6 gill rakers on the upper limb and 11 to 15 on the lower limb of the first gill arch respectively.

The overall body colour of the fish is pearly white becoming yellowish posteriorly. The body in front and at the top has about 8 streaks or lines directed obliquely upwards and backwards. This set of streaks is met at an angle by another set of about 12 lines running downwards and backwards. A broad black bar running through the eye is prominent. There are further two black bars, one a blackish crescentic bar, running along the bases of the dorsal, caudal

Chaetodon vagabundus

and anal fins, and the other which has a narrow white edge on the anal fin. The caudal fin is yellow in colour.

The species occurs throughout tropical Indo-West Pacific region. The fish inhabits shallow coral rocky reefs down to 30 metres depth and is usually encountered in pairs. The diet of this species includes small crabs, worms and other small invertebrates. The vagabond butterfly fish attains a total length of 23 centimeters.

Silver Moony

The body of this diamond-shaped *Monodactylus argenteus* is strongly composed and very deep being 1.3 to 1.5 times of its standard length. The mouth is small and terminal bears very fine villiform teeth in each jaw. The dorsal and anal fins are symmetrical with dorsal fin comprising 7 or 8 spines and 26 to 30 soft rays and the anal fin with 3 spines is small and inconspicuous. The pectoral fins are short and rounded and have 16 rays in each. The large caudal fin is truncate to slightly emarginated. The body is covered with small deciduous ctenoid scales, which are arranged in 52 to 60 series along the lateral line. Scales also extends out to the fins. There are 25 to 28 gill rakers on the first gill arch with 6 to 8 rakers occurring on the upper limb and 19 to 23 on the lower limb of the gill arch.

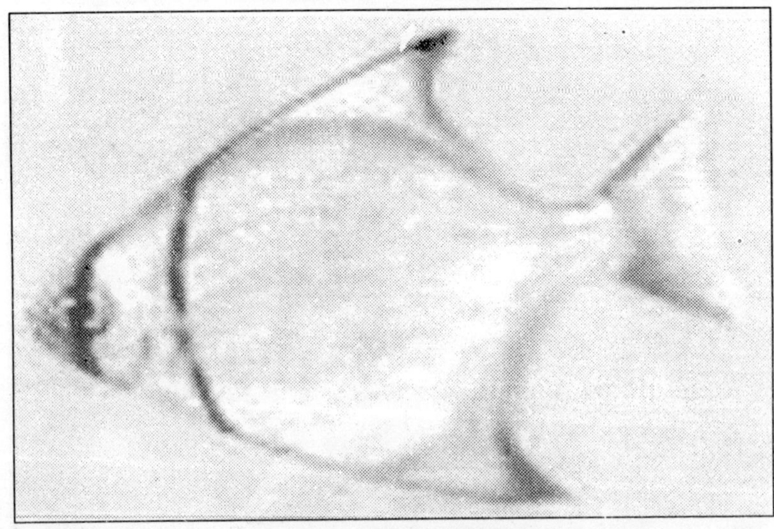

Monodactylus argenteus

The fish is silvery to silvery grey in colour with its fins having yellowish tips. The dorsal and anal fin lobes are dusky to black. Young specimens have two dusky bars accross the head which fade with age.

The fish is widespread in the tropical Indo-Pacific and widely distributed in Omani waters occurring along the entire coastline except in the Arabian Gulf. The fish is gregarious and is commonly found in small but dense schools in estuaries, warfs and shallow reefs. It can tolerate a wide fluctuations in salinity and may occur in freshwater also. It mainly feeds in the water column and its diet consists of planktonic organisms and detritus. It attains sexual maturity at about 15 to 18 cm in length and lay a few numbers of large eggs. It uses estuaries as a nursery area. They grow upto 25 cm in total length.

Blue and Gold Fusilier

The body of *Caesio caerulaurea* is elongate, slender and slightly compressed. Dorsal and ventral profile are equally convex. Anterior region of head and lower region of preoperculum and operculum are scaleless. Interorbital space is slightly concave, width about 1.3 times of eye diameter. Eyes are moderately large, adipose eyelid not well-developed, covering only orbital margin of eye. Eye diameter is

Caesio caerulaurea

greater than snout length. Snout slightly blunt. Small terminal mouth with thin lips. A series of minute conical teeth are present in both jaws. There is no teeth on palatine and vomer. Maxilla reaching below the anterior margin of pupil. Head and body is covered with small ctenoid scales. Lateral line runs horizontally. Dorsal and anal fins almost completely scaled. Pelvic fins are with auxillary scales. Dorsal fin with minute first spine. Caudal fin is deeply forked.

Upper half of head and body is bright metallic blue and lower half pinkish. Golden horizontal band runs from head to caudal fin base, just above the lateral line. A black blotch is present at upper base of pectoral fin. Broad black band is also present on each lobe of caudal fin.

The species is found in Indo-Pacific region including Malaysia. They are very common in coral reef areas, usually seen in large schools. Prefers swimming at the surface of water or under floating objects and boats. Feeds mainly on zooplankton, crustaceans and small fishes. Common size is about 20 cm, but maximum size is 35 cm. Usually caught in purse-seines and traps.

Convict Surgeon Fish

The body of *Acanthurus lineatus* is deep and compressed, body depth is 2.2 times in standard length. Mouth is small with spatulate teeth, close-set, with denticulate edges. Caudal peduncle with a lancet-like spine folded into deep grooves on each side. Caudal spine

Acanthurus iineatus

is long and venomous. Stomach is thin-walled. Caudal fin is deeply lunate with filamentous upper and lower rays.

Body and head with alternating blue and yellow black-edge stripes except for lower one-fourth, which is bluish-white. Dorsal fin with pale blue and yellowish stripe. Anal fin is grey, yellow basally with light blue edge. Caudal fin is blackish with a grey crescent centro-posteriorly, front with bluish-white edge and black margins. Pectoral fins pale, pelvic-fins orange-yellow, whitish along lateral margin and black sub-marginally. Found in Indo-Pacific, and western Indian Ocean. The fish inhabits inshore waters feeding on benthic algae on coral reefs or rocky areas. Known to be an aggressive territorial fish. Common size is 25 cm and maximum length is 30 cm.

Blue Striped Surgeon Fish

The body of *Acanthurus triostegus* is high and compressed, body depth 1.8 times of standard length. Dorsal profile of head before eye is concave and above eye is convex. Mouth is small, teeth spatulate, close-set with denticulated edges. Dorsal fin is continuous, intermembrane is without notch. Caudal peduncle with a lancet-like spine which folds into a deep groove. Caudal fin is truncate to slightly emarginated. The fish is whitish-green dorsally and white venterally. 5 dark vertical bars from nape passes through eye downward. There is also one dark vertical bar on caudal peduncle. Found in Indo-Pacific, Japan and South China Sea.

Acanthurus triostegus

The fish inhabits in inshore waters and feeds on algae. Dwells either singly or in large groups. Spawning occurs at dusk from 12 days before to 2 days after full moon. Eggs are pelagic and spherical with a single oil droplet. Average diameter of the egg is 0.68 mm. Eggs hatch after 26 hours. Juveniles live in very shallow waters of tide pools or reef flats and grow at about 12 mm per month. Common length is 17 cm and maximum length of the species is 24 cm.

Bibliography

Appukuttan K. K. Studies on the development stages of Hammerhead shark *Sphyrna blochii* from the Gulf of Mannar, Indian Journal of Fisheries, Vol. 25, Nos 1 and 2, 1978.

Annam V. P. and S. K. Dharma Raja. Trends in the catch of Silverbellies by mechanised boats in Tamil Nadu during 1971-75, Indian Journal of Fisheries, Vol 28, Nos 1 and 2, 1981.

Appa Rao T. On some aspects of fishery and biology of sardines of Waltair area, Indian Journal of Fisheries, Vol 28, Nos, 1 and 2, 1981.

Appa Rao T. and B. Krishnamoorthi. Diurnal variation in the catches of demersal fishes in the north-west region of Bay of Bengal during 1959-60, Indian Journal of Fisheries, Vol. 29, Nos 1 and 2, 1982.

Ameer Hamsa K.M.S. and G. Arumugam. A record of the snake mackerel, *Gempylus serpans* Cuvier, from Gulf of Mannar, Indian Journal of Fisheries, Vol. 29, Nos 1 and 2, 1982.

Appa Rao T. Length-weight relationship in *Pennahia macropthalmus* (Bleeker) and *Johnius carutta* (Bloch), Indian Journal of Fisheries, Vol. 29 Nos 1 and 2, 1982.

Appukuttan, K. K., P.N. Radhakrishnan Nair, and K. K. Kunhikoya. Studies on the fishery and growth rate of oceanic skipjack

Katsuwonus pelamis (Linnaeus) of Minicoy Island from 1966 to 1969, Indian Journal of Fisheries, Vol. 24, Nos 1 and 2, 1977.

Aravindakshan, M. and J. P. Karbhari. *Acetes* shrimp fishery of Bombay coast. Marine Fisheries Information Service, No. 80, 1988, CMFRI, Cochin.

Aravindakshan, M. and J. P. Karbhari. Some aspects on the fishery and biology of periscope shrimp from Bombay waters, Marine Information Service No. 83, CMFRI, Cochin ICAR, 1988.

Ah Kow Tham. The role of planktonology in fisheries development, Proc. Symp. on marine and fresh water plankton in the Indo-Pacific. FAO, 1954.

Bapat S. V. A preliminary study of the pelagic fish eggs and larvae of the Gulf of Mannar and the Palk Bay, Indian Journal of Fisheries, Vol. 2, 1955.

Bal D. V. and S. K. Banerji. A survey of the sea fisheries of India., Indian Journal of Fisheries, Vol. 2, 1955.

Balan V. The sail fish fishery off Calicut during 1974-75 and 1975-76 Indian Journal of Fisheries, Vol. 25, Nos 1 and 2, 1978.

BOBP. Report of the consultation on stock assessment for small-scale fisheries in the Bay of Bengal, Proc. Vol I, 1980.

BOBP. Status paper on coastal fishery resources, Malaysia, Report of consultation, of stock assessment for small-scale fisheries in the Bay of Bengal, Proc. Vol. 2, 1980.

Bhatia Udom and Somsak Chullasorn. Status paper on coastal fishery resources, Thailand, Report of consultation on stock assessment for small scale fisheries in the Bay of Bengal, Proc. Vol. 2, BOBP, 1980.

Biswas K. P. Corals of tropical Oceans, Daya Publishing House, Delhi, 2008.

Biswas K. P. Inshore demersal fisheries off Orissa coast, Fishery Technology, Vol. 16. No. 2, July, 1979.

Biswas, K. P. A text book of fish, fisheries and technology, Narendra Publishing House, Delhi, 1990.

Collette Bruce B. and Cornelia E. Nauen. Scombrids of the world, FAO Fishries Synopsis No 125 Vol. 2, 1983.

Chakraborty, S. K. Growth and mortality of bronze croaker, *Otolithoides biauritus* (Cantor) from Mumbai waters, Indian J. Fish (53(1), 85-89, Jan-Mar, 2006.

Dan S. S. and P. Majumder. Length-weight relationship in cat fish *Tachysurus tenuispinis* (Day) Indian Journal of Fisheries, Vol. 25, Nos 1 and 2, 1978.

Devadoss P. Maturation and breeding habit of *Dasyatis imbricatus* (Schneider) at Porto NOVO, Indian Journal of Fisheries, Vol. 25, Nos 1 and 2, 1978.

Devadoss P. A preliminary study on the Batoid fishery of Cuddalore with a note on the biology, Indian Journal of Fisheries, Vol. 25, Nos 1 and 2, 1978.

Dan S. S. Mortality rates and yield per recruit of cat fish *Tachysurus tenuispinis* (Day) Indian Journal of Fisheries, Vol. 28, Nos 1 and 2, 1981.

Devaraj M. Age and growth of three species of Seerfishes, *Scomberomorus commerson, S. guttatus* and *S. lineolatus*, Indian Journal of Fisheries Vol. 28, Nos 1 and 2, 1981.

De D. K. and P. M. Mitra. Studies on the fecundity of *Setipinna taty* (Val) and *Trichiurus pantului* (Gupta) of the Hooghly estuarine system, Indian Journal of Fisheries, Vol. 28, Nos 1 and 2, 1981.

Dorairaj K. and G. Nandakumar. Observations on the Indian short-finned eel, *Anguilla bicolor bicolor* Mc Clelland, caught for the first time at sea during spawning migration, Indian Journal of Fisheries, Vol. 29, Nos 1 and 2, 1982.

Devadoss P. On the embryonic stage of the mottled ray *Aetomylus maculatus* (Gray) with a note on its breeding season. Indian Journal of Fisheries, Vol. 29, Nos 1 and 2, 1982.

Druzhinin A. D. A morphometric study of *Rastrelliger* spp from the Mergui Archipelago Burma, Proc. IPFC, 13th session, FAO, Thailand, 1970.

Druzhinin A. D. Indian mackerel, *Rastrelliger* spp in Burma waters Proc. IPFC 13th session, FAO, Thailand, 1970.

Dalal S. G. A critique of Indian Ocean Fishery estimates, Seafood Export Journal, Vol XIX No. 5, 1987.

Devasundaram M. Peter. Scale study of *Mugil cephalus*, Linnaeus of Chilka Lake, J. Madras. Univ. B XXII No. 1., 1952.

Dan S. S. Intraovarian studies and fecundity in *Nemipterus japonicus* (Bloch) Indian Journal of Fisheries, Vol. 24, Nos 1 and 2, 1977.

Devadoss, P., P. K. Mahadevan Pillai and P. Natarajan. Observations on some aspects of the biology and fishery of *Psettodes erumei* (Bloch) at Porto Novo, Indian Journal of Fisheries, Vol. 24, Nos 1 and 2, 1977.

Dan S. S. Maturity, spawning and fecundity of cat fish, *Tachysurus tenuispinis* (Day), Indian Journal of Fisheries, Vol. 24, Nos 1 and 2, 1977.

Dhulkhed M. H. Determination of mean length of oil sardine at different ages based on annual size-frequency distribution, Indian Journal of Fisheries, Vol. 24, Nos 1 and 2, 1977.

Dhulkhed, M. H. and C. G. Annigeri. Marine Fish calendar, Karwar, Marine Fisheries Information Service, No. 88, 1988, CMFRI, Cochin.

Eales J. G. 1968. The eel fisheries of eastern Canada, Fishery Research Board of Canada, Bull 166, 1968.

Fotedar, R. K. and Y. D. Savaria. Unusual fishery for oil sardines along the west Saurashtra Coast, Marine Fisheries Information Service, No. 80, 1988, CMFRI, Cochin.

Ganapati P. N. and V. S. R. Murthy. Preliminary observations on the hydrography and inshore plankton in the Bay of Bengal off Visakhapatnam coast, Indian Journal of Fisheries, Vol. 2, 1955.

Gandhi V. Studies on the biometry of *Pennahia aneus* Bloch, Indian Journal of Fisheries, Vol. 29, Nos 1 and 2, 1982.

Girijavallabhan K. G. Occurrence of eggs of Myctophidae in the shelf region off Madras coast, Indian Journal of Fisheries, Vol. 29, Nos 1 and 2, 1982.

Girijavallabhan K. G. and J. C. Gnanamuttu. Occurrence of eggs of *Stolephorus bataviensis* in the fishing grounds off Madras, Indian Journal of Fisheries, Vol. 29, Nos 1 and 2, 1982.

Ghosh A. N. and T. D. Nangpal. On the winter breeding of *Hilsa ilisha* (Ham) in the Ganga river system, Proc. IPFC, 13th Session, FAO, Thailand, 1970.

Garibaldi Luca and Luca Limongelli. Trends in oceanic captures and clustering of large marine ecosystem FAO Fisheries Technical Paper, 435, 2003.

Ganapati P. N. and V. S. R. Murthy. Salinity and temperature variations of the surface waters off the Visakhapatnam coast, Andhra Univ. Ser. No. 49, Mem in Oceangr. Vol. I, 1954.

Gordon Ann. The by-catch from Indian shrimp trawlers in the Bay of Bengal, Bay of Bengal Programme, BOBP/WP/68, Madras, India, 1991.

Hussain Syed Makhdoom and J. Ali Khan. Fishes of the family Cynoglossidae from Pakistan coast, Indian Journal of Fisheries, Vol. 28, Nos 1 and 2, 1981.

Hoda S. M. Samsul. Maturation and spawning of the anchovy, *Thryssa mystax* in the northern Arabian Sea, Indian Journal of Fisheries, Vol. 29, Nos 1 and 2, 1982.

Hapuarachchi K. P. and D. S. Jayakody. Status paper on coastal fishery resources, Sri Lanka, Report of consultation on stock assessment for small-scale fisheries in the Bay of Bengal, Proc. Vol. 2, BOBP, 1980.

Indra T. J. A note on the occurrence of *Nemipterus metopias* (Bleeker) from Madras coast, Indian Journal of Fisheries, Vol. 28, Nos 1 and 2, 1981.

Irani K. K. Time ripe for profitable fishing ventrues, Proc. Symp. On the need for techno-economic survey of deep sea fishing resources, 1973.

Jones S. and V. Rayappa Pantulu. On some Ophichthyid larvae from the Indian coastal waters. Indian Journal of Fisheries, Vol. 2, 1955.

Jones S. and K. H. Sujansingani. The Hilsa fishery of the Chilka Lake, J. Bom. Nat. Hist. Soc., Vol. 50, No. 2, 1951.

Jones S. Notes on eggs, larvae and juveniles of fishes from Indian waters V. *Euthynnus affinis* (Cantor), Indian Journal of Fisheries, Vol. 7 No. 1, 1960.

Joseph, K. M. Salient observations on the results of fishery resource survey during 1983-84, FSI Bull.; 13/84, 1984.

James and P.S.B.R and M. Badrudeen. Biology and fishery of Silverbelly *Leiognathus dussumieri* (Val) from Gulf of Mannar, Indian Journal of Fisheries, Vol. 28, Nos 1 and 2, 1981.

Krishna Menon M. Notes on the bionomics and fishery of the prawn, *Metapenaeus dobsoni*, Miers, on the south-west coast of India, Indian Journal of Fisheries, Vol. 2, 1955.

Krishna Menon M. The life history and bionomics of the Indian penaeid prawn *Metapenaeus dobsoni*, Miers, Indian Journal of Fisheries, Vol. 2, 1955.

Krishnamurthy K. N. and M. Kaliyamurthy. Studies on the age and growth of sand whiting *Sellago sihama* (Forskal) from Pulicat Lake with observations on its biology and fishery, Indian Journal of Fisheries, Vol. 25, Nos 1 and 2, 1978.

Kulkarni G. M. and T. S. Balasubramaniam. On the occurrence of the deep sea snake fish *Acanthocepola limbata* (Cuvier) in Karwar waters, Indian Journal of Fisheries, Vol. 25, Nos 1 and 2, 1978.

Kunjipalu K. K. and A. C. Kuttappan. Note on an abnormal catch of Devil rays *Dicerobatis eregoodoo* (Day) in gill nets off Veraval, Indian Journal of Fisheries, Vol. 25, Nos 1 and 2, 1978.

Krishnapillai S. Species composition of trawler landings at Sassoon Dock, Bombay, during 1971, Indian Journal of Fisheries, Vol. 29, Nos 1 and 2, 1982.

Khan Mohammad Zafar. A probable nursery ground of silver pomfret, *Pampus argenteus* (Eupher) off Nawabunder (Gujrat), Indian Journal of Fisheries, Vol. 29, Nos 1 and 2, 1982.

Kunjipalu K. K. and P. George Mathai. Catch of a whale shark, *Rhincodon typus* (Smith) off north-west coast of India (Veraval), Fishery Technology, Vol. XIII, No. 2, 1976.

Karbhari, J. P., J. R. Dias and M. Aravinda Kshan on the capture of a giant sized, Indian threadfin, *Polynemus indicus* (Shaw) at Satpati, Moharashtra coast, Marine Fisheries Information Service, No. 81, 1988, CMFRI, Cochin.

Kumaran, M. T. M.Yohannan, and Feroz Khan. Marine Fish calendar Calicut, Marine Fisheries Information Service, No. 81, 1988, CMFRI, Cochin.

Kagwade, P. V., S. K. Chakraborty, Mohamed Zafar Khan and J. D. Sarang. Marine Fish calendar, Bombay, Marine Fisheries Information Service, No. 89, 1988, CMFRI, Cochin.

Kasinathan, C. Bumper catch of "Kalaru" from Dhanuskodi and Palk Bay, Rameswaram Island, Marine Fisheries Information Service No. 89, 1988, CMFRI, Cochin.

Kathirvel M. and V. Selvaraj. Nursery ground for early juveniles of tiger prawn in Kovalam backwater near Madras, Marine Fisheries Information Service No. 85 CIMFRI, Cochin, ICAR, 1988.

Lal Mohan R. S. An illustrated synopsis of the fishes of the family sciaenidae of India, Indian Journal of Fisheries, Vol. 28, Nos 1 and 2, 1981.

La. Fond, E. C. Environmental factors affecting the vertical temperature structure of the upper layers of the sea. Andhra Univ. Mem. in Oceanography, Vol. I, 1954.

La. Fond, E. C., and C. Borreswara Rao Rotary currents in Bay of Bengal, Andhra Univ. Mem. Oceanogr. Vol. I, 1954.

La. Fond, E. C. On upwelling and sinking off the east coast of India, Andhra Univ. Mem. Oceanogr. Vol I, 1954.

Lazarus S. Observations on the food and feeding habits of *Sardinella gibbosa* from Vizhinjam, Indian Journal of Fisheries, Vol. 24, Nos 1 and 2, 1977.

Lal Mohan R. S. A nomograph for the various characters of the sciaenid fish, *Pennahia aneus* (Bloch), Indian Journal of Fisheries Vol. 24, Nos 1 and 2, 1977.

Lipton, A. P., T. Appa Rao, S. G. Raje, C. Gopal, Ranjit Singh, P. B. Thumber and H. K. Dhokia,. Marine Fish calendar Veraval, Marine Fisheries Information Service, No. 86, 1988, CMFRI, Cochin.

Livingston, P., M. Sivadas and M. Badrudeen. Marine Fish calendar Mandapam, Marine Fisheries Information Service, No. 90, 1988, CMFRI, Cochin.

Luther, G. Oil Sardine, an emerging new fishery resource along the east coast, Marine Fisheries Information Service, No. 88, 1988, CMFRI, Cochin.

Mohamed K. H. Preliminary observations on the biology and fisheries of the Threadfin, *Polydactylus indicus*, Shaw in the Bombay and Saurashtra waters, Indian Journal of Fisheries, Vol 2, 1955.

Mojumder P. Maturity and spawning of the cat fish *Tachysurus thalassinus* (Ruppel) off Waltair coast, Indian Journal of Fisheries, Vol 25, Nos 1 and 2, 1978.

Murugan, M, M. K. Sahu, M. Srinivasan, Kamala Devi, and L. Kannan. Marine ornamental fishes of the Little Andaman waters, Fishing Chimes, Vol. 28, No. 8, 2008.

Murty V. Sriramchandra. *Nemipterus mesoprion* (Bleeker, 1853) (Nemipteridae Pisces) a new record from the seas around India, Indian Journal of Fisheries, Vol. 25, Nos 1 and 2, 1978.

Murty V. Sriramchandra. Observations on some aspects of biology of Threadfinbream, *Nemipterus mesoprion* (Bleeker) from Kakinada Indian Journal of Fisheries, Vol. 28, Nos 1 and 2, 1981.

Manisseri Mary K. and C. Manimaran. On the fishery of the Indian white prawn, *Penaeus indicus* along the Tinnevelly coast, Tamil Nadu, Indian Journal of Fisheries, Vol. 28, Nos 1 and 2, 1981.

Mohan Madan and G. Gopakumar. On the occurrence of *Nemipterus metopias* (Bleeker) from the south-west coast of India, Indian Journal of Fisheries, Vol. 28, Nos 1 and 2, 1981.

Manisseri Mary K. On the fishery of juveniles of *Penaeus semisulcatus* along the Tinnevelly coast, Tamil Nadu, Indian Journal of Fisheries, Vol 29 Nos 1 and 2, 1982.

Muthiah C. Study on the biology of *Johnieops vogleri* (Bleeker) of Bombay waters, Indian Journal of Fisheries, Vol. 29, Nos, 1 and 2, 1982.

Misra K. S. An aid to the identification of the fishes of India, Burma and Ceylon, Elasmobranchii and Holocephali, Records of the Indian Museum, Vol. XLIX, Part I, 1952.

Maldeniya R. and S. L. Suraweera. Exploratory fishing for large pelagic species in Sri Lanka, BOBP Rep. 47, 1991.

Mohiuddin M. S. Golam Kabira, Md. Mokammal Hossain and Liaquat Ali. Status paper on coastal fishery resources, Bangladesh. Report of the consultation on stock assessment for small-scale fisheries in the Bay of Bengal, Proc. Vol 2, BOBP, 1980.

Mojumder P. Length-frequency studies on cat fish, *Tachysurus thalassinus* at Waltair, during 1964-1970, Indian Journal of Fisheries, Vol. 24, Nos 1 and 2, 1977.

Menon M. Devidas. Findings of the investigations of the pelagic fishery project, Diversified fishing in Indian waters, Bull. No2, IFP, 1977.

Menon M. Devidas and K. M. Joseph. Development of Kalava (Rock cod) fishing off south-west coast of India Diversified fishing in Indian waters, Bull. No2, IFP, 1977.

Madhavan P., T. S. Unnikrishnan Nair and K. K. Balachandran A review on oil sardine, Fishery Technology Vol-XI, No. 2, 1974.

Menon, N. Gopinath and K. Balachandran. Marine fish calendar, Visakhapatnam, Marine Fisheries Information Service, No. 80, 1988, CMFRI, Cochin.

Menon, N. Gopinath and K. Balachandran. Marine fish calendar, Vizhinjam, Marine Fisheries Information Service, No. 87, 1988, CMFRI, Cochin.

Menon, N. Gopinath and K. Balachandran. Marine Fish calendar Cochin, Marine Fisheries Information Service, No. 82, 1988, CMFRI, Cochin.

Murty, V. Sriramachandra, M. K. Bandhopadhyay and P. Ramalingam. Marine Fish calendar, Kakinada, Marine Fisheries Information Service, No. 83, 1988, CMFRI, Cochin.

Mahadevan, S., P. Sam Bennet, K.M.S. Ameer Hamsa and H. Mohamad Kasim. Marine Fish calendar, Tuticorin, Marine Fisheries Information Service, No. 84, 1988, CMFRI, Cochin.

Mathew K. J., P. A. Thomas, Rani Mary George, K. G. Girijavallabham, T. S. Naomi, K. R. Nair, Geetha Antony, G. S. Bhat and M. Selvaraj. Phytoplankton blooms along the Indian coasts,–some highlights, Marine Fisheries Information Service, No. 84, 1988, CMFRI, Cochin.

Mohanty, S. K. P. V. Rao and S. Choudhury. Ganjam coast in Orissa can support good trawl fishery. Sea food Export Journal, Vol XI, No. 3, 1979.

Nair R. Velappan. Studies on the revival of the Indian Oil sardine fishery. Indian Journal of Fisheries, Vol. 2, 1955.

Nair K. N. V. and K. M. Joseph. Important observations on Deep sea fish resources made during 1983-84, Fishery Survey of India, Bull. No. 13, 1984.

Nair, K. V. Somasekharan. Maturity and spawning of *Johnius (Johnieops) sina* (Cuvier) at Calicut during 1969-72, Indian Journal of Fisheries, Vol. 24, Nos 1 and 2, 1977.

Nair R. Velappan. Studies on the life history, bionomics and fishery of the white sardine, *Kowala coval* (Cuv.), Indian Journal of Fisheries, Vol. 2, 1955.

Nanda Kumar G. Observations on the fishery of Banana prawn along the north Kanara coast with notes on its schooling behaviour and migration, Marine Fisheries Information Service, No 81, 1988, CMFRI, Cochin.

Ninan, T. V., S. P. Basu, and P. K. Bhargava. Observations on the demersal fishery resources along Andhra Pradesh coast, FSI Bull., 13/84, 1984.

Nakai Z. On the methodology of marine plankton collection, with a suggested classification. Proc. Symp, on marine and fresh water plankton in the Indo-Pacific, FAO, 1954.

Nurul Amin S. M., M. A. Rahaman, G. C. Haldar, M. A. Mazid and D. Milton. Population dynamics and stock assessment of Hilsa Shad, *Tenualosa ilisha* in Bangladesh, Asian Fisheries Science 15 (2002), 123-128.

Nagabhushanam R. and G. K. Kulkarni. Reproductive biology of the female of a penaeid prawn *Parapenaeopsis hardwickii* (Miers), Indian J. Fisheries, Vol. 29, Nos 1 and 2, 1982, ICAR.

Olsen Steinar. Role of acoustics in pelagic fishing, Symp. On the need for a techno-economic survey of the deep-sea fishing resources, AFC, 1973.

Prabhu M. S. Some aspects of the biology of the ribbon fish *Trichiurus haumela* (Forskal), Indian Journal of Fisheries, Vol. 2, 1955.

Pillay T.V.R. A preliminary biometric study of certain populations of Hilsa, *Hilsa ilisha* (Ham), Indian Journal of Fisheries, Vol. 2, 1955.

Prasanna Kumari B. and S. K. Dharmaraja. On the pomfret fishery of India with special reference to the catch statistics of Maharashtra and Gujarat coasts, Indian Journal of Fisheries, Vol. 25, Nos 1 and 2, 1978.

Pillai P. K. Mahadevan. A preliminary study on the catfish fishery off Blangad on the south-west coast of India, Indian Journal of Fisheries, Vol. 25; Nos 1 and 2, 1978.

Pillai S. Krishna and A. A. Jayaprakash. Occurrence of juveniles of the Indian mackerel, *Rastrelliger kanagurta* (Cuvier) in Bombay waters, Indian Journal of Fisheries Vol. 25, Nos 1 and 2, 1978.

Patel B. H. and Ibrahim A. Balapatel. Some observations on the prawn fishery of Gulf of Khambhat, Gujarat, Indian Journal of Fisheries, Vol. 29, Nos, 1 and 2, 1982.

Pati, S. and D. K. Pati. The role of railfall on the hilsa fishery along the Orissa coast, Indian Journal of Fisheries, Vol. 29, Nos 1 and 2, 1982.

Pandit C. G. and Sundar Lal Hora. The probable role of the Hilsa fish, *Hilsa ilisha* (Ham) in maintaining cholera endimicity in India, Indian Journal of Medical Sciences, Vol. 5, No. 8, 1951.

Pajot G. Capture of flying fish; findings of BOBP trials and studies, BOBP Publication No. 41, 1991.

Patnaik, S. A contribution to the fishery and biology of Chilka Shal, *Eleutheronema tetradactylum* (Shaw) Proc. Indian National Science Academy, Vol. 36, B, No. I, 1970.

Perumal, M. C., P. S. Joy and V. Narayana Pillai. Observations on the trawl fish catches in the shallow-waters off the south-west coast of India, 1969-1972. Seafood Export Journal, Vol. VI, No. 12, 1974.

Pandey A. K. and G. S. Sandhu. Encyclopaedia of fishes and Fisheries of India. Vol-VII Anmol Publications, Darya Ganj, New Delhi–110002, 1992.

Prabhu M. S. and M. H. Dhulkhed. On the occurrence of small-sized oil sardine, *Sardinella longiceps* (Val.) Current. Sci. 36 (15), August 5, 1967.

Pillai P. P., G. Gopa Kumar and K. K. Kunhikoya. Marine Fish calendar, Minicoy, Marine Fisheries Information Service, No. 90, 1988, CMFRI, Cochin.

Phillip, K. P., B. Premchand, G. K. Avhad and P. J. Joseph. A note on the deep sea demersal resources of Karnataka–North Kerala coast, FSI Bull., 13/84, 1984.

Royce W. F. Preliminary report on a comparison of the stocks of yellowfin tuna. Indian Journal of Fisheries, Vol. 2, 1955.

Ramamurthy S., G. G. Annigeri and N. S. Kurup. Resource assessment of the penaeid prawn *Metapenaeus dobsoni* (Miers) along the Mangalore coast, Indian Journal of Fisheries, Vol. 25, Nos 1 and 2, 1978.

Rajan K. N., K. K. Sukumaran and S. Krishna Pillai. On "dol" net prawn fishery of Bombay during 1966-76, Indian Journal of Fisheries, Vol. 29, Nos 1 and 2, 1982.

Ravindranath K. The Krishna estuarine complex with reference to its shrimp and prawn fishery, Indian J. Fisheries, Vol 29, Nos 1 and 2, 1982, ICAR.

Rao, Sudhakara, G. Prawn fishery by the "Big trawlers" along the north east coast, Marine Fisheries Information Service, No. 87, 1988, CMFRI, Cochin.

Ramamurthy S, J. C. Gnanamuthu, E. Vivekanandan, P. Ramadoss and S. Chandrasekhar. Marine Fish calendar, Madras, Marine Fisheries Information Service, No. 85, 1988, CMFRI, Cochin.

Ramaswamy V., S. A. Rajesh Kumar, S. Deepak Jamy, C. Arun Kumar, R. V. Arjun, B. V. Praveesh, S. Hakeem Mohammed, S. V. Raj, S. Martin Raina and K. Jenny Sharlet; . Toxic rainbow blankets in the Arabian Sea, J. Indian Ocean Studies Vol. 13, No 3, December, 2005.

Rengaswamy V. S. Studies on length frequency of oilsardine at Calicut during 1970-71 to 73-74, Indian Journal of Fisheries, Vol. 24, Nos 1 and 2, 1977.

Ramanathan N. P. Vijaya, V. Ramaiyan and R. Natarajan. On the biology of the large-scaled Tongue sole *Cynoglossus macrolepidotus* (Bleeker), Indian Journal of Fisheries, Vol 24, Nos 1 and 2, 1977.

Sekharan K. V. Observations on the *Choodae* fishery of Mandapam area. Indian Journal of Fisheries, Vol. 2, 1955.

Seshappa G. and B. S. Bhimachar. Studies on the fishery and biology of the Malabar sole, *Cynoglossus semifasciatus*, Day, Indian Journal of Fisheries, Vol. 2, 1955.

Sarvaiya R. T. Prawn fisheries of Kutch with special reference to Sukhper and Lakhpat, Indian Journal of Fisheries, Vol. 25, Nos 1 and 2, 1978.

Sreenivasan P. V. Observations on the fishery and biology of *Megalaspis cordyla* (Linnaeus) at Vizhinjam, Indian Journal of Fisheries Vol. 25, Nos 1 and 2, 1978.

Seshappa G. Some observations on the size distribution and the occurrence of growth rings in the scales of three species of *Cynoglossus* at Calicut, Indian Journal of Fisheries, Vol. 25, Nos, 1 and 2, 1978.

Subba Rao K. Venkata. Food and feeding of Lizard fishes (*Sourida* spp) from north western part of Bay of Bengal, Indian Journal of Fisheries, Vol. 28, Nos 1 and 2, 1981.

Sukumaran K. K. and K. N. Rajan. Studies on the fishery and biology of *Parapenaeopsis hardwickii* (Mieri) from Bombay area, Indian Journal of Fisheries, Vol. 28, Nos 1 and 2, 1981.

Seshappa G. and B. K. Chakrapani. Length-weight relationship of *Cynoglossus lida* (Bleeker) Indian Journal of Fisheries, Vol. 28, Nos 1 and 2, 1981.

Sreenivasan P. V. Length-weight relationship in *Decapterus dayi* (Wakiya), Indian Journal of Fisheries, Vol. 28, Nos 1 and 2, 1981.

Subba Rao K. Venkata. Studies on populations of *Saurida tumbil* (Bloch) from Indian waters, Indian Journal of Fisheries, Vol 29. Nos 1 and 2, 1982.

Suseelan C. and M. Kathirvel. A study on the prawns of Ashtamudi backwaters in Kerala with special reference to penaeids, India Journal of Fisheries, Vol. 29, Nos 1 and 2, 1982.

Shreenivasan P. V. Age and growth of the scad, *Decapterus dayi*. Wakiya, Indian Journal of Fisheries, Vol. 29, Nos 1 and 2, 1982.

Stequert B. and F. Marsac. Tropical tuna–surface fisheries in the Indian Ocean, FAO Fisheries Technical paper, 282, 1989.

Silas, E. G., T. Jacob, K. C. George and M. J. George. Status paper on coastal fishery resources along the east coast, India, Report of the consultation on stock assessment for small-scale fisheries in the Bay of Bengal, Proc. Vol 2, BOBP, 1980.

Seshappa G. Observations on the physical and biological features of the inshore sea bottom along the Malabar Coast, Proc. Nat. Ins. Sci India, Vol XIX, No 2, 1953.

Subba Rao K. Venkata. Systematics and comparative osteology of Indian Lizard fishes (*Saurida* spp), Indian Journal of Fisheries, Vol. 24, Nos 1 and 2, 1977.

Sarojini K. K. Biology and fisheries of the grey mullets of Bengal, II. Biology of *Mugil cunnesius* (Val), Indian Journal of Fisheries, Vol V, No 1, 1958.

Sarojini K. K. The food and feeding habits of the grey mullets, *Mugil parsia* (Hamilton) and *M. speigleri* (Bleeker), Indian Journal of Fisheries, Vol I, 1954.

Sukumaran, K. K., Ali C. Gupta, Uma S. Bhat, D. Nagaraja, H. Ramachandra, O. Thippeswamy and Y. Munyappa. Monsoon prawn fishery by "Mata Bala" along the Mangalore coast–a case study, Marine Fisheries Information Service, No. 82, 1988 CMFRI, Cochin.

Srinivasarengan, S. Unusual landing of cat fish, *Tachysurus dussumieri* at Madras, Marine Fisheries Information Service, No. 89, 1988, CMFRI, Cochin.

Somvanshi, V. S. and P. K. Bhar. A note on the demersal fishery resources survey of Gulf of Mannar, FSI Bull., 13/84, 1984.

Subramanian P., S. Sambasivan, and K. Krishnamurty. A survey of natural communities of juveniles of the penaeid prawns proc. of the first national symposium on shrimp farming MPEDA, Cochin, 1980.

SSFP, South Asia, FAO General description of marine small-scale fisheries, Kerala, RAS/77/044 Working paper, No. 30, 1980.

Sudarsan D. Results of exploratory survey around the Andaman Islands, Bull. No. 7, Exploratory Fisheries Project, Govt. of India.

Thomas P. A. and M. M. Kunju. On an unusual catch of Ghol, *Pseudosciaena diacanthus* off Goa, Indian Journal of Fisheries, Vol. 25, Nos 1 and 2, 1978.

Tiews K. On the biology of Anchovies (*Stolephorus lacepede*) in Philippine waters, Proc. IPFC, 13th Session, FAO, Thailand, 1970.

Tiews K. On the biology of Round scads (*Decapterus,* Bleeker) in Philippine waters, Proc. IPFC, 13th Session, FAO, Thailand, 1970.

Venkataraman G. M. Badrudeen and R. Thiagarajan. Population dynamics of Silverbelly *Leiognathus jonesi* in Palk Bay, Indian Journal of Fisheries, Vol. 28, Nos 1 and 2, 1981.

Vinci G. K. Threadfin bream (*Nemipterus*) resources along the Kerala coast with notes on biology of *Nemipterus japonicus,* Journal of Fisheries, Vol. 29, Nos 1 and 2, 1982.

Verghese, C. P. Introduction of purse seine fishing in Indian coasts, operations from 36' and 57' vessels for Sardine and Mackerel, Diversified fishing in Indian waters, Bull. No. 2, IFP, 1977.

Venkataraman K. Coral reefs in India, National Biodiversity Authority, Chennai, India, 2006.

Varghese K. K., and V. Shivaji. Some observations on the tuna resources of Indian Ocean FSI/Bull/13/84, 1984.

Vijayaraghavan P. Life-history and feeding habits of the spotted seer *Scomberomorus guttatus* (Bloch Schneider), Indian Journal of Fisheries, Vol.–II, No. 2, October, 1955.

Yohannan T. M. Population dynamics of Indian mackerel based on data from Mangalore during 1967-1975, Indian Journal of Fisheries, Vol. 29, Nos 1 and 2, 1982.

Yohannan T. M. Studies on the mackerel fishery of Mangalore area during 1969-73, Indian Journal of Fisheries, Vol. 24, Nos 1 and 2, 1977.

Yoshida Howard O. Synopsis of biological data on tunas of the genus *Euthynnus* FAO Fisheries Synopsis No 122, NOAA Technical Report NMFS Circular 429, October, 1979.

Yamaha Motor Co. 1981 Eels are eaten when they have grown 1000 times, Fishery Journal No 14, 1981.

Index

U

V

W

X

Z